$$
\begin{array}{r}
\overset{2}{5},976 \\
2,753 \\
385 \\
4 \\
+\quad 56 \\
\hline
4
\end{array}
$$

BARRON'S
E-Z
ARITHMETIC

Edward Williams, B.A., M.A.
Former Assistant Principal-Supervision
Mathematics Department
Washington Irving High School, New York

Katie Prindle, B.A., M.S.
Mathematics Teacher
Bemus Point Schools, Bemus Point, New York

Edited by
Eugene J. Farley, Ed.D.
Consultant in Education

BARRON'S

Better Grades or Your Money Back!

As a leader in educational publishing, Barron's has helped millions of students reach their academic goals. Our E-Z series of books is designed to help students master a variety of subjects. We are so confident that completing all the review material and exercises in this book will help you, that if your grades don't improve within 30 days, we will give you a full refund.

To qualify for a refund, simply return the book within 90 days of purchase and include your store receipt. Refunds will not include sales tax or postage. Offer available only to U.S. residents. Void where prohibited. Send books to **Barron's Educational Series, Inc., Attn: Customer Service** at the address on this page.

All inquiries should be addressed to:
Barron's Educational Series, Inc.
250 Wireless Boulevard
Hauppauge, New York 11788
www.barronseduc.com

Library of Congress Control Number: 2011007730

ISBN: 978-0-7641-4466-0

Library of Congress Cataloging-in-Publication Data
Williams, Edward, 1926–
  E-Z arithmetic / Edward Williams, Katie Prindle; edited by Eugene J. Farley.—5th ed.
    p. cm.
  Previously published: Arithmetic the easy way.
  Includes bibliographical references and index.
  ISBN 978-0-7641-4466-0
  1. Arithmetic–Problems, exercises, etc. I. Prindle, Katie. II. Farley, Eugene J.
III. Williams, Edward, 1926– Arithmetic the easy way. IV. Title. V. Title: Easy arithmetic.
  QA139.W55   2011
  513.076—dc22
                                        2011007730

PRINTED IN THE UNITED STATES OF AMERICA
9 8 7 6 5 4 3 2 1

CONTENTS

PREFACE

his book will help you to improve your skills in arithmetic—the basic or beginning area of mathematics. You use arithmetic in many ways in everyday life, such as in dealing with household and personal income and expenses, buying and selling, taxes, banking, and measurement. As you deal with more advanced problems in your job, your home, or your contacts with stores or other business firms, you need more advanced skills.

You picked up this book because you wanted help. However, you probably know and remember much of this material. Maybe some of it has been forgotten because you have not used it recently. Your interests, previous schooling, work, and daily activities develop your skills. Our purpose is to provide the explanations and practice. We also show you how to study topics of mathematics in a logical and practical order. You should try the test at the beginning of each chapter to see if you understand the ideas well or need to review them. The book is set up so you can move at your own pace.

Barron's E-Z Arithmetic will help you move up to high school level work. Topics near the end of the book will be more advanced, but you will be ready for them. Many others have used these skills to move ahead to better jobs or training. Your plans for yourself are important; we want to help you make them come true.

I would like to acknowledge the assistance of the many people who helped to shape this book. My thanks go to Dr. Farley, who contributed his knowledge of adult education techniques; to Lester Schlumpf, for his guidance in presenting accurate, easily understood mathematical problems and explanations; and to Linda Bucholtz Ross, whose illustrations have brightened the text and made mathematics a truly enjoyable subject.

Edward Williams

INTRODUCTION

Barron's E-Z Arithmetic reviews and explains basic mathematics. For those of you who want to improve your chances for a better job or to be better informed when you buy something or do business with someone, mathematical skills are important. With this book, you can move from beginning level mathematics to high school level, at your own speed. You can use this text on your own, or as a student in an adult education class.

WHAT THIS BOOK CAN DO

E-Z Arithmetic will try to:

1. Review and explain the basic ideas and skills of beginning mathematics.
2. Give many examples to practice the ideas and skills.
3. Increase your confidence to try more advanced material.

You could summarize the purposes of this book as *building your skills and confidence*.

HOW THE BOOK IS ORGANIZED

We emphasize basic mathematics—addition, subtraction, multiplication, and division. Each of these is applied to whole numbers, fractions, and decimals. The ideas and skills are related to problems of everyday life at work or at home.

Each chapter begins with a test so that you can check whether you know and understand the ideas. This is called a *pretest* because it comes *before* the chapter work. If you do well on the pretest, you can save time by skipping the material in that chapter. If you find you need some help, you have the advantage of knowing which ideas are not clear to you. Then you can read the chapter with a specific purpose. Each topic and each method will be carefully explained. Practice examples are given to help you learn.

After studying each chapter you can check how well you have learned by taking the *chapter test*, so named because it tests the material presented within the chapter. If you do well on it, you can feel confident about moving to the next chapter. You may find some problems that you cannot do. To help you, there will be a section toward the end of the chapter which guides you to the parts you need to study again.

The standard for doing well varies for each test. You can progress as fast as you can study and understand. Answers are given so that you may check whether you did the problems correctly. Obviously, there is no value to looking at the answer unless you have worked the problem. You want to know *how* to do the problems and to learn the technique so that you can use that knowledge on similar problems.

For additional interest and a little relaxation, there are some mathematical games and puzzles scattered throughout the book. These also give you practice in the basic mathematical skills.

HOW TO USE THIS BOOK

Each individual studies in his or her own way. However, the book is planned in such a way that the following suggestions should work best for you:

1. Read the brief introduction to the topic at the beginning of a chapter.
2. Take the pretest. Check your answers to decide whether
   a. You need to study that chapter, or
   b. You can move to the next chapter.
3. If you are not sure, study the material in the chapter by
   a. Reading the explanations and
   b. Working the practice examples.
      (Note: Practice examples are necessary to gain good control over the material. Do not skip the practice until you are sure you do the problems well.)
4. Take the chapter test at the end of the chapter to see how well you have learned.
5. Stress explanations and methods.
   (Note: You are urged to spend most of your time on **explanations.** *The* **how and why** *of mathematics, or the* **methods** *and the* **reasons,** *are the most important parts. Once you understand the* **reason** *for something and the* **method** *or way to do it, you can apply the knowledge to many other problems.)*

Although you are encouraged to work on your own, you are not limited to that method. Check with the board of education or a community college in your area to find adult education classes that you may attend. Working in a class has certain advantages—you can get help from a teacher and you can discuss the problems with others. The main point is to ask questions and to discover the places that offer help. Usually help is available free or at a very low cost.

WHY THIS IS IMPORTANT

Only you can decide how important education and skills are to you. Total lifetime earnings can be increased by getting more training. This is called the *economic* value of education. There is no question that better jobs, chances for promotion, and new kinds of training are related to learning skills. Employers try to select workers who gain new skills easily and rapidly. Advancement or promotion on your present job or to a different kind of job often depends upon the skills that will be covered in this book.

Personal values cannot be figured in terms of money or job, but they are of great importance. You gain in self-respect and confidence when you know that your skills have improved. Perhaps you just want to prove to yourself or others that you can be better. There is no point in regretting your past failures to achieve. What counts now are the changes you want to make. If you want to concentrate on the past, this book cannot help you. If you want to think of the present and your future, help is available.

Perhaps you have been away from learning for a while. You may even feel that it is too much effort to study again. Let us not talk about reasons that keep you from trying to do your best. The important action now is to apply your energy to renewing habits and skills. Each skill you learn helps you to learn more. The habit of study leads to success and your success will encourage you to try further. You have already shown faith in yourself by reading this book.

Once you master the basic skills of mathematics, reading interpretation, and English usage, you will want to *show* that you know them. The goal of this book is to help you raise your level of skill so that you will be ready for high school level work. When you have finished the materials in this book, *or* if you find these materials are very easy for you, try the next higher level, *E-Z Math*.

LET'S GET STARTED

It is a good idea to talk about the future. We can benefit from setting a goal or planning ahead. However, our concern for now is to get started on a review of arithmetic. Chapter 1 starts the refresher course by discussing numbers and number systems.

REFERENCE TABLES

United States System of Weights and Measures

UNITS OF LENGTH

12 inches (in.) = 1 foot (ft.)
3 feet (ft.) = 1 yard (yd.)
1,760 yards (yd.) = 1 mile (mi.)

UNITS OF CAPACITY

16 fluid ounces (fl. oz.) = 1 pint (pt.)
2 pints (pt.) = 1 quart (qt.)
4 quarts (qt.) = 1 gallon (gal.)

UNITS OF WEIGHT

16 ounces (oz.) = 1 pound (lb.)
2,000 pounds (lb.) = 1 ton (T.)

UNITS OF TIME

60 seconds (s.) = 1 minute (min.)
60 minutes (min.) = 1 hour (h.)
24 hours (h.) = 1 day (da.)
7 days (da.) = 1 week (wk.)
52 weeks (wk.) = 1 year (yr.)

We Can Work Together

Everyone has the need to count and to use numbers. We get change in a store and count the money to be sure we have not been cheated. We read the thermometer to see how cold or warm it is. Our pay envelopes or checks, prices, amounts of food or materials, and many other examples are part of our everyday life.

Counting requires a system of making marks or using numbers. This chapter will show that a number system is also the basis for all the mathematics that follows. We refer to some old number systems but explain mostly the number system that is commonly used today.

A PLAN TO SAVE TIME

- You have just read a general idea of this chapter.
- See how much you know by taking the pretest.
- Check your answers with those listed at the end of the chapter.
- Review sections of the chapter that gave you trouble.
- Study the entire chapter if necessary to be sure you have a mastery of these skills and problems.

Pretest 1—See What You Know and Remember

Do as many of these questions or problems as you can. Some may be harder than others for you. Write your answers in the spaces provided.

1. Write the different number symbols used in our number system. _____

2. Our number system is called a *decimal* system because it is based on: _____

3. In a system of marking or tallying, how would you show *six* of something? _____

4. If a large stone equals *ten* of something and a small stone equals *one* of something, what is the total number represented here? _____

5. How many *digits* are in each of these numerals?
    a. 6   b. 29   _____ _____
    c. 8   d. 341   _____ _____

6. What *place* is shown by the underlined number
    a. in the numeral 6<u>3</u>?   b. in the numeral 5<u>9</u>?   _____ . _____

7. In the following numerals, what *place* is shown by the underlined number?
    a. <u>4</u>27   b. <u>7</u>61,598   _____ _____
    c. <u>1</u>,835   d. <u>9</u>6,240   _____ _____
    e. <u>6</u>,952,064   _____

8. Write the numerals indicated by each of the following:
    a. 4 hundreds, 2 tens, 3 ones. _____
    b. 2 thousands, 0 hundreds, 3 tens, 5 ones. _____
    c. 5 thousands, 9 tens, 1 one. _____
    d. 1 ten-thousand, 8 thousands, 3 hundreds, 6 tens, 7 ones. _____
    e. 1 million, 0 hundred-thousands, 0 ten-thousands, 2 thousands, 5 hundreds, 9 tens, 0 ones. _____

9. The numbers below follow a certain pattern. Fill in the two blanks on each line to continue the pattern.

   a. 1   10   100   1,000   _____ _____

   b. 30   300   3,000   30,000   _____ _____

   c. 600,000   60,000   6,000   600   _____ _____

   d. 1,500,000   150,000   15,000   1,500   _____ _____

10. In the Roman numeral system the symbol M represents 1,000, the symbol D represents 500, C is 100, L is 50, X is 10, V is five, and I is one. Write the following in Roman numerals.

    a. 7   b. 28   _____ _____

    c. 61   d. 620   _____ _____

    e. 1,163   _____

    Write each of the following Roman numerals in our number system.

    f. II   g. XVIII   _____ _____

    h. CCLVI   i. MDCCLXXVI   _____ _____

Now turn to page 417 to check your answers. Add up all that you had correct. Count by the number of separate answers, not by the number of questions. In this pretest there were 10 questions, but 37 separate answers.

A Score of	Means That You
34–37	Did very well. You can move to Chapter 2.
29–33	Know this material except for a few points. Read the sections about the ones you missed.
24–28	Need to check carefully on the sections you missed.
0–23	Need to work with this chapter to refresh your memory and improve your skills.

Questions	Are Covered in Sections
1, 2	1.1
3, 4	1.2
5	1.3
6–9	1.4 and 1.5
10	1.6

1.1 Number Symbols We Use

A *symbol* is a sign or device that stands for something else. We all know that a red traffic light means *stop*. The American flag represents the original thirteen states and the fifty states of today. A red cross  means a hospital or medical help. Many other symbols are used to advertise products, tell the kinds of stores, show places on road maps, illustrate ideas in cartoons or pictures, and so forth. Two hands clasped together show cooperation. Two hands of the same person gripped overhead show victory. A fist means power or protest.

Number symbols represent an amount or quantity. For example, consider the number of toes an average person has on one foot. You could answer five, or if you spoke Spanish you could say *cinco*. Or you could write it as 5. The number or the amount of toes could also be shown as ⅢⅢ. Thus, the idea or *concept* of 5 can be shown in different ways. This will not apply to toes, but the *concept* of 5 can be shown as $3 + 2$, 1×5, $6 - 1$, $20 \div 4$, or by many other *numerals*. A *numeral* is a symbol used to denote a number. Thus, 5 is a numeral representing the number five.

In the same manner we can show any other amount or number by using only ten symbols: 0, 1, 2, 3, 4, 5, 6, 7, 8, 9. These symbols can be arranged to show any number that is needed. We must thank the Hindus for developing the symbols. The Arabs get credit for bringing the ideas to the Europeans, who brought them to our country. As a result, this number system is known as the Hindu-Arabic system. Because there are *ten* symbols and all numbers that we use are based on *ten*, we also call this a *decimal* system (*decimal* comes from the Roman word *decem*, which means *ten*). Later in this chapter we shall show more about the Roman and some other number systems.

1.2 Counting and Tallying

Many, many years ago, man kept a record of his flock of sheep by the use of number symbols. For each sheep he sent into the meadow, he made a notch in a stick. This notch was called a tally. There would be as many tallies or notches in his stick as there were sheep feeding in the meadow.

Man also used pebbles to count with. He kept a count of his sheep by placing one pebble in a pile for each sheep. The number of pebbles in the pile represented the number of sheep. This pile of pebbles represented six sheep:

For each method used, there were as many notches in a stick or pebbles in a pile as the number of things counted.

Each *one* of the pebbles or notches matches or corresponds with *one* of the objects. Therefore, we call this method of counting a *one-to-one correspondence*.

Years later, man devised a method to represent several objects by using a single symbol. He might have used a rock to replace each pile of five pebbles. We do the same today when we record data by tallying. We mark each *one* by a single stroke like I. We keep recording by single strokes until we have four strokes, IIII. The fifth mark is written as IHI, thereby renaming that picture, five.

Imagine that you achieved a score of 29 on Pretest 1. This would be written as IHI IHI IHI IHI IHI IIII. How would you write your score using tally marks? Write it here. _____

_____

Written numerals developed when man traveled and traded his goods. He had a need for a more permanent record.

1.3 One-Digit, Two-Digit, and Larger Numbers

You know that the number symbols can be used to represent any number. When only *one* of the symbols is used, as in these examples,

    a. 3   b. 9   c. 6   d. 4

we have a *single-digit* or a *one-digit* number. A *digit* is a single-number symbol.

A *two-digit* number would be similar to any of the following examples:

    a. 21   b. 68   c. 59   d. 47

You see that two number symbols are put together to represent a larger amount. The same idea follows with *three-digit* numbers such as 613 or *four-digit* numbers such as 5,782 and so on as high as you want to go. However, the digits have different meaning according to their *place*, as you will learn in the next section.

It is interesting to note that *digit* comes from a Roman word which means *finger*. This shows a close connection between old (and still-used) methods of counting and the terms we use today.

TERMS YOU SHOULD REMEMBER

Symbol   A sign representing something else.
Decimal System   A system of counting by 10.
Tally   An account kept by notches.
Digit   Any one of the Arabic numerals.
Numeral   A symbol used to denote a number.

1.4 Each Place Has a Value

When you look at the numeral 2, you know it means *two* of something. How about the 2 in the numeral 23? It also means two of something. However, there is a big difference in the *somethings* that are represented.

Each digit takes on a different value, depending on the place it occupies. Thus the digit 2 stands for 2 tens and the 3 stands for 3 ones. This is shown in the following diagram.

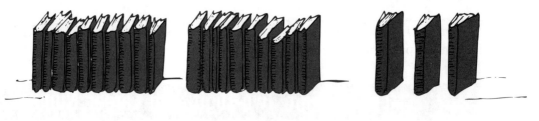

The number 23 has 2 tens and 3 ones =

TENS	ONES
2	3

You see that each digit in a number has a different value depending on its *place* in the number.

If you had a three-digit number like 537, the 5 occupies a new place called the hundreds place. Thus:

537 has 5 hundreds, 3 tens, and 7 ones =

HUNDREDS	TENS	ONES
5	3	7

See how place value permits you to write a number as large as you please without adding new symbols to the ten we have already.

PRACTICE EXERCISE 1

Fill in the blanks for each of the following numbers.

1. 37 has _____ tens and _____ ones.

2. 73 has _____ tens and _____ ones.

3. 239 has _____ hundreds and _____ tens and _____ ones.

4. 932 has _____ hundreds and _____ tens and _____ ones.

5. 444 has _____ hundreds and _____ tens and _____ ones.

Look at this numeral: 6,937

The digit 6 is in the fourth place to the left. This place tells how many *thousands* there are. Using a comma between the hundreds and thousands place makes the number easier to read. Commas are used after each group of three digits starting from the right. The comma in 6937 is placed between the 6 and the 9: 6,937. Thus, 6,937 has 6 thousands, 9 hundreds, 3 tens, and 7 ones.

$$6,937 = \begin{array}{c|c|c|c} \text{THOUSANDS} & \text{HUNDREDS} & \text{TENS} & \text{ONES} \\ \hline 6 & 9 & 3 & 7 \end{array}$$

If you regrouped the same four digits in a different way like 7,693, you would have a different numeral and hence an entirely different number. This is illustrated in the following diagram:

6,937 has 6 thousands, 9 hundreds, 3 tens, and 7 ones.

THOUSANDS (1,000) (1,000) (1,000) (1,000) (1,000) (1,000)

HUNDREDS (100) (100) (100) (100) (100) (100) (100) (100) (100)

TENS (10) (10) (10)

ONES (1) (1) (1) (1) (1) (1) (1)

7,693 has 7 thousands, 6 hundreds, 9 tens, and 3 ones.

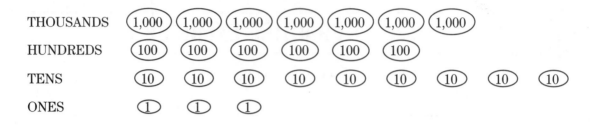

THOUSANDS	(1,000) (1,000) (1,000) (1,000) (1,000) (1,000) (1,000)
HUNDREDS	(100) (100) (100) (100) (100) (100)
TENS	(10) (10) (10) (10) (10) (10) (10) (10) (10)
ONES	(1) (1) (1)

The names for some of the other places are shown in the diagram below, using the number 3,237,471:

MILLIONS	HUNDRED-THOUSANDS	TEN-THOUSANDS	THOUSANDS
3	2	3	7

HUNDREDS	TENS	ONES
4	7	1

The number 3,237,471 has 3 millions, 2 hundred-thousands, 3 ten-thousands, 7 thousands, 4 hundreds, 7 tens, and 1 one. It is read as three million, two hundred thirty-seven thousand, four hundred seventy-one. Notice that the second comma is placed after the second group of three digits, or between the millions and the hundred-thousands.

PRACTICE EXERCISE 2

Work this exercise carefully, and see if you can get all the answers correct.

1. In the numeral 578
    a. what does the 8 mean?
    b. what does the 7 mean?
    c. what does the 5 mean?
    _____
    _____
    _____

Fill in the missing blanks for each of these numbers:

2. 3,721 has _____ thousands, _____ hundreds, _____ tens, and _____ one.

3. 15,526 has _____ ten-thousand, _____ thousands, _____ hundreds, _____ tens, and _____ ones.

4. 737,985 has _____ hundred-thousands, _____ ten-thousands, _____ thousands, _____ hundreds, _____ tens, and _____ ones.

Write each of the following in numbers.

5. forty-six   _____

6. three hundred ninety-eight   _____

7. four thousand, two hundred nineteen   _____

8. twenty-one thousand, four hundred thirty-seven   _____

9. six hundred twenty-three thousand, one hundred twenty-one   _____

Write the numeral that means:

10. 5 tens and 3 ones.   _____

11. 3 hundreds, 4 tens, and 6 ones.   _____

Write each of the following numbers in written words.

12. 3,693   _____

13. 726   _____

14. 459,822   _____

15. 8,333,333 _____

16. Write the numeral, using all the digits 8, 6, 9, and 5, which represents:

   a. the largest number.   _____

   b. the smallest number.   _____

17. The four numbers below follow a pattern. Write the next two numbers that follow the same pattern.

a.	1	10	100	1,000	_____ _____
b.	30	300	3,000	30,000	_____ _____
c.	100,000	10,000	1,000	100	_____ _____
d.	600,000	60,000	6,000	600	_____ _____

1.5 Even Nothing Means Something

Zero is used as a *place holder*. If you want to write the number six hundred, you place the digit 6 in the hundreds place (the third place to the left) and follow it with zeros in the tens and ones places. Thus 600 is the numeral for the number six hundred. Without the use of the zeros, there would be no way to show that the digit 6 was in the hundreds place.

The symbol *zero* allows you to write a number as large as you wish. Years ago, before the invention of zero as a place holder people wrote 66 to mean many different numbers: sixty-six, six hundred sixty, six hundred six, and so on. With the invention of zero as a place holder, guessing the actual number is no longer necessary.

66 has 6 tens and 6 ones
606 has 6 hundreds, 0 tens, and 6 ones
660 has 6 hundreds, 6 tens, and 0 ones

PRACTICE EXERCISE 3

1.   Write the numeral that means

   a. 4 tens and 0 ones.   _____

   b. 9 hundreds, 8 tens, and 0 ones.   _____

   c. 6 hundreds, 0 tens, and 1 one.   _____

   d. 7 thousands, 3 hundreds, and 4 ones.   _____

   e. 5 thousands and 6 tens.   _____

2. Write as numerals

   a. Two thousand, thirty-two.   _____

   b. Three thousand, sixty.   _____

   c. Six hundred ninety.   _____

   d. Forty thousand, three hundred ten.   _____

   e. Ten thousand.   _____

1.6 Old Number Systems Still Influence Us

The movie *Chinatown* was released in MCMLXXIV. Do you know the year that represents?

Those numerals are Roman numerals, which formed the Roman number system. You still see these numbers used to name the hours on the faces of some clocks, to record dates on the cornerstone of public buildings, or to represent the year a movie (such as *Chinatown*) has been released. These are but a few places you see these numerals. Perhaps you could name many more places.

Each letter in the *Roman system* names a number. The basic Roman symbols and their values are listed in the following table.

Our Numeral	Roman Numeral Symbols
1	I
5	V
10	X
50	L
100	C
500	D
1,000	M

In this system, new numbers are formed by both addition and subtraction. For example,

$$\text{VIIII} = 5 + 4 = 9 \quad \text{or} \quad \text{IX} = 10 - 1 = 9$$

The four strokes after the five (*V*) means that four is to be *added* to the five; the stroke before the ten (*X*) means that 1 is to be *subtracted* from the ten. In each case, the result is 9.

Each form is acceptable but we usually look for the one which has fewer symbols. In this case IX is preferable.

Here are a few examples of Roman numerals illustrating the use of addition and subtraction to form new numbers.

Roman Numeral	Explanation	Hindu-Arabic* Numeral
XL	50 – 10. The X precedes the L so subtract ten from fifty.	40
LXXII	50 + 2 tens + 2 ones. L is fifty, XX is twenty, and II = 2 ones.	72
CCLXVI	200 + 50 + 10 + 5 + 1. CC = 200, L = 50, X = 10, V = 5, and I = 1.	266
DCXLI	500 + 100 + 50 – 10 + 1. D = 500, C = 100, XL = 50 – 10 = 40 and I = 1.	641
MDCXLIV	1,000 + 500 + 100 + 50 – 10 + 5 – 1. M = 1,000, D = 500, C = 100, XL = 50 – 10 = 40, IV = 5 – 1 = 4.	1,644

*System we use

In all number systems you are able to write a number of any size. Why is it to your advantage to use the Hindu-Arabic system rather than the older methods described?

Writing the same number each way will illustrate the *simplicity* of our number system. Take the number, one thousand five hundred forty-three. In Roman numerals it is written MDXLIII; and in our present system it is written 1,543.

As you see, fewer symbols are used to write a number in our Hindu-Arabic system in most cases than in the other system. This is especially true when writing large numbers. A simple number like 8, written as VIII in the Roman system, can be expressed as a single symbol in our decimal system.

Other systems can be formed using a different number of digits. The Babylonians had a *base sixty system* which probably helped to create 60 minutes in an hour, 60 seconds in a minute, 360° (° means degrees) in a circle, and other measurements.

PRACTICE EXERCISE 4

Show how well you understand Roman numerals by solving all these problems correctly. Use the chart on page 11 to figure out those you do not know.

1. What Roman numerals represent these numbers?

    a. 67 _____   f. 873 _____

    b. 105 _____   g. 1,974 _____

    c. 345 _____   h. 1,031 _____

    d. 508 _____   i. 1,541 _____

    e. 101 _____   j. 2,865 _____

2. The following numerals are written in Roman numerals. Express these numbers in our decimal system.

    a. LXII _____   f. MCL _____

    b. XLIV _____   g. CMLXI _____

    c. CXLIX _____   h. CDX _____

    d. CCLXVII _____   i. MCMLXXIII _____

    e. DCXLI _____   j. MMCMIX _____

3. The date on the cornerstone of a building reads MCMXII. What year does this represent?   _____

TERMS YOU SHOULD REMEMBER

Numeral   Symbol used to denote a number.
Place Value   The value of the symbol 1, according to its place in the numeral (in 213, 1 has the value of ten).
Base   The foundation on which the number system rests (decimal system—base 10).

REVIEW OF IMPORTANT IDEAS

Some of the most important ideas in Chapter 1 were:

 Numbers play a very important role in our life.

 A numeral is a name for a number.

 Each symbol in the numeral is the name of a number.

 The position of the symbol in the numeral determines the value of the symbol.

Example: 742 = 7 hundreds, 4 tens, and 2 ones

 The symbol *zero* represented by 0, holds a place that would otherwise be unoccupied.

Example: 502 = 5 hundreds, 0 tens, and 2 ones

 The number of digits in a numeral determines the relative size of the number.

Example: A three-place numeral is larger than a two-place numeral.

CHECK WHAT YOU HAVE LEARNED

The following test lets you see how well you have learned the ideas in Chapter 1. We call it a *chapter* test because it tests how well you understand the material presented in the chapter. You noted that we called the test at the beginning a *pre*test because it came *before* the chapter. You can call them before and after tests, or whatever you like. The main point is to take the test to show yourself that you have learned the material.

In counting up your answers, remember that there were 31 separate answers in this test.

A Score of	Means That You
27–31	Did very well. You can move to Chapter 2.
24–26	Know this material except for a few points. Reread the sections about the ones you missed.
19–23	Need to check carefully on the sections you missed.
0–18	Need to review the chapter again to refresh your memory and improve your skills.

Questions	Are Covered in Sections
1, 2	1.1
3, 4	1.2
5	1.3
6–8	1.4 and 1.5
9	1.6

Chapter Test 1

1. Write the different number symbols used in our number system.   _____

2. Our number system is called a decimal system because it is based on the number __?__   _____

3. In a system of tallying, how would you indicate five of something?   _____

4. If ⊞ represents five of something and I represents one of some-thing, what is the total number represented here? ⊞ ⊞ ⊞ ⊞ II   _____

5. How many digits are in each of these numerals?
   a. 38   _____
   b. 7   _____
   c. 101   _____
   d. 27   _____

6. What place is shown by the underlined number?
   a. <u>3</u>8   _____
   b. <u>3</u>,822   _____
   c. <u>8</u>3,213   _____
   d. <u>8</u>   _____
   e. <u>9</u>91   _____

7. Fill in the blanks for each of the following numbers.
   a. 38 = _____ tens and _____ ones.
   b. 3,822 = _____ thousands, _____ hundreds, _____ tens, and _____ ones.
   c. 83,213 = _____ ten-thousands, _____ thousands, _____ hundreds, _____ ten, and _____ ones.
   d. 8 = _____ ones.
   e. 991 = _____ hundreds, _____ tens, and _____ one.

8. Write the numerals indicated by each of the following.
   a. 5 hundreds, 3 tens, and 2 ones.   _____
   b. 3 thousands, 0 hundreds, 3 tens, and 5 ones.   _____
   c. 5 thousands, 9 hundreds, and 1 one.   _____
   d. 1 ten-thousand, 8 thousands, 3 hundreds, 2 tens, and 5 ones.   _____
   e. 2 millions, 0 hundred-thousands, 0 ten-thousands, 2 thousands, 5 hundreds, and 9 ones.   _____

9.   In the Roman Numeral system, the symbol M represents 1,000,
      D = 500, C = 100, L = 50, X = 10, V = 5, and I = 1.

      Write the following in Roman numerals.

      a. 16   _____

      b. 97   _____

      c. 495   _____

      d. 1,974   _____

      Write each of the following Roman numerals in our decimal
      system.

      e. XVIII   _____

      f. XL   _____

      g. MDCLXVII   _____

      h. MCMLXXIV   _____

YOU ARE ON YOUR WAY

You have made a good start. With a sound understanding of our number system you can
learn to work with numbers. Chapter 2 will discuss addition, one of the basic operations
with numbers.

   Each individual must work out his or her own way to study and learn. Along the way
we shall try to give ideas and methods that could help. Before you start Chapter 2, think
about these points:

    You study because you want to learn, not because somebody tells you to
                        study.

    We try to *suggest*, rather than tell. That leaves the choice up to you, as it
                        should be for an adult.

    A study suggestion—start with a goal. Have a reason for doing the study
                        that is important to you. The goal or reason will pull you through the
                        times when you get discouraged.

MATHEMATICAL MAZE

This maze contains *seven* mathematical words. They are written horizontally →, vertically ↕, or diagonally ↗ ↘. When you locate a word in the maze, draw a ring around it. The word DIGIT has already been done for you.

```
S   A   N   B   C   X   D
R   Y   U   Z   Y   H   E
P   O   M   S   R   L   C
L   J   E   B   T   T   I
A   K   R   A   O   A   M
C   I   A   S   F   L   A
E   T   L   E   E   L   L
D   I   G   I   T   Y   O
```

WORDS OF:

4 letters	6 letters
BASE	SYMBOL

5 letters	7 letters
PLACE	NUMERAL
DIGIT	DECIMAL
TALLY	

The solution is on page 419.

It All Adds Up!

The Totality of Things

In the last chapter you learned about number systems. Chapter 2 starts to talk about things you can *do* with numbers. In other words, this discusses how to *use* numbers or how to *operate* with them.

The topic is *addition*, which is a basic operation or use of numbers. You know you can figure the total of two or more numbers by adding them together. No matter how large or how long an addition example may seem, it can be done by following certain basic steps. Check your skill on this topic in the pretest which follows.

Pretest 2A—See What You Know and Remember

Work these exercises carefully, doing as many problems as you can. Write the answers below each question.

1. $\begin{array}{r} 4 \\ +3 \\ \hline \end{array}$

2. $\begin{array}{r} 6 \\ +0 \\ \hline \end{array}$

3. $\begin{array}{r} 34 \\ +\ 4 \\ \hline \end{array}$

4. $\begin{array}{r} 42 \\ +25 \\ \hline \end{array}$

5. $\begin{array}{r} 2 \\ 3 \\ +4 \\ \hline \end{array}$

6. $\begin{array}{r} 242 \\ +325 \\ \hline \end{array}$

7. $\begin{array}{r} 257 \\ +\ 32 \\ \hline \end{array}$

8. $7 + 2 =$

9. $\begin{array}{r} 123 \\ 12 \\ 210 \\ +\ 24 \\ \hline \end{array}$

10. $\begin{array}{r} 31 \\ 12 \\ +23 \\ \hline \end{array}$

11.   Find the sum of 341, 2, 20, 401, and 133.

12.   How many clams would I have if I dug up 12 on Monday, 21 on Tuesday, 10 on Wednesday, and 14 on Thursday?

13.   There were 221 qt. of milk sold during the first hour, 233 qt. during the second hour, 320 qt. during the third hour, and only 121 during the fourth hour. What was the total number of quarts sold during the four-hour period?

14.   There are 4,032 people working for the Transit Authority on the night shift and 25,146 people on the day shift. What is the total number of people employed?

15.   In the blank in each example, insert one of these symbols to make a true statement:

>   which means *more than*
=   which means *the same*
<   which means *less than*

a. 4_____7   b. 6_____4   c. 7_____7

d. 15_____13   e. 1_____3

     Now turn to the back of the book to check your answers. Add up all that you had correct. Count by the number of separate answers, not by the number of questions. In this pretest there were 15 questions but 19 separate answers.

A Score of	*Means That You*
18–19	Did very well. You can move to the second part of this chapter, "The Sum of Things," beginning on page 37.
15–17	Know this material except for a few points. Read the sections about the ones you missed.
12–14	Need to check carefully on the sections you missed.
0–11	Need to work with this part of the chapter to refresh your memory and improve your skills.

Questions	*Are Covered in Sections*
1, 4, 6, 8	2.2
15	2.3
2	2.4
3, 7	2.5
5, 9–11	2.6
12–14	2.7

2.1 Whole Numbers

The numbers we generally use are called *whole numbers*. This term is used because there are other examples which we will learn later that are parts of a whole. They are called *fractions* and *decimals*. For now and the next few chapters we are working with whole numbers. These are the numbers:

$$0, 1, 2, 3, 4, 5, 6, 7, 8, 9$$

and all the higher numbers made by combining two or more of these symbols.

2.2 The Language of Addition

EXAMPLE

Mr. Lewis paid $20 to the baby-sitter on Thursday evening and $16 to the same sitter on Friday evening. How much money did he give to the baby-sitter for the two nights of work?

SOLUTION

To find out what the baby sitter earned for the two nights, add $20 and $16. Written in column form, the two *addends* look like this: ─────────┐

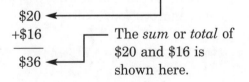

To show that you must add, use a *plus sign*.

```
   $20  ◄────────┘
  +$16
  ─────     ┌─ The *sum* or *total* of
   $36  ◄───┘   $20 and $16 is
               shown here.
```

You'll notice that addition has a language. Or you could say there are terms you must know to understand addition.

TERMS YOU SHOULD REMEMBER

Add   To join to another.
Addend   One of the quantities that is to be added to another.
Sum   The result obtained when two or more numbers are added.
Plus   The sign (+) to denote addition.
Column   A single vertical file.

PRACTICE EXERCISE 5

Find the sum of each of the following problems. Show how well you can do by getting all 36 correct.

1. 3 +6	2. 1 +1	3. 2 +3	4. 7 +2	5. 1 +3	6. 5 +1
7. 3 +4	8. 5 +2	9. 2 +2	10. 1 +5	11. 5 +3	12. 1 +7
13. 3 +2	14. 4 +2	15. 6 +2	16. 3 +3	17. 4 +5	18. 8 +1
19. 3 +5	20. 4 +1	21. 7 +1	22. 5 +4	23. 6 +1	24. 2 +1
25. 1 +8	26. 2 +7	27. 4 +4	28. 1 +4	29. 2 +4	30. 6 +3
31. 2 +5	32. 1 +2	33. 4 +3	34. 2 +6	35. 3 +1	36. 1 +6

EXAMPLE

Mrs. Patrick is employed as a domestic worker. She earned $56 on Monday and $42 on Tuesday. How much money did she earn altogether?

SOLUTION
To find out what Mrs. Patrick earned you must add her earnings for the two days. Add $56 and $42 like this:

Problem	Step 1	Step 2
$56 + $42	$5⌐6⌐ + $4⌊2⌋ ⌊8⌋	$⌐5⌐6 + $⌊4⌋2 ⌊9⌋8
	Add the *ones* place first: 6 + 2 = 8	Add the *tens* place next: 5 + 4 = 9

The sum of $56 + $42 = $98.

PRACTICE EXERCISE 6

Add each of the following.

1. 35
   +23
   ___

2. 12
   +47
   ___

3. 25
   +24
   ___

4. 53
   +25
   ___

5. 71
   +11
   ___

6. 83
   +15
   ___

7. 33
   +25
   ___

8. 71
   +28
   ___

9. 65
   +24
   ___

10. 56
   +31
   ___

EXAMPLE

Union dues are collected each month from Mrs. Ryan's paycheck. If she paid $112 in January and $123 in February, what was the total money deducted from her salary for the two months?

SOLUTION

Once again you must add $112 and $123 to find the total deducted.

Problem	Step 1	Step 2	Step 3
$112 +$123	$ 1 1 2 + $ 1 2 3 5	$ 1 1 2 + $ 1 2 3 3 5	$ 1 1 2 + $ 1 2 3 $ 2 3 5
	Add the *ones* place: 2 + 3 = 5	Add the *tens* place: 1 + 2 = 3	Add the *hundreds* place: 1 + 1 = 2

The sum of $112 + $123 = $235.

PRACTICE EXERCISE 7

Add.

1.	235 +342	2.	346 +621	3.	712 +273	4.	823 +166	5.	526 +253

6.	511 +345	7.	387 +412	8.	452 +311	9.	313 +316	10.	621 +271

This same method is applied when you add numbers of more than three places so try the next practice exercise to test your skill.

PRACTICE EXERCISE 8

Add.

1.	3,437	2.	35,432	3.	123,456	4.	35,382	5.	5,383,112
	+2,521		+23,517		+345,212		+23,412		+3,112,583

Numbers are written horizontally sometimes to save space but it is easier to rewrite the problem in column form and add as you did before.

Problem	Step 1	Step 2	Step 3
33 + 24 =	33 +24	3 3 + 2 4 7	3 3 + 2 4 5 7
	Rewrite the problem in column form.	Add the *ones* place: 3 + 4 = 7	Add the *tens* place: 3 + 2 = 5

Thus: 33 + 24 = 57.

PRACTICE EXERCISE 9

Rewrite each of the following problems in column form first and then add.

1. 41 + 45 =   2. 42 + 32 =

3. 64 + 11 =   4. 73 + 23 =

5. 85 + 13 =   6. 21 + 31 =

7. 12 + 71 =   8. 66 + 23 =

9. 52 + 24 =   10. 41 + 53 =

2.3 Help from the Number Line

It is sometimes convenient to represent *numbers* by points on a line. For example, draw the line below, and assign the value zero to any point you wish. Then, along the line at any convenient distance from zero, represent the number one. Using the same distance from zero to one as your measuring unit, mark off the points associated with two, three, four . . . (the three dots [ellipsis] indicate that the line may go on indefinitely). There is no end to the line or numbers on the line. A line such as this is called a *number line*.

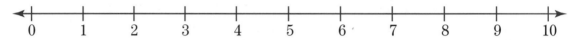

Using the number line, you can find the number of units between 3 and 7. Start at 3 and move to the right from *A* to *B* as shown below.

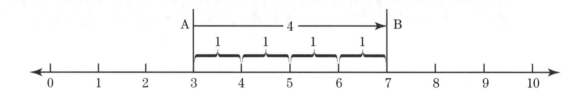

Thus you see that there are *four* units between 3 and 7.

You can also see that any number to the *right* of another has the greater numerical value. Thus 7, which is to the right of 3, is the greater number. This is indicated by a new symbol, >, which means *greater than*. Since 7 is *greater than* 3, this statement is written as 7 > 3.

Similarly, 9 is *greater than* 7 because 9 is to the right of 7. This statement is written as 9 > 7.

These are examples of *inequalities*. An *inequality* is a statement that two amounts are not equal.

Any number to the left of another has a smaller numerical value. Because 5 is to the left of 8, 5 is the smaller number. This is indicated by the symbol <, *less than*. Since 5 is *less than* 8, it is written as 5 < 8.

The number 8 is to the left of 10, so 8 is *less than* 10. This is written: 8 < 10. These are also examples of inequalities.

A number can be equal to, greater than, or less than another number. Once again, the symbols are:

greater than	>
equal to	=
less than	<

PRACTICE EXERCISE 10

In the parentheses between the numbers insert the symbol >, <, or = which will make a true statement. The first one has been done for you.

1. 7 (>) 6
    7 is greater than 6

2. 5 (   ) 2

3. 3 (   ) 8

4. 6 (   ) 6

5. 1 (   ) 2

6. 10 (   ) 10

7. 5 (   ) 7

8. 9 (   ) 3

9. 8 (   ) 8

10. 4 (   ) 5

To summarize, let a represent one number and b represent a different number.

>   If a is *larger* than b, we write $a > b$.
>   If a is the same as or *equal* to b, we write $a = b$.
>   If a is *less* than b, we write $a < b$.

PRACTICE EXERCISE 11

Try this exercise to see how well you understand these ideas.

1. Using the horizontal line below, draw a number line labeling it from 0 to 10.

$\longleftarrow \hspace{6cm} \longrightarrow$

2. What number is 3 units to the right of 3? _____

3. What number is 1 unit to the right of 8? _____

4. Count 3 units to the right of 2. What number corresponds to this point? _____

5. Count 2 units to the left of 8. What number corresponds to this point? _____

6. What number is 4 units to the right of 8? (The number line may be extended, if necessary.)   _____

7. Which is the larger number, 6 or 5? Write this, using the inequality symbol.   _____

8. The temperature on Thursday morning was 45° and in the afternoon was 52°. How many degrees did it rise? (*Hint:* Picture the number line in your head when working with large numbers.)   _____

9. The price of an automobile rose from $16,482 to $16,491. How much money has it risen?   _____

Adding whole numbers can also be shown on the number line. Let's look again at Mr. Lewis's baby-sitting problem in Section 2.2. He paid $20 one night and $16 the next night. To find out how much was paid, we added the two amounts of money together. Adding $20 and $16 on the number line can be shown in the following way.

To begin, represent 20 by a line segment starting at 0 and extending 20 units to the right of 0. Remember 0 is our starting point.

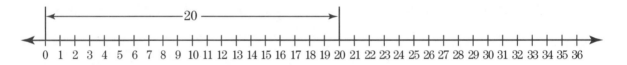

Represent 16 by a line segment starting at 20, the end of the first line segment, and extending 16 units to the right.

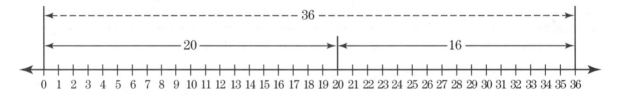

The second segment ends at a point marked 36 on the number line. This point can also be reached with a single line segment starting at 0 and extending 36 units to the right. Thus the dotted segment represents the sum of the other two segments. The sum of 20 and 16, written as 20 + 16, is equivalent to 36. Thus:

$$\$20 + \$16 = \$36.$$

PRACTICE EXERCISE 12

Using the number line given for each exercise, illustrate the addition indicated by each problem. First represent each portion; then use a dotted line to show the single line segment. Refer to the example just shown.

1. 2 + 5

    ← | | | | | | | | | | | →
    0   1   2   3   4   5   6   7   8   9   10

2. 3 + 2

    ← | | | | | | | | | | | →
    0   1   2   3   4   5   6   7   8   9   10

3. 3 + 6

    ← | | | | | | | | | | | →
    0   1   2   3   4   5   6   7   8   9   10

2.4 Nothing Is Still Important

Adding a one-place number and zero is done in this way:

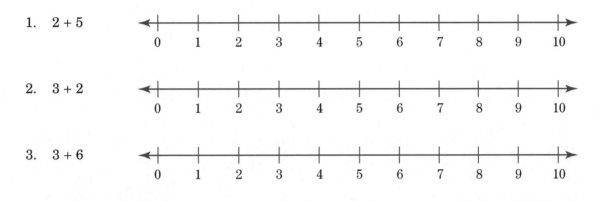

Problem	Step 1
3 +0	3 +0
	Zero added to any number results in the number itself: $3 + 0 = 3$

PRACTICE EXERCISE 13

Add.

1. 8 +0	2. 4 +0	3. 0 +7	4. 5 +0	5. 3 +0
6. 0 +9	7. 0 +3	8. 1 +0	9. 6 +0	10. 0 +7
11. 0 +6	12. 0 +8	13. 0 +1	14. 0 +2	15. 9 +0
16. 2 +0	17. 0 +5	18. 0 +4	19. 0 +0	

EXAMPLE

A candy bar cost 75¢ before it increased 10¢ in price. What is its present cost?

SOLUTION
You add 75¢ and 10¢:

Problem	Step 1	Step 2
75¢ +10¢	7 5 ¢ + 1 0 ¢ 5 ¢	7 5 ¢ + 1 0 ¢ 8 5 ¢
	Add the *ones* place: $5 + 0 = 5$	Add the *tens* place: $7 + 1 = 8$

Thus: 75¢ + 10¢ = 85¢.

PRACTICE EXERCISE 14

Add these problems.

1.   53
  +20
	2. 408
	  +151
		3. 360
		  +127
			4. 532
			  +230
				5. 646
				  +200

6.   3,003
  +2,402
	7. 30,000
	  +23,452
		8. 3,553,222
		  +2,220,037
			9. 6,903
			  +1,002
				10. 403,246
				  +190,030

2.5 Unequal Place Addition

EXAMPLE

At the end of the sixth frame in a bowling game, Jim's score is 142 pins. In the seventh frame he scores 6 pins more. How many pins does he now have to his credit?

SOLUTION

Add 142 and 6 to find his score. This requires you to add a one-place number to a three-place number. To do this, line up the 2 (the ones column in 142) and the 6 (the ones column in 6). It would appear like this:

Problem	Step 1	Step 2	Step 3

142
+ 6
	14$\boxed{2}$
	+ $\boxed{6}$
	    $\boxed{8}$
		1$\boxed{4}$2
		+ $\boxed{0}$6
		   $\boxed{4}$8
			$\boxed{1}$42
			+$\boxed{0}$06
			$\boxed{1}$48

	Add: 2 + 6 = 8	Add: 4 + 0 = 4	Add: 1 + 0 = 1

Thus: 142 + 6 = 148.

PRACTICE EXERCISE 15

Find these sums.

1. 35 + 4	2. 308 + 41	3. 546 + 52	4. 4,332 + 146

5. 6,721 + 57	6. 387 + 12	7. 43 + 6

Write in column form before adding.

8. $483 + 12 =$ 9. $5,432 + 27 =$ 10. $3,432,703 + 4,221 =$

2.6 No Addition Problem Is Long

No matter how long a column of numbers may be, the addition process basically involves two numbers being added at one time. For example: you really have a series of small problems of addition.

Step 1: $6 + 1 = 7$

Step 2: $7 + 2 = 9$

PRACTICE EXERCISE 16

Write each of these examples in column form and then add.

1. $3 + 2 + 4 =$ 2. $5 + 0 + 3 =$ 3. $2 + 2 + 3 + 4 =$

4. $2 + 1 + 4 + 1 =$ 5. $6 + 0 + 1 + 0 + 2 =$

Numbers with more than a ones place are added the same way. Look at this problem.

Problem	Step 1	Step 2
43 22 +14	4⌐3⌐ 2⌐2⌐ + 1⌐4⌐ ⌐9⌐	⌐4⌐3 ⌐2⌐2 +⌐1⌐4 ⌐7⌐9
	Add the *ones* place: 3 + 2 + 4 = 9	Add the *tens* place: 4 + 2 + 1 = 7

Thus: 43 + 22 + 14 = 79.

2.7 Word Problems

Word problems introduce new ideas in this book. Whenever possible, this type of question will be used to show you the wording involved in the particular operation.

GENERAL INSTRUCTIONS

- Read the problem carefully. Reread it again to be sure you understand it.
- Discover the information that you are told or what is *given*.
- Find out what question you are being asked.
- Determine the *method* you need to work the problem. Should you add, subtract, multiply, divide?
- Finally, *check* to see if your answer fits. Did you answer the question asked?

KEY WORDS OR PHRASES FOR PROBLEMS SOLVED BY ADDING

- *Sum* or *total*—"What is the *sum* or *total* money deducted from his salary for the two months?"
- *Increased*—"An item *increased* 10¢ in price."
- *More than*—"I have 5¢ *more than* I had yesterday."
- *Added to*—"Sixteen pens were *added to* the 142 I had before."

PRACTICE EXERCISE 17

Search for the key words and solve the problems.

1. On Tuesday 2 bottles of milk were bought, on Wednesday 3 bottles, and on Thursday 1 bottle. How many bottles were purchased altogether?

2. There were 43 cans of vegetables sold one day and 22 the next day in the L & Z store. What was the total number of cans sold during the two-day period?

3. Membership in the museum was increased $10 from the previous rate of $65. What is the new membership cost?

4. The New Mexico Giants added 3 more runs to the 4 they had already. How many runs do they have?

5. It is 4 blocks to the bowling alley, 3 more to the pizza parlor, and 2 more to the candy store. How many blocks is it to the candy store?

6. Last year there were 103 people on the day shift and only 46 on the night shift. How many people were employed altogether?

7. No more than 8 items purchased permits you to use the express lane in the super-market. If I buy 3 bottles of soda, 1 bar of butter, 2 cans of tomato juice, 1 can of cleanser, and 2 packages of frozen vegetables, may I use the express lane?

8. Tickets for illegal parking are issued each day at the town hall parking area. What was the total number of tickets issued for the first three days in August if 41, 23, and 14 summonses were given?

9. Your earnings were $34,324 last year and your wife's earnings were $35,225. What were your combined earnings?

10. In a neighborhood street fair $5,035 was collected on Saturday and $1,703 more on Sunday. What was the total money collected by Sunday?

TERMS YOU SHOULD REMEMBER

Equal   Having the same value.
Unequal   Not having the same value.
Number Line   Points on a line, used to represent our number system.

REVIEW OF IMPORTANT IDEAS

Some of the most important ideas in this part of Chapter 2 were:

 The numbers we are adding are called addends.

 The symbol + means to add.

 The same place values are lined up in columns before adding.

 Numbers are added by pairs only.

LET'S CHECK HOW WELL YOU'RE DOING

This section of the chapter moved from simple to more advanced addition. You probably found much of this coming back to you from the last time you used it. Chapter test 2A lets you check on your understanding.

In counting up your answers, remember that there are 19 separate answers in this test.

A Score of	Means That You
18–19	Did very well. You can move to the second part of this chapter, "The Sum of Things."
15–17	Know this material except for a few points. Reread the sections about the ones you missed.
12–14	Need to check carefully on the sections you missed.
0–11	Need to review this part of the chapter again to refresh your memory and improve your skills.

Questions	Are Covered in Section
1, 4, 6, 8	2.2
15	2.3
2	2.4
3, 7	2.5
5, 9–11	2.6
12–14	2.7

Chapter Test 2A

Write your answers below each problem.

1.
$$\begin{array}{r} 5 \\ +3 \\ \hline \end{array}$$

2.
$$\begin{array}{r} 4 \\ +0 \\ \hline \end{array}$$

3.
$$\begin{array}{r} 43 \\ +\ 3 \\ \hline \end{array}$$

4.
$$\begin{array}{r} 55 \\ +23 \\ \hline \end{array}$$

5.
$$\begin{array}{r} 4 \\ 1 \\ +3 \\ \hline \end{array}$$

6.
$$\begin{array}{r} 523 \\ +245 \\ \hline \end{array}$$

7.
$$\begin{array}{r} 457 \\ +\ 41 \\ \hline \end{array}$$

8. $6 + 3 =$

9.
$$\begin{array}{r} 213 \\ 14 \\ 110 \\ +\ 52 \\ \hline \end{array}$$

10.
$$\begin{array}{r} 41 \\ 14 \\ +22 \\ \hline \end{array}$$

11. Find the sum of 143, 2, 30, 401, and 123.

12. Twelve oysters were found on the beach on Monday, 15 on Tuesday, 20 on Wednesday, and 21 on Thursday. What was the total of all oysters found?

13. There were 222 cups of coffee sold during the first hour, 233 cups in the second hour, 403 cups during the third hour, and only 120 cups during the fourth hour. What was the total number of cups of coffee sold during this four-hour period?

14. There were 43,622 elementary school teachers and 6,257 secondary school teachers in the year 1969. What was the total number of teachers employed that year?

15. In the blank, place the symbol >, =, or < to make a true statement.

a. 6 ____ 4   b. 8 ____ 3   c. 1 ____ 10

d. 9 ____ 9   e. 13 ____ 15

THERE'S MORE TO COME

You have finished the first part of adding whole numbers. You haven't completed all the ideas in addition yet. There is more to come.

The Sum of Things

Numbers come in many sizes. You have to learn to add many numbers together and you have to learn to add numbers that have large digits. Each type of example in this half of the chapter uses all that you have learned so far in Chapter 2, as well as some new ideas. Check your skill on these new ideas in addition in the pretest which follows.

Pretest 2B–See What You Know and Remember

Work these exercises carefully, doing as many problems as you can. Write the answers below each question.

1.	2.	3.	4.	5.
9 +7	62 +53	38 +23	75 + 5	275 +351

6. 3,253
    +2,926

7. 8
    7
    6
    9
    1
    4
    +6

8. 5,687
    +3,445

9. 2,011
    +1,989

10. 3,456
    2,235
    4,197
    1,469
    +2,106

11.   5,976
       2,753
         385
           4
     + 56
     ─────────

12.   Mr. Carrasco worked overtime for the post office during December. His wages during that time period were $444, $516, $648, and $471. What were his total earnings?

13.   Find the sum of 5,897; 290; 1,035; 87; 491; and 34.

14.   A committee from work raised money for the Cancer Fund. Each department contributed some money. How much money was donated if the sums collected were $435; $2,861; $3,085; $762; and $1,530?

15.   Find the number of feet of fencing needed to enclose a triangular plot of ground whose sides are 73 ft., 69 ft., and 91 ft.

Now turn to the back of the book to check your answers. Add up all that you had correct.

A Score of	Means That You
14–15	Did very well. You can move to Chapter 3.
12–13	Know this material except for a few points. Read the sections about the ones you missed.
10–11	Need to check carefully on the sections you missed.
0–9	Need to work with this part of the chapter to refresh your memory and improve your skills.

Questions	Are Covered in Section
1	2.8
2–6, 8–10	2.9
7	2.10
11, 13	2.11
12, 14	2.13
15	2.14

2.8 Every Number Has Its Place

And every place has a value. You will recall that in Chapter 1 we mentioned that the numeral 23 has 2 tens and 3 ones. Sometimes we find that in adding two one-place addends we get a sum greater than 10. What happens then is shown in the following example, where the 1 is *carried* over into the tens place.

Problem

$$
\begin{array}{r|r}
 & 7 \\
+ & 6 \\
\hline
1 & 3 \\
\end{array}
$$

Another way of looking at it—To see why $7 + 6 = 13$, maybe this approach appeals to you. Think of $7 + 6$ as:

$$7 + 3 + 3 \text{ or}$$
$$10 + 3 = 13$$

Our number system is *base ten* and this method may speed up your work and increase your accuracy.

Add $9 + 7$. Think of it as $9 + 1 + 6$. This gives you $10 + 6 = 16$.

PRACTICE EXERCISE 18

The following exercise provides you with single addend addition. Remember, practice makes perfect, so work this exercise carefully and try to get all 45 problems correct. Each sum for these problems is 10 or more than 10.

1. $\begin{array}{r} 6 \\ +8 \\ \hline \end{array}$	2. $\begin{array}{r} 5 \\ +9 \\ \hline \end{array}$	3. $\begin{array}{r} 5 \\ +6 \\ \hline \end{array}$	4. $\begin{array}{r} 9 \\ +2 \\ \hline \end{array}$	5. $\begin{array}{r} 8 \\ +7 \\ \hline \end{array}$
6. $\begin{array}{r} 6 \\ +9 \\ \hline \end{array}$	7. $\begin{array}{r} 6 \\ +5 \\ \hline \end{array}$	8. $\begin{array}{r} 7 \\ +4 \\ \hline \end{array}$	9. $\begin{array}{r} 8 \\ +5 \\ \hline \end{array}$	10. $\begin{array}{r} 6 \\ +7 \\ \hline \end{array}$
11. $\begin{array}{r} 3 \\ +9 \\ \hline \end{array}$	12. $\begin{array}{r} 9 \\ +9 \\ \hline \end{array}$	13. $\begin{array}{r} 7 \\ +6 \\ \hline \end{array}$	14. $\begin{array}{r} 4 \\ +8 \\ \hline \end{array}$	15. $\begin{array}{r} 9 \\ +5 \\ \hline \end{array}$
16. $\begin{array}{r} 9 \\ +4 \\ \hline \end{array}$	17. $\begin{array}{r} 8 \\ +6 \\ \hline \end{array}$	18. $\begin{array}{r} 7 \\ +8 \\ \hline \end{array}$	19. $\begin{array}{r} 8 \\ +9 \\ \hline \end{array}$	20. $\begin{array}{r} 3 \\ +8 \\ \hline \end{array}$

21. 9 +7	22. 8 +8	23. 3 +7	24. 4 +6	25. 8 +4
26. 9 +6	27. 9 +3	28. 5 +7	29. 9 +1	30. 7 +9
31. 4 +7	32. 9 +8	33. 5 +8	34. 2 +8	35. 6 +6
36. 5 +5	37. 4 +9	38. 1 +9	39. 8 +3	40. 7 +3
41. 7 +7	42. 8 +2	43. 7 +5	44. 2 +9	45. 6 +4

2.9 Place Value and Column Addition

If you can work with number combinations and place value, you can do any addition problem. For example: suppose we had to find the sum of 85 and 73.

When the sum of numbers in *any* column, ones-place, tens-place, etc., is greater than 10, we carry to the next column as illustrated below.

Problem	Step 1	Step 2
85 +73	8|5 + 7|3 |8	8|5 + |7|3 1|5|8
	Add the *ones* place: 5 + 3 = 8	Add the *tens* place: 8 + 7 = 15 Place the 5 in the *tens* place and carry the 1 to the *hundreds* place.

Thus: 85 + 73 = 158.

PRACTICE EXERCISE 19

In these examples you are asked to carry to the next column. Be sure you understand the preceding example before you begin. Add.

1. 63
  +61
	2. 93
	  +54
		3. 723
		  +536
			4. 803
			  +392

5. 423 + 704 =	6. 9,438
	  +2,511
		7. 85,423
		  +72,541
			8. 923,003
			  +450,892

9. 8,432,000 + 4,527,000 =   10. 909
  +990

EXAMPLE

You worked 8 hours more than your usual 35 hours last week. What was the total number of hours you worked?

SOLUTION

The word *total*, which is one of our clue words, says that we must add. Add: 35 + 8.

Problem	Step 1	Step 2
35 + 8	1 3 5 + 8 3	1 3 5 + 8 4 3
	Add: 5 + 8 = 13 Place the 3 in the *ones* column and carry the 1 to the *tens* column.	Add the 1 you carried to the 3 in the *tens* column. 1 + 3 = 4

Thus: 35 + 8 = 43.

PRACTICE EXERCISE 20

Find the sum for each of these exercises. You are asked to carry into the tens column as in the previous illustration.

1.	79 + 5	2.	88 + 9	3.	76 + 8	4.	78 + 3	5.	87 + 7
6.	68 + 9	7.	58 + 7	8.	85 + 9	9.	49 + 6	10.	39 + 7
11.	22 + 9	12.	56 + 6	13.	47 + 6	14.	29 + 8	15.	58 + 4
16.	48 + 5	17.	13 + 9	18.	79 + 9	19.	57 + 3	20.	29 + 7

Numbers can be carried to any column. Look at the two following examples before beginning the next practice exercise.

Problem	Step 1	Step 2	Step 3
453 +284	45**3** +28**4** **7**	1 4**5**3 +2**8**4 **3**7	1 **4**53 +**2**84 **7**37
	Add: 3 + 4 = 7	Add: 5 + 8 = 13 Place the 3 in the *tens* place. Carry the 1 to the *hundreds* place.	Add: 1 + 4 + 2 = 7

Thus: 453 + 284 = 737.

Problem	Step 1	Step 2	Step 3	Step 4
5,931 +2,345	5 9 3 |1| +2 3 4 |5| ‾‾‾‾‾‾‾ |6|	5 9 |3|1 +2 3 |4|5 ‾‾‾‾‾‾‾ |7|6	1 5 |9|3 1 +2 |3|4 5 ‾‾‾‾‾‾‾ |2|7 6	1 |5|9 3 1 +|2|3 4 5 ‾‾‾‾‾‾‾ |8|2 7 6
	Add: 1 + 5 = 6	Add: 3 + 4 = 7	Add: 9 + 3 = 12 Place the 2 in the *hundreds* place and carry the 1 to the *thousands* place.	Add: 1 + 5 + 2 = 8

Thus: 5,931
       +2,345
       ‾‾‾‾‾‾
        8,276.

PRACTICE EXERCISE 21

Add these problems.

1.   48
      +33
      ‾‾‾

2.   757
      +272
      ‾‾‾‾

3.   5,432
      +3,707
      ‾‾‾‾‾

4.   39,000
      +19,873
      ‾‾‾‾‾‾

5.   459,321
      +190,005
      ‾‾‾‾‾‾‾

6.   67
      +58
      ‾‾‾

7.   296
      +130
      ‾‾‾‾

8.   6,737
      +2,922
      ‾‾‾‾‾

9. $23,450
      + 18,320
      ‾‾‾‾‾‾

10.   3,450,000
      +2,720,000
      ‾‾‾‾‾‾‾

Sometimes it is necessary to carry in more than one column. Look at this example:

EXAMPLE

Shop A produced 148 articles yesterday while Shop B made 156 of the same articles. How many did they make altogether?

SOLUTION
The words *how many* indicate that you must add because you have two *parts* of the total, 148 and 156.

Problem	Step 1	Step 2	Step 3
148 +156	1 148 +156 4	11 148 +156 04	1 14 8 +15 6 30 4
	Add: 8 + 6 = 14 Carry the 1.	Add: 1 + 4 + 5 = 10 Carry the 1.	Add: 1 + 1 + 1 = 3

Thus: 148 + 156 = 304.

PRACTICE EXERCISE 22

Add the following numbers.

1. 98
    +56

2. 732
    +481

3. 3,932
    +1,587

4. 53,493
    +37,507

5. 573,258
    +273,852

6. 77
    +66

7. 856
    +375

8. 4,803
    +5,219

9. 33,337
    +27,082

10. 7,832,000
    +5,989,000

2.10 Longer Columns of Numbers

Adding longer columns of numbers sometimes results in carrying more than a one into the next column. Look at this problem:

Problem

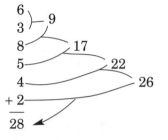

Step 1: 6 + 3 = 9
Step 2: 9 + 8 = 17
Step 3: 17 + 5 = 22
Step 4: 22 + 4 = 26
Step 5: 26 + 2 = 28

PRACTICE EXERCISE 23

Find the sum of each of the following problems.

1.	2	2.	9	3.	4	4.	1	5.	9
	2		7		1		8		9
	5		6		9		0		7
	+8		+2		+1		+3		+6

6.	8	7.	3	8.	1	9.	5	10.	7
	8		8		1		9		8
	2		5		7		1		4
	4		4		1		3		4
	+3		+2		+1		+3		+5

2.11 When the Columns Are Broken

Up to the present moment, you have added groups of numbers that contain the same number of digits in each addend. What happens if you are asked to find the sum of a group of addends like 56 + 137 + 9 + 7,638 + 589? Write each addend so that the numbers with the same place value lie in the same column. The 6 in 56, the 7 in 137, the 9, the 8 in 7,638, and the 9 in 589 are all numbers in the ones place, so they are lined up in column form like this:

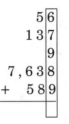

This is called *broken-column addition*, and once the numbers are arranged correctly, the procedure for adding the numbers is the same as before.

Problem	Step 1	Step 2	Step 3	Step 4
56 137 9 7,638 + 589	3 5 6 1 3 7 9 7 6 3 8 + 5 8 9 9	2 3 5 6 1 3 7 9 7 6 3 8 + 5 8 9 2 9	1 2 5 6 1 3 7 9 7 6 3 8 + 5 8 9 4 2 9	1 5 6 1 3 7 9 7 6 3 8 + 5 8 9 8 4 2 9
	Add: 6 + 7 + 9 + 8 + 9 = 39 Write the 9 and carry 3.	Add: 3 + 5 + 3 + 3 + 8 = 22 Write the 2 and carry 2.	Add: 2 + 1 + 6 + 5 = 14 Write the 4 and carry 1.	Add: 1 + 7 = 8

Thus: 56 + 137 + 9 + 7,638 + 589 = 8,429.

PRACTICE EXERCISE 24

Find the sum of each of the following problems. Use a calculator to check your answers after you have done the addition.

1.	2.	3.	4.
70	635	45	8
5	28	399	125
8,769	1,038	+10,037	3,456
+ 127	+ 77		+ 29

5.	6.	7.	8.
4	7	94	7,756
3,456	31	958	2,398
907	3,987	9	5
10,203	53	+6,738	420
+ 56	+ 7		+1,006

In questions 9 and 10, rewrite the problem in column form before adding.

9.   Find the sum of 3,257; 4,009; 138; 4; 67; and 3,470.

10.   Add 1; 1,002; 208; 4,001; 20,009.

2.12 Know the Combinations

You can see that skill and speed in addition depend very much on knowing how numbers combine. There are basic number combinations using all the symbols from 0 to 9. These are well worth practicing to gain skill because they save so much time when you know them.

Any number can be added to itself or to any of the other numbers. For example, a problem can be 0 + 0, or 0 + each of the other nine numbers. Or it can be 1 + 0, or 1 + each of the other nine numbers. Thus there are *ten* combinations for each number, or one hundred altogether. However, once you know these hundred combinations you can unlock any addition problem.

PRACTICE EXERCISE 25

This time we suggest that you do not write the answers in the book. Write your answers on a separate piece of paper so you can use this exercise for practice as many times as you wish. Don't write the problem. Place your paper under the first row and write each answer. Then fold back the paper and place it under the next row. Write the answers, fold back the paper, and so forth, until you have all twenty rows. The idea is to see how you can improve by quickly recognizing the answers. Keep your own record on this.

This group of examples is a review. Repeat them as often as you need so they become second nature to you. Remember, practice makes perfect.

1.　0　+1	2.　3　+3	3.　4　+2	4.　7　+0	5.　8　+5
6.　1　+1	7.　6　+2	8.　5　+6	9.　9　+9	10.　2　+2
11.　1　+5	12.　4　+4	13.　3　+0	14.　0　+7	15.　9　+0
16.　5　+0	17.　7　+2	18.　8　+6	19.　2　+7	20.　6　+7
21.　3　+4	22.　0　+2	23.　5　+8	24.　1　+2	25.　4　+6

26. 6 +3	27. 7 +4	28. 2 +1	29. 8 +4	30. 9 +8
31. 4 +8	32. 5 +1	33. 1 +3	34. 3 +4	35. 6 +8
36. 2 +8	37. 0 +0	38. 8 +7	39. 7 +6	40. 9 +1
41. 6 +4	42. 9 +7	43. 8 +3	44. 5 +7	45. 0 +8
46. 3 +1	47. 2 +6	48. 1 +6	49. 4 +3	50. 7 +8
51. 5 +5	52. 1 +4	53. 9 +2	54. 8 +8	55. 2 +3
56. 7 +9	57. 3 +5	58. 4 +5	59. 6 +9	60. 0 +6
61. 8 +2	62. 0 +3	63. 9 +6	64. 2 +0	65. 1 +9
66. 6 +0	67. 5 +9	68. 3 +8	69. 7 +7	70. 4 +7

71. 7   72. 4   73. 2   74. 6   75. 3
    +5   +0   +5   +6   +7
    ——   ——   ——   ——   ——

76. 0   77. 1   78. 8   79. 5   80. 9
    +9   +8   +9   +2   +3
    ——   ——   ——   ——   ——

81. 8   82. 2   83. 6   84. 5   85. 7
    +0   +9   +1   +4   +3
    ——   ——   ——   ——   ——

86. 4   87. 9   88. 3   89. 0   90. 1
    +9   +5   +9   +5   +7
    ——   ——   ——   ——   ——

91. 2   92. 4   93. 0   94. 9   95. 3
    +4   +1   +4   +4   +2
    ——   ——   ——   ——   ——

96. 6   97. 7   98. 8   99. 1   100. 5
    +5   +1   +1   +0   +3
    ——   ——   ——   ——   ——

2.13 More Difficult Word Problems Requiring Addition

EXAMPLE

The Cordero family are members of a traveling circus. They travel quite a bit. Last summer the family started in Buffalo, New York, played in many cities, and finished in Atlanta, Georgia. This is their path.

From	To	Mileage
Buffalo, NY	Pittsburgh, PA	217
Pittsburgh, PA	Cincinnati, OH	284
Cincinnati, OH	Des Moines, IA	576
Des Moines, IA	St. Paul, MN	253
St. Paul, MN	Milwaukoo, WI	330
Milwaukee, WI	Omaha, NE	491
Omaha, NE	Tulsa, OK	385
Tulsa, OK	Houston, TX	510
Houston, TX	Dallas, TX	245
Dallas, TX	Memphis, TN	472
Memphis, TN	Louisville, KY	389
Louisville, KY	Atlanta, GA	432

Mr. Cordero wanted to know the total number of miles his automobile would travel on this trip. He had to find the *total* mileage. To find the total, he had to add that long list of numbers. Can you help him?

SOLUTION

```
          6 4
Add:     2 1 7
         2 8 4
         5 7 6
         2 5 3
         3 3 0
         4 9 1
         3 8 5
         5 1 0
         2 4 5
         4 7 2
         3 8 9
    +    4 3 2
       ─────────
       4 , 5 8 4  mi.
```

Step 1: The sum of the ones column is 44; carry the 4.
Step 2: The sum of the tens column is 68; carry the 6.
Step 3: The sum of the hundreds column is 45.

That's right; the Cordero family traveled many miles across this land to work. Most people don't travel half that amount going to and from work. Do you have any idea how far you travel each year going to and from work?

Did you notice that the key word *total* appeared in this problem, too? Recall that this word sometimes indicates that you must add in order to find the answer to the problem. Go back to Section 2.7 to review the key words that indicate addition.

PRACTICE EXERCISE 26

This exercise gives you an opportunity to practice solving word problems along with the basic position facts. Key words in two examples have been underlined for you. Check your solutions with a calculator before looking up the answers.

1.   If you purchased 6 bottles of soda yesterday and 7 more today, how many bottles have you purchased in the last two days?

2.   In five games of cards, Daniel scored the following number of points: 2, 6, 9, 1, and 5. What was his total number of points scored?

3.   During a six-week period, your salary checks vary each payday. *How much* would you earn altogether if your checks were $615, $643, $626, $619, $637, and $625?

4. In a one-week period, an automobile company produced 4,325 sedans, 2,786 station wagons, and 1,938 trucks. *How many* vehicles were produced?

5. Your street association collects money from 15 families to purchase trees. If the amounts collected are $2, $5, $5, $3, $2, $1, $5, $7, $10, $5, $3, $2, $2, $5, and $1, what is the total amount of money collected?

6. The A & M Auto Company rented 35 cars more this week than the 178 they rented last week. What was the total number of cars rented this week?

7. Cigarette sales last week were 3,423 cartons. This week they increased by 789 cartons. What was the total number of cartons of cigarettes sold this week?

8. Your electric bill states the number of kilowatt hours of electricity you use each month. If you use 232 kilowatt hours in January, 246 in February, 207 in March, 198 in April, and 163 in May, what is the total number of kilowatt hours of electricity used during this five-month period?

9. If you study 3 hours today, 2 more tomorrow, 4 on Saturday, and 5 on Sunday, how many hours will you spend studying mathematics this week?

10. The A.I.M. Medical Clinic saw 162 patients on Monday, 146 on Tuesday, 87 on Wednesday, 243 on Thursday, and 109 on Friday. How many patients visited the clinic this week?

2.14 More Problems with Addition

Mathematics has many branches. One of these is geometry. All around us are various geometric shapes. You probably recognize rectangles, squares, triangles, and circles. Geometry deals with these shapes, as well as with the measurement of lines, angles, and surfaces. Surely there have been times when you used a ruler and measured the length of lines. Once the measurements were made, you might have added them as illustrated in the following problem.

EXAMPLE

Grace Badilla wanted to make a vegetable garden in the shape of a rectangle in the rear of her house. It could be 21 ft. long and 13 ft. wide. To protect the vegetables she wanted to enclose the garden with a wire fence. How many feet of fencing did she require?

SOLUTION

We'll begin this problem with a diagram of the garden:

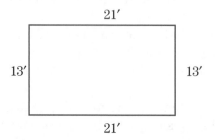

The rectangle, as you recall, has four sides, and each pair of opposite sides are the same length. Thus there are two lengths of 21 ft. and two widths of 13 ft. Since the fencing will be placed around the entire edge of the rectangle, the problem is simply to find the sum:

$$21 \text{ ft.} + 21 \text{ ft.} + 13 \text{ ft.} + 13 \text{ ft.}$$

The sum of all the sides of a rectangle or geometric shape is given a special name. It is called the *perimeter*. We can write this word in a simplified form by using the letter *P*. We can also use an abbreviation (') for *feet*.

The problem in its simplest form now becomes:

$$P = 21' + 21' + 13' + 13'$$

Finding the sum of the four addends, we get:

$$P = 68'$$

Grace now knew she had to purchase at least 68 ft. of fencing to enclose her garden.

The layout or format of the problem may cause you to wonder why you can't just add all the sides. You can, but you will want to learn a new idea which will be very helpful later when you study more advanced mathematics.

$P = 21' + 21' + 13' + 13'$ is called a *statement of equality or* an *equation*. It is a good idea to learn this form because you will use it often.

Perimeters of other geometric shapes can also be found.

EXAMPLE

Find the perimeter of the triangle drawn below. The symbol (″) means inches.

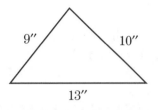

9″   10″

13″

SOLUTION

Perimeter = 9 in. + 10 in. + 13 in.

or

$P = 9″ + 10″ + 13″$

$P = 32″$

PRACTICE EXERCISE 27

In each of the following, first write the perimeter in the form of an equation, as illustrated in (a) of the first problem. Then solve the problem.

1. Find the perimeter for each of these:

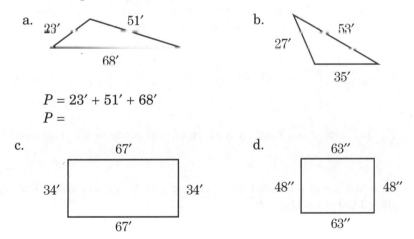

a.   23′   51′
      68′

$P = 23′ + 51′ + 68′$
$P =$

b.   27′   53′
      35′

c.   67′
   34′   34′
      67′

d.   63″
   48″   48″
      63″

2.   Find the amount of fencing required to enclose a *triangular* plot of ground measuring 27 ft., 54 ft., and 69 ft.

   a. Draw a diagram.

   b. Find the perimeter.

3.   How many feet of fencing are required to enclose a *rectangular*-shaped garden whose dimensions are 47 ft. by 35 ft.?

TERMS YOU SHOULD REMEMBER

Place value   The location of the symbol in the numeral signifying its value.
Triangle   A three-sided geometric figure.
Perimeter   The sum of all the sides of a geometric figure.

REVIEW OF IMPORTANT IDEAS

 The tens number in a sum greater than 10 must be carried to the next column on its left.

 Geometry is a branch of mathematics dealing with shapes and with measurements of lines, angles, and surfaces.

 Rectangles (▢) and triangles (◺) are just two of the many shapes.

 The perimeter is found by adding *all* the sides of the geometric figure.

 An equation is a statement of equality of two quantities. The sign = is placed between them.

LET'S CHECK HOW WELL YOU'RE DOING

This half of the chapter completed all ideas associated with adding whole numbers. Most of this material was probably familiar to you. The chapter test lets you check on your understanding. Try these examples.

A Score of	Means That You
14–15	Did very well. You can move to Chapter 3.
12–13	Know this material except for a few points. Reread the sections about the ones you missed.
10–11	Need to check carefully on the sections you missed.
0–9	Need to review this part of the chapter again to refresh your memory and improve your skills.

Questions	Are Covered in Section
1	2.8
2–6, 8–10	2.9
7	2.10
11, 13	2.11
12, 14	2.13
15	2.14

Chapter Test 2B

Write your answers below each question. Use a calculator to check your answers after you have completed the entire chapter test.

| 1. | 8
+5 | 2. | 73
+41 | 3. | 48
+27 | 4. | 65
+ 5 |

5.   382
   +146

6.   4,813
   +2,625

7.
  9
  5
  3
  3
  6
  8
 +7

8.   4,956
   +3,687

9.   4,099
  +2,901

10.
  2,624
  3,540
  4,748
  2,794
 +1,960

11.
  9,754
  5,860
    17
  1,063
 + 297

12. The first seven people who purchase rolls in the F & R Bakery store on Tuesday morning buy 6, 8, 12, 4, 8, 5, and 18 rolls. What is the total number of rolls purchased?

13. Find the sum of 1,089; 2; 674; 80; and 6,872.

14. During seven workdays you produced 286, 340, 197, 432, 263, 307, and 169 parts per day. How many parts did you complete at the end of that time?

15. Mr. O'Connor wanted to install molding on the ceiling of a 13 ft. by 8 ft. rectangular-shaped room. How many feet of molding are required?

HOLD IT!

Now that you have had lots of practice in addition, go back to Section 2.12 and see if you can improve your score on "Know the Combinations." Can you make a perfect score?

PUZZLE TIME

Try your hand with this cross-number puzzle which reviews the *addition* of *whole numbers*. Do it as you would a regular crossword puzzle. The answers down and across must all check. Two answers, 2 Across and 1 Down, have been filled in for you. The solution appears on page 426.

ACROSS

2. $9 + 8 =$
5. $6 + 7 =$
6. $59 + 44 =$
7. $39 + 98 =$
8. $8 + 7 =$
9. $20,373 + 4,842 =$
11. $93 + 5 =$
12. $683 + 752 =$
14. $56 + 89 + 67 =$
17. $107 + 343 =$
18. $752 + 398 =$
20. $69 + 8 =$
22. $872 + 759 + 10,832 =$
25. $74 + 7 =$
26. $548 + 197 =$
28. $97 + 734 + 9 =$
29. $37 + 9 =$
30. $65 + 6 =$

DOWN

1. $1,686 + 2,672 =$
2. $5,872 + 2,762 + 1,868 =$
3. $497 + 235 =$
4. $79 + 4 =$
5. $98 + 21 =$
6. $8 + 4 =$
7. $75 + 79 =$
10. $827 + 308 =$
13. $25 + 8 =$
15. $957 + 745 =$
16. $17 + 4 =$
17. $23,750 + 20,425 + 1,466 =$
19. $345 + 170 =$
20. $197 + 4,862 + 2,784 =$
21. $47 + 82 + 545 + 42 =$
23. $239 + 229 + 19 =$
24. $26 + 4 =$
27. $28 + 19 =$

Subtraction– Take It Away!

Subtraction is the second basic operation you need in working with numbers. Finding the difference between two amounts is a subtraction problem. If one can of beans costs 89¢ and another costs 78¢, you want to know the *difference* or how much you would save with the second. If your total purchases cost $3.37 and you have $5.00, you subtract to find the change.

You will see as you work with subtraction that it is not a new idea. In fact, subtraction is the opposite or the *inverse* of addition. The main ideas of subtraction are covered in Pretest 3.

Pretest 3—See What You Know and Remember

Show how well you can do with these examples. Write your answers below each question.

1.   8
   −5
	2. 73
	   −51
		3. 256
		   − 34
			4. 456
			   −255
				5. 366
				   −150

6.   96
   − 3
	7. 15
	   − 9
		8. 73
		   − 8
			9. 325
			   −171
				10. 812
				   −567

11.   3,069
   −2,154
	12. 2,001 − 1,983 =		13. 4,000
			   − 208	

14. What is the difference between 89¢ and 78¢?

15. On the first day of a two-day charity drive $42,607 was collected. If, on the second day, $6,118 less than the first day was collected, how much was collected the second day?

16. How much change do you receive from a ten dollar bill after a purchase of 4 dollars 82 cents?

17. How much less is 212 than 632?

18. The value of a house decreased $1,150 from its original purchase price of $109,345. What is its present value?

19. If two sides of a triangle measure 74 ft. and 58 ft., what is the length of the third side if the perimeter of the triangle is 191 ft.?

20. Show the result of subtracting 5 from 8 on the number line below.

```
◄───┼────┼────┼────┼────┼────┼────┼────┼────┼────┼────►
    0    1    2    3    4    5    6    7    8    9    10
```

Check your answers by turning to the back of the book. Add all that you had correct.

A Score of	Means That You
18–20	Did very well. You can move to Chapter 4.
16–17	Know the material except for a few points. Read the sections about the ones you missed.
13–15	Need to check carefully on the sections you missed.
0–12	Need to work with this chapter to refresh your memory and improve your skills.

Questions	Are Covered in Section
1, 2, 4, 5	3.2
20	3.3
3, 6	3.4
14, 17	3.5
16	3.6
7	3.7
8–13	3.8
15, 18, 19	3.9

3.1 What Happens in Subtraction?

If you have 8 coins and I "take away" 5 of them, what do you have left? This is a problem in subtraction.

You have 8 coins.

I "take away" 5 coins.

You have 3 coins left.

You write the number that you "take away" with a minus sign to its left, under the amount you had when you started. Taking away is another name for subtraction and the minus sign (–) is used to indicate the operation subtraction. Thus:

$$\begin{array}{r} 8 \text{ coins} \\ -5 \text{ coins} \\ \hline 3 \text{ coins} \end{array}$$

Subtraction is not a new operation for you. It is the inverse of addition. Webster defines *inverse* as the *direct opposite*, and subtraction is the opposite operation of addition. Instead of *adding on* we are *taking away*.

What does this mean? If we subtract 5 from 8, we write it vertically like this:

$$\begin{array}{r} 8 \\ -5 \\ \hline ? \end{array}$$

or horizontally like this: 8 – 5 = ? Since subtraction is the opposite of addition, we mean 5 + ? = 8. The examples:

$$\begin{array}{rcr} 8 & \text{or} & 5 \\ -5 & & +? \\ \hline ? & & 8 \end{array}$$

mean the same thing. The missing number is 3:

$$\begin{array}{cc} 8 & \text{or} & 5 \\ -5 & & +3 \\ \hline 3 & & 8 \end{array}$$

3.2 The Language of Subtraction

EXAMPLE

Mrs. Maxson paid $130 to the Day Care Center last week and only $127 this week to pay for the care of her daughter. How much less did it cost her this week than last week?

SOLUTION

To find out how much less the care cost, you subtract $127 from $130. The answer to subtraction is called the difference.

Written in column form, the problem would look like this:

$$\begin{array}{r} 130 \\ -127 \\ \hline 3 \end{array}$$

The *minus* sign indicates subtraction →   3   difference

You notice that subtraction also has a language. These are terms that you must know to understand subtraction.

TERMS YOU SHOULD REMEMBER

Subtract   To take away.
Minus   The sign (−) of subtraction.
Difference   That which is left after a part has been taken away.

You can *check* your answer in subtraction by adding the answer (difference) to the number you subtracted and the result should be the number you are subtracting from.

$$
\begin{array}{cc}
\textit{Example} & \textit{Check} \\
130 & 127 \\
\underline{-127} & \underline{+\ \ 3} \quad \text{difference} \\
3 & 130
\end{array}
$$

Before you start the practice exercise, let's discuss a few basic facts you are already aware of.

You know that zero taken away from a number will make no change in the number.

$$
\begin{array}{ccc}
4 & 7 & 9 \\
\underline{-0} & \underline{-0} & \underline{-0} \\
4 & 7 & 9
\end{array}
$$

Similarly, *one* taken away from a number will leave the number *one* lower.

$$
\begin{array}{ccc}
6 & 8 & 3 \\
\underline{-1} & \underline{-1} & \underline{-1} \\
5 & 7 & 2
\end{array}
$$

Subtracting the same quantity from itself will **always** result in a remainder of 0.

$$
\begin{array}{cc}
9 & 6 \\
\underline{-9} & \underline{-6} \\
0 & 0
\end{array}
$$

You can also figure quickly examples taking away two, three, etc. The practice exercise that follows has all possible combinations of single-digit examples. Write your answers on a separate piece of paper so you can use this exercise for practice as many times as you wish. Don't write the problem. Place your paper under the first row and write each answer. Then fold back the paper and place it under the next row. Write the answers, fold back the paper, and so forth, until you have completed the exercise. The idea is to see how you can improve by recognizing the answers more and more quickly. Keep your own record on this.

PRACTICE EXERCISE 28

Subtract and check by adding. Try to get all correct.

1. 6 −1	2. 4 −4	3. 5 −0	4. 8 −2	5. 8 −3
6. 1 −1	7. 6 −4	8. 3 −2	9. 6 −5	10. 0 −0
11. 4 −3	12. 9 −2	13. 1 −0	14. 7 −4	15. 7 −6
16. 7 −1	17. 7 −7	18. 6 −0	19. 9 −8	20. 9 −9
21. 2 −1	22. 7 −5	23. 8 −4	24. 4 −2	25. 7 −0
26. 3 −1	27. 8 −5	28. 2 −0	29. 5 −3	30. 8 −8
31. 8 −7	32. 7 −2	33. 6 −6	34. 3 −0	35. 6 −3
36. 9 −5	37. 8 −0	38. 9 −6	39. 9 −4	40. 4 −1
41. 9 −3	42. 9 −7	43. 5 −2	44. 8 −1	

This last group of examples is written horizontally. Write each of them in column form before subtraction.

45. $3 - 3 =$ 46. $9 - 0 =$ 47. $6 - 2 =$ 48. $9 - 1 =$

49. $5 - 5 =$ 50. $5 - 1 =$ 51. $8 - 6 =$ 52. $7 - 3 =$

53. $5 - 4 =$ 54. $4 - 0 =$ 55. $2 - 2 =$

EXAMPLE

Raul and Brianna are planning to go to dinner before the high school dance. Dinner for two at a local family restaurant, including tax and tip, is roughly $35. Dinner for two at a French restaurant is roughly $66. If they choose to go to the family restaurant, they will have enough money to go to the arcade after the dance. How much *less* will Brianna and Raul pay by going to the family restaurant?

SOLUTION
To find the difference in the prices of these meals you must subtract the two costs of the meals.

Problem	Step 1	Step 2	Check
$66 − 35	$6⌐6⌐ − 3⌊5⌋ ⌊1⌋	$⌐6⌐6 − ⌊3⌋5 $⌊3⌋1	$35 + 31 $66
	Subtract the *ones* column first: $6 - 5 = 1$	Subtract the *tens* column: $6 - 3 = 3$	

Thus: The difference of $66 − $35 = $31.

PRACTICE EXERCISE 29

Subtract the following examples. Check by adding.

1.   35
    −23

2.   47
    −12

3.   25
    −14

4.   55
    −23

5.   74
    −11

6.	88	7.	36	8.	78	9.	65	10.	56
	−15		−13		−21		−24		−31

The same method is used when subtracting numbers with more than two places. You begin subtracting in the ones column and move to the left. Be sure the numbers are written in column form with digits having the same place value under one another.

Subtractions that result in zero are meaningful. Remember $101 is not the same as $11, so be sure you place the zero in the answer when it is needed.

EXAMPLE

Quarterly union dues of $123 and $112 were deducted from two of Mr. Godwin's checks. By how much was the second deduction decreased?

SOLUTION

Once again we must subtract $112 from $123 to find the decrease.

Problem	Step 1	Step 2	Step 3	Check
$123 − 112	$ 1 2 3 − 1 1 2 1	$ 1 2 3 − 1 1 2 1 1	$ 1 2 3 − 1 1 2 0 1 1	$112 + 011 $123
	Subtract the *ones* place: 3 − 2 = 1	Subtract the *tens* place: 2 − 1 = 1	Subtract the *hundreds* place: 1 − 1 = 0	

The difference between $123 and $112 = $011. The zero in this case precedes the whole number and shows there are no hundreds. When zero is the initial digit in a number it is customary not to write it. The result is $11.

PRACTICE EXERCISE 30

Subtract. Check by adding. Be sure zero is included in your answer when needed.

1.	68	2.	629	3.	2,985	4.	24,679	5.	623,481
	−43		−304		−1,383		−22,569		−212,401

6.	39	7.	843	8.	3,497	9.	38,437	10.	$4,385,000
	−25		−641		−3,100		−16,202		− 3,152,000

11.	437	12.	8,543	13.	63,439	14.	$3,795	15.	847,356
	−212		−2,123		−63,212		− 3,550		−615,301

3.3 Help from the Number Line

Subtraction can also be done on the number line. For example, $8 - 5 = ?$ can be shown as the reverse operation of addition. Draw a line from 0 to 8 units representing the amount you are starting with. From 8, draw a line 5 units long extending in the *reverse* direction. This represents the amount you are subtracting. The answer is the segment extending from 0 to 3 units.

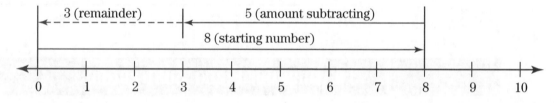

or $8 - 5 = 3$.

Can we subtract a larger number from a smaller one like $5 - 8 = ?$ Represent 5 as a line extending from 0 to 5 units and then reverse the direction of the number you are taking away by drawing a line 8 units to the left. This will extend beyond the zero of the number line:

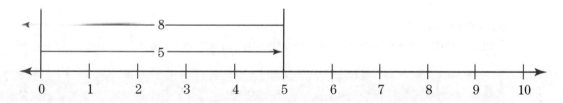

There is no positive number associated with points to the left of zero. You see that an answer to this example is impossible if you use a number line which begins at zero and extends only to the right as far as you desire.

Check to see if you understand this idea by doing the practice exercise which follows.

PRACTICE EXERCISE 31

Using the given number line for each problem, illustrate the subtraction indicated. Write the problem in column form before you begin.

1. 9 – 3 = ?

$\longleftarrow$ |———|———|———|———|———|———|———|———|———|$\longrightarrow$
  1   2   3   4   5   6   7   8   9   10

2. 8 – 6 = ?

$\longleftarrow$ |———|———|———|———|———|———|———|———|———|$\longrightarrow$
  1   2   3   4   5   6   7   8   9   10

3. 10 – 3 = ?

$\longleftarrow$ |———|———|———|———|———|———|———|———|———|$\longrightarrow$
  1   2   3   4   5   6   7   8   9   10

4. 7 – 3 = ?

$\longleftarrow$ |———|———|———|———|———|———|———|———|———|$\longrightarrow$
  1   2   3   4   5   6   7   8   9   10

5. 5 – 2 = ?

$\longleftarrow$ |———|———|———|———|———|———|———|———|———|$\longrightarrow$
  1   2   3   4   5   6   7   8   9   10

3.4 Unequal Place Subtraction

EXAMPLE

Edna saved $46 from the selling price of $758 when purchasing a sofa for cash. What was the cost of the sofa?

SOLUTION

You are subtracting a two-place number from a three-place number: $758 – $46 = ?

Problem	Step 1	Step 2	Step 3	Check
$758 – 46	$ 7 5 **8** – 4 **6** **2**	$ 7 **5** 8 – **4** 6 **1** 2	$ **7** 5 8 – 4 6 **7** 1 2	$46 +712 $758
	Subtract: 8 – 6 = 2	Subtract: 5 – 4 = 1	Subtract: 7 – 0 = 7 Consider 46 as 046.	

Thus: $758 – $46 = $712.

PRACTICE EXERCISE 32

Subtract. Check by adding.

| 1. | 35
– 4 | 2. | 368
– 41 | 3. | 546
– 32 | 4. | 4,332
– 120 | 5. | 6,747
– 43 |

Write in column form before subtracting.

6. $483 - 12 =$

7. $5,439 - 27 =$

8. $387 - 12 =$

9. $48 - 3 =$

10. $\$432,000 - \$12,000 =$

3.5 Word Problems

Before you begin, go back to "General Instructions" in Chapter 2, Section 2.7, to review the methods used to solve any word problem.

KEY WORD FOR PROBLEM SOLVING BY SUBTRACTING

Certain key words indicate that the problem will be solved by subtracting.

- *Decreased*—"What is the *decreased* cost of the sofa?"
- *Less*—"How much *less* is it from Boston than from Provincetown?"
- *Difference*—"What is the *difference* in price?"
- *Fewer*—"How many *fewer* jokes did he tell?"
- *Remaining*—"How many people are *remaining* for the weekend?"

PRACTICE EXERCISE 33

Do these problems, after you search out key words or phrases.

1. You had $48 before you went to the movies with the family. After spending $45, how many dollars remained?

2. One new camera can be purchased for $173 while another brand costs $151. How much less money does the cheaper camera cost?

3. A paintbrush which usually costs $29 is on sale this week for $21. What is the difference in price?

4. Last week the fishing boats caught 338,400 lb. of fish. This week the poundage dropped to 231,200. How many fewer pounds of fish were caught this week?

5. Mrs. Catablanco earned $48,764 in 2004. She earned only $46,560 in 2005. By how much did her salary decrease?

6. By how much do the houses shown above differ in price?

7. The town baseball team won 31 games this season while playing a total of 45 games. How many games did they lose?

8. The boat *Widgeon* now costs $20,075 but the price will rise to $20,995 next month. How much will you save by buying it now?

9. A pair of earrings sells for $42 while a necklace costs $53. How much more does the necklace cost than the earrings?

10. On Monday 43 cans of vegetables were on the shelf of the L & Z Store. By Wednesday, only 22 cans remained. How many were sold during that time?

3.6 Subtraction by Addition

Subtraction can be performed by thinking of it as a related addition problem. This is especially true when you purchase an item in a store and the salesperson gives you change from your bill. Suppose your purchase amounted to $8.78 and you paid with a $10.00 bill.

Instead of subtracting $8.78 from $10.00, some people would treat this as an addition example. They would keep adding to $8.78 until they got to $10.00. They would say, for example, $8.79, $8.80, $8.90, $9.00, $10.00—meaning that they would give you 1¢, 1¢, 10¢, 10¢, and then a $1.00 bill, or a total of $1.22 change. This is illustrated below.

8 dollars 78 cents + 1 cent = $8.79
8 dollars 79 cents + 1 cent = $8.80
8 dollars 80 cents + 10 cents = $8.90
8 dollars 90 cents + 10 cents = $9.00
9 dollars + 1 dollar = $10.00

This is sometimes a simpler way of subtracting.

EXAMPLE

How much change would you get from a 20 dollar bill after a purchase of 13 dollars and 59 cents?

SOLUTION
20 dollars – 13 dollars 59 cents = $6.41
Adding from 13 dollars 59 cents to 20 dollars you get

13 dollars 59 cents + 1 cent = $13.60
13 dollars 60 cents + 5 cents = $13.65
13 dollars 65 cents + 10 cents = $13.75
13 dollars 75 cents + 25 cents = $14.00
14 dollars + 1 dollar = $15.00
15 dollars + 5 dollars = $20.00

PRACTICE EXERCISE 34

Subtract by adding. The first one has been done for you.

1.   5 dollars – 3 dollars 74 cents = 1 dollar 26 cents

   <u> 0 </u> $5; <u> 1 </u> $1; <u> 0 </u> 50¢; <u> 1 </u> 25¢; <u> 0 </u> 10¢; <u> 0 </u> 5¢; <u> 1 </u> 1¢

2.   2 dollars – 1 dollar 59 cents =

   _____ $5; _____ $1; _____ 50¢; _____ 25¢; _____ 10¢; _____ 5¢; _____ 1¢

3.   10 dollars – 9 dollars 95 cents =

   _____ $5; _____ $1; _____ 50¢; _____ 25¢; _____ 10¢; _____ 5¢; _____ 1¢

4.   20 dollars – 11 dollars 43 cents =

   _____ $5; _____ $1; _____ 50¢; _____ 25¢; _____ 10¢; _____ 5¢; _____ 1¢

5.   7 dollars – 6 dollars 67 cents =

   _____ $5; _____ $1; _____ 50¢; _____ 25¢; _____ 10¢; _____ 5¢; _____ 1¢

3.7 Numbers Can Be Exchanged

EXAMPLE

You spend $8 from a total of $13. How much do you have left?

SOLUTION
You must subtract $8 from $13 or $13 – 8 = ?

Problem	Step 1	Step 2	Check
$13 − 8	0 13 $ 1 3 − 8	0 13 $ 1 3 − 8 5	$8 + 5 $13
	In the *ones* column you can't subtract 8 from 3, so *exchange* 1 ten in the top number for 10 ones. You have 0 tens and 13 ones.	Subtract: 13 − 8 = 5	

Thus: $13 − 8 = 5.

PRACTICE EXERCISE 35

Subtract these problems. Check by adding. As was suggested in Section 3.2, write your answers on a separate piece of paper, folding back the paper after each row, so that you can review this exercise as many times as you wish. Do not write in the book.

1. 12
  − 9
	2. 16
	  − 7
		3. 12
		  − 8
			4. 14
			  − 6
				5. 18
				  − 9

6. 11
  − 6
	7. 12
	  − 7
		8. 13
		  − 5
			9. 17
			  − 8
				10. 11
				  − 8

11. 15
  − 9
	12. 14
	  − 7
		13. 10
		  − 5
			14. 16
			  − 8
				15. 11
				  − 7

16. 14
  − 5
	17. 11
	  − 4
		18. 11
		  − 9
			19. 10
			  − 7
				20. 12
				  − 4

21.	10 – 9	22.	10 – 6	23.	10 – 4	24.	11 – 2	25.	10 – 3
26.	16 – 9	27.	13 – 8	28.	15 – 6	29.	10 – 2	30.	10 – 8
31.	14 – 8	32.	11 – 3	33.	11 – 5	34.	14 – 9	35.	13 – 7
36.	13 – 6	37.	17 – 9	38.	12 – 5	39.	12 – 3	40.	12 – 6
41.	13 – 4	42.	15 – 7	43.	10 – 1	44.	15 – 8	45.	13 – 9

3.8 Exchanging in Subtraction

The 45 examples you just completed along with the 55 examples in Practice Exercise 28 comprise the basic 100 combinations you need to know to do all subtraction examples. Be sure you know them well before continuing.

EXAMPLE

The high temperature for today was 85° and the low was 66°. What is the difference in the two temperatures?

SOLUTION
Subtract: 85° – 66° = ?

Problem	Step 1	Step 2	Step 3	Check
85 −66	7 15 8 5 − 6 6	7 15 8 5 − 6 6 9	7 15 8 5 − 6 6 1 9	66 +19 85
	You can't subtract 6 from 5, so *exchange* 1 ten in the top number for 10 ones. Instead of 8 tens and 5 ones you now have 7 tens and 15 ones.	Subtract: 15 − 6 = 9	Subtract: 7 − 6 = 1	

Thus: 85° − 66° = 19°.

PRACTICE EXERCISE 36

Subtract. Follow the method just illustrated. The first two examples have been partially completed for you.

1.   8 16
   9 6
−3 8
2.   3 12
   4 2
−2 5
3.   62
−39
4.   45
−29
5.   64
−48

6.   37
−19
7.   82
−57
8.   51
−16
9.   64
−29
10.   56
−17

EXAMPLE

Mary deposited $923 in the bank and wrote checks for $758. How much money has she left in her checking account?

SOLUTION

To find out what remains in her checking account we must subtract $758 from $923. Subtract: $923 − 758 = ?

Problem	Step 1	Step 2	Step 3	Check
$923 − 758	$\begin{array}{r} 1\ \fbox{13} \\ \$9\,2\!\!/\,\fbox{3\!\!/} \\ -\ 7\,5\,\fbox{8} \\ \hline \fbox{5} \end{array}$	$\begin{array}{r} 8\ \fbox{11}\ 13 \\ \$9\!\!/\,\fbox{2\!\!/}\,3\!\!/ \\ -\ 7\,\fbox{5}\,8 \\ \hline \fbox{6}\,5 \end{array}$	$\begin{array}{r} \fbox{8}\ 11\ 13 \\ \$\fbox{9\!\!/}\,2\!\!/\,3\!\!/ \\ -\ \fbox{7}\,5\,8 \\ \hline \$\fbox{1}\,6\,5 \end{array}$	$758 + 165 $923
	Exchange 1 ten in the top number for 10 ones since you can't subtract 8 from 3. You now have 1 ten and 13 ones instead of 2 tens and 3 ones. Subtract: 13 − 8 = 5	In the *tens* column you can't subtract 5 from 1, so exchange 1 hundred in the top number for 10 tens. You now have 8 hundreds and 11 tens rather than 9 hundreds and 1 ten. Subtract: 11 − 5 = 6	Subtract: 8 − 7 = 1	

Thus: $923 − $758 = $165.

Once you have mastered the idea of *exchanging* you can do all subtraction examples having any number of places. Look at the following examples before continuing to the practice exercise.

Problem	Step 1	Step 2	Step 3	Step 4	Check
3,814 −1,859	0 14 3 8 1̸ 4̸ − 1 8 5 9 ───── 5	7 10 14 3 8 1̸ 4̸ − 1 8 5 9 ───── 5 5	2 17 10 14 3̸ 8̸ 1̸ 4̸ − 1 8 5 9 ───── 9 5 5	2 17 10 14 3̸ 8̸ 1̸ 4̸ − 1 8 5 9 ───── 1 9 5 5	1,859 +1,955 ───── 3,814
	In the *ones* column, you can't subtract 9 from 4, so *exchange* 1 ten in the top number for 10 ones. You now have 0 tens and 14 ones rather than 1 ten and 4 ones. Subtract: $14 - 9 = 5$	In the *tens* column, you can't subtract 5 from 0, so *exchange* 1 hundred in the top number for 10 tens. You now have 7 hundreds and 10 tens rather than 8 hundreds and 0 tens. Subtract: $10 - 5 = 5$	You can't subtract 8 from 7, so exchange 1 thousand in the top number for 10 hundreds. You now have 2 thousands and 17 hundreds. Subtract: $17 - 8 = 9$		

Thus: $3,814 - 1,859 = 1,955$.

Problem	Step 1	Step 2	Step 3	Check
704 −398	6 9 14 7̷ 0̷ 4̷ − 3 9 8 ———— 6	6 9 14 7̷ 0̷ 4̷ − 3 9 8 ———— 0 6	6 9 14 7̷ 0̷ 4̷ − 3 9 8 ———— 3 0 6	398 +306 ———— 704
	In the *ones* column, you can't subtract 8 from 4, so exchange 1 ten in the top number for 10 ones. Instead of 70 tens you now have 69 tens and 14 ones. Subtract: 14 − 8 = 6	In the *tens* column, subtract: 9 − 9 = 0	Subtract: 6 − 3 = 3	

Problem	Step 1	Step 2	Step 3	Check
400 −259	3 9 10 4̷ 0̷ 0̷ − 2 5 9 ———— 1	3 9 10 4̷ 0̷ 0̷ − 2 5 9 ———— 4 1	3 9 10 4̷ 0̷ 0̷ − 2 5 9 ———— 1 4 1	259 +141 ———— 400
	You can't subtract 9 from 0, so exchange 1 ten in the top number for 10 ones. Instead of 40 tens you now have 39 tens and 10 ones. Subtract: 10 − 9 = 1	Subtract: 9 − 5 = 4	Subtract: 3 − 2 = 1	

PRACTICE EXERCISE 37

In each of the following examples illustrate the method of exchanging in subtraction.

1.   6 tens and 3 ones = 5 tens and _____ ones

2.   2 tens and 0 ones = 1 ten and _____ ones

3.   8 tens and 4 ones = 7 tens and _____ ones

4.   3 hundreds and 4 tens = 2 hundreds and _____ tens

5.   6 hundreds and 2 tens = 5 hundreds and _____ tens

Find the difference in each of the following problems. Check by adding.

6.	383 – 56	7.	996 – 48	8.	535 –286	9.	684 –455	10.	209 – 98
11.	200 – 98	12.	100 – 37	13.	1,690 – 956	14.	6,075 –2,921	15.	1,023 – 394

Write each of the following problems in column form before subtracting.

16.   5,303 – 4,729 =   17.   5,970 – 5,796 =   18.   9,895 – 5,599 =

19.   7,290 – 329 =   20.   915 – 92 =

REVIEW OF IMPORTANT IDEAS

Some important ideas covered in Chapter 3 so far are

 Subtraction is "taking away."

 The symbol – means to subtract.

 Digits having the same place values are lined up in columns before subtracting.

 Numbers are subtracted by pairs only.

 We *exchange* when it is impossible to subtract in any column.

3.9 More Difficult Word Problems Requiring Subtraction

Before we attempt to solve any more problems in mathematics, let's look again at some special hints that may help us become better problem solvers.

1.   Develop habits of careful reading in math. These habits result from reading critically (with questioning mind) and from practice. This book gives much practice in reading directions, explanations, and problems.

2.   Do not be alarmed if your reading rate in mathematics is slow—it should be. Often you must reread to be sure you understand. Directions should be read slowly so that you know what is expected. Many people skip over directions quickly because they are eager to start on the problems. But what a great waste it is when work is done incorrectly because the directions were ignored or read carelessly. Directions often include sample problems to show how the work is done and how the answer is to be chosen.

3.   Read mathematics with thought and questions. As you read, ask yourself:

   1.   What is given? (facts in problem)
   2.   What is unknown? (question to be answered)
   3.   How should this be solved? (method to be used to find the answer)

This is a much different style of reading from what you might use when you look through a newspaper or magazine. In mathematics, directions, explanations, and problems are compressed into very few words. Each word is therefore very important and cannot be overlooked by speedy reading.

Try to keep some of these suggestions in mind when you approach this problem and all others that follow.

EXAMPLE

Mary earned $435 last week for 5 days of work. This week her paycheck was $462 since she worked some overtime hours. How much more did she earn this week than last week?

SOLUTION
1. What is given?
   You are told Mary's two earnings, $435 for last week and $462 this week.
2. What is unknown?
   You are asked to find out how much more she earned this week than last week.
3. How should this be solved?
   The problem asks for the *difference* in the two paychecks, so subtract $435 from $462.

$$\begin{array}{r} \$462 \\ -\ 435 \\ \hline \$\ 27 \end{array}$$

Mary earned $27 more this week than last week.

PRACTICE EXERCISE 38

1. You buy a wallet for $38 and give the cashier a $50 bill. How much change do you get?

2. A man weighed 218 lb. before he dieted. If he lost 39 lb., how much does he now weigh?

3. A lawn mower usually sells for $498. Its sale price is $449. How much would you save buying it during the sale?

4. The local paper announced that 7,500 lb. of haddock were caught this week and 3,600 lb. last week. How many more pounds were caught this week?

5. Mr. Smith buys a new house for $155,125 and sells his old house for $147,750. How much more does his new house cost?

6. A cashier is given $85 at the start of her shift to help her make change. At the end of her shift, she counts her money and finds that it totals $1,804. How much money did she collect during her shift?

7. Two sides of a triangle are 27 ft. and 32 ft. Find the length of the third side if the perimeter is 100 ft. (*Hint:* First add the two known sides together before subtracting the sum from the perimeter.)

8. The distance from Omaha to Tulsa to Houston is 510 mi. If the distance from Omaha to Tulsa is 385 mi., how far is it from Tulsa to Houston?

9. An automobile company produced 8,634 cars in February. How many cars did it produce in March if the monthly output was 796 cars fewer than in February?

10. A dance was held in a social hall on two successive Saturday nights. If the attendance on the first night was 240 couples and on the second night was 229 couples, how many more couples appeared on the first night?

LET'S SEE HOW YOUR SKILL IN SUBTRACTION HAS IMPROVED

This chapter proceeded from simple to more complex subtraction examples. You have probably found your past skills returning and understand the problems better than you did before. The chapter test will help check your progress.

A Score of	Means That You
18–20	Did very well. Move on to Chapter 4.
16–17	Know the material except for a few points. Reread the section about the ones you missed.
13–15	Need to check carefully the sections you missed.
0–12	Need to review the chapter again to refresh your memory and improve your skills.

Chapter Test 3

Write your answer in the spaces provided. Use a calculator to check your answers after you have completed the entire chapter test.

1. 9
   −2

2. 94
   −73

3. 563
   −361

4. 478
   −120

5. 78
   − 5

6. 365
   − 43

7. 13
   − 7

8. 65
   − 9

9. 436
   −282

10. 743
    −485

11. 8,076
    −2,341

12. 3,004 − 2,877 =

13. 5,000
    − 309

14. What is the difference between 63¢ and 96¢?

15. If a dress manufacturer ships 2,390 dresses as a partial order of 3,265 dresses, how many more must he ship?

16. How much change do you receive from a 20-dollar bill if your purchases total 13 dollars 56 cents?

17. From 1,812 subtract 345.

18. If 101 ft. of wire is needed to enclose a triangular plot of ground, find the length of the third side if the sum of the other two sides is 66 ft.

19.   This house has been reduced in price to $249,995. How much money can be saved buying the house at the new price?

$254,000

20.   Show the result of subtracting 3 from 7 on the number line below.

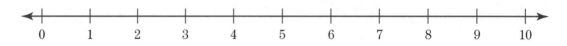

HOLD IT!

Now that you have reviewed all the combinations in subtraction, go back to Practice Exercises 28 and 35 and see how well you can now do on the 100 combinations. Can you make a perfect score?

PUZZLE TIME

Did you enjoy the other cross-number puzzle? Here is one which will review *subtraction of whole numbers*. The solution is on page 430.

ACROSS

1. $469 - 234 =$
2. $78 - 19 =$
4. $532 - 278 =$
6. $481 - 206 =$
8. $4,610 - 2,845 =$
10. $300 - 162 =$
12. $82 - 64 =$
13. $106 - 89 =$
16. $704 - 589 =$
18. $2,811 - 1,993 =$
22. $201 - 185 =$
23. $100 - 48 =$
26. $482 - 293 =$
27. $23,051 - 15,838 =$
29. $2,000 - 1,367 =$
30. $2,005 - 1,755 =$
31. $83 - 6 =$
32. $602 - 197 =$

DOWN

1. $790 - 589 =$
3. $1,071 - 143 =$
4. $8,970 - 6,390 =$
5. $6,460 - 2,347 =$
7. $100 - 48 =$
9. $250 - 189 =$
11. $500 - 463 =$
14. $63 + 37 - 9 =$
15. $57 - 19 =$
17. $81 - 23 =$
19. $71 - 57 =$
20. $4,040 - 2,348 =$
21. $11,068 - 2,498 =$
22. $97 - 79 =$
24. $100 - 78 =$
25. $71 - 35 =$
26. $340 - 203 =$
28. $1,827 - 1,432 =$

Multiplication– Remember Those Times?

nother basic operation in mathematics is multiplication. We use multiplication when we want to find the price of more than one item. The cost of 2 coffees at 93¢ a cup is two times 93¢. The cost of 3 doughnuts at 55¢ a doughnut is three times 55¢.

At first glance you would say it looks like addition. You are right. Multiplication is really repeated addition, but by a faster method. For example, 2 times 93 means 93 + 93. In an example such as 2 times 93, adding and multiplying are equally fast. In an example such as 8 times 28, addition is a slow process compared to multiplication. The larger the numbers, the more useful multiplication becomes.

Pretest 4–See What You Know and Remember

This test gives you a chance to see how good you are in multiplication. Do all your work in the spaces provided.

1. $\begin{array}{r} 7 \\ \times 5 \\ \hline \end{array}$	2. $\begin{array}{r} 94 \\ \times 2 \\ \hline \end{array}$	3. $\begin{array}{r} 212 \\ \times 4 \\ \hline \end{array}$	4. $\begin{array}{r} 73 \\ \times 3 \\ \hline \end{array}$	5. $\begin{array}{r} 402 \\ \times 6 \\ \hline \end{array}$
6. $\begin{array}{r} 330 \\ \times 2 \\ \hline \end{array}$	7. $\begin{array}{r} 35 \\ \times 12 \\ \hline \end{array}$	8. $\begin{array}{r} 43 \\ \times 34 \\ \hline \end{array}$	9. $\begin{array}{r} 56 \\ \times 47 \\ \hline \end{array}$	10. $\begin{array}{r} 306 \\ \times 93 \\ \hline \end{array}$
11. $\begin{array}{r} 510 \\ \times 70 \\ \hline \end{array}$	12. $\begin{array}{r} 703 \\ \times 30 \\ \hline \end{array}$	13. $\begin{array}{r} 500 \\ \times 200 \\ \hline \end{array}$	14. $\begin{array}{r} 378 \\ \times 98 \\ \hline \end{array}$	

15.   How many cents do 6 cans of juice at 93¢ a can cost?

16.   Find the product of 3×3 on the number line.

17.   A weekly newspaper costs 75¢. If you purchase the paper each week for 50 weeks, what will be your total cost in cents?

18.   A family wishes to purchase MP3 players and cell phones from a local electronics store. Each MP3 player is priced at $169, and the cell phones are priced at $99 apiece. What would be the cost of 2 MP3 players and 4 cell phones?

19.   A car travels 9 mi. for each gallon of gasoline. If you purchase 16 gal. of gasoline, how many miles can you travel?

20.   A dress manufacturer makes and sells dresses for $82 each. If he packs one to a box and ships 1,032 boxes, how much money will he receive?

21.   Martin is covering his rectangular kitchen floor with linoleum. How many square feet of material must he purchase if his room measures 6 ft. by 13 ft.?

22.   Using the formula for the perimeter of a rectangle, $P = 2L + 2W$, find P if $L = 14$ in. and $W = 8$ in.

Check your answers by turning to the back of the book. Add up all that you had correct.

A Score of	Means That You
19–22	Did very well. You can move to Chapter 5.
17–18	Know this material except for a few points. Read the sections about the ones you missed.
14–16	Need to check carefully on the sections you missed.
0–13	Need to work with this chapter to refresh your memory and improve your skills.

Questions	Are Found in Section
16	4.2
1	4.4
2–6	4.5
15, 18, 19	4.6
7–14	4.7
21, 22	4.8
17, 20	4.9

4.1 Multiplication Is Repeated Addition

You may be wondering why we said that *multiplication is repeated addition*. Let's look into the meaning of this statement.

When you add the same term a given number of times—for example, "Add 6 three times"—the problem is written as

$$
\begin{array}{r}
6 \\
6 \\
+6 \\
\hline
18
\end{array}
$$

Since 6 is an addend *three* times, you can write it as a *multiplication* example:

$$
\begin{array}{r}
6 \\
\times 3 \\
\hline
18
\end{array}
$$

This is pictured as 3 groups of 6 items:

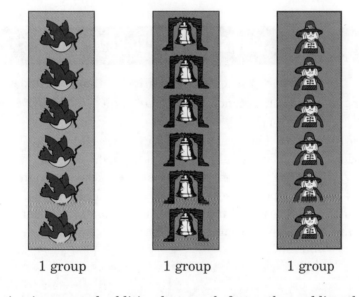

1 group   1 group   1 group

Multiplication is repeated addition but much faster than adding the same number many times. The symbol × is called the *times* sign, so whenever you see that sign you know to perform the operation of multiplication.

There is a language for multiplication too. Look at this example:

$$
\begin{array}{r}
6 \leftarrow \text{multiplicand} \\
\times 3 \leftarrow \text{multiplier} \\
\hline
18 \leftarrow \text{product}
\end{array}
$$

The *times* sign tells us to multiply ───────→

If you were to say, "Add 3 six times," you would write it as $3 + 3 + 3 + 3 + 3 + 3$. Since 3 is the addend six times, you can write it as a multiplication example:

$$3 \leftarrow \text{multiplicand}$$
$$\times 6 \leftarrow \text{multiplier}$$
$$\overline{18} \leftarrow \text{product}$$

This is pictured as 6 groups of 3 items.

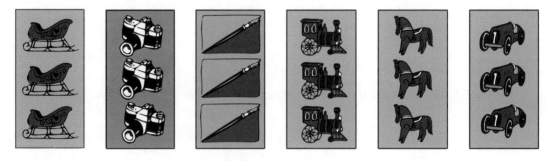

Therefore, $3 \times 6 = 6 \times 3$, since each product is 18. *Changing the order of the numbers doesn't change the value of their product.*

TERMS YOU SHOULD REMEMBER

Multiply   To perform repeated addition of the same addend using as many of the addends as there are units in another number (the multiplier).

Multiplication   The act of multiplying.

Multiplicand   The number to be multiplied.

Multiplier   The number which tells how many times an addend is to be used in repeated addition.

Product   The result of multiplying two or more numbers together.

Times   The sign $\times$ of multiplication.

4.2 Help from the Number Line

One number can be added to another and one number can be subtracted from another. This joining of two things is called a *binary* operation. Since you also multiply two numbers together at a time, *multiplication* is a binary operation.

Multiplication can be shown on the number line. To multiply 3×2, represent this as a repeated addition of $2 + 2 + 2$ on the number line. Beginning at 0, draw a line segment 2 units to the right. From that point continue with another line segment also 2 units in length. This now brings you to a value corresponding to the point 4. Complete the addition of the three terms at the point which is 6 units to the right of 0. Thus the product of $3 \times 2 = 6$.

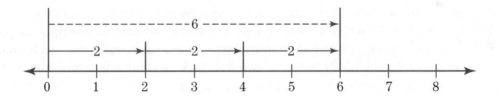

Multiplying 2×3 is pictured on the number line below. It is the repeated addition of 3 ＋ 3. As you see, the final result is 6, the same as 3×2. This further illustrates that changing the order of the numbers doesn't change the value of the product.

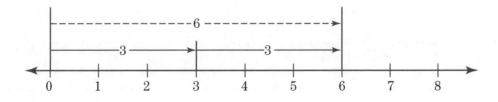

PRACTICE EXERCISE 39

Illustrate each of the following multiplication problems on a number line and find the product.

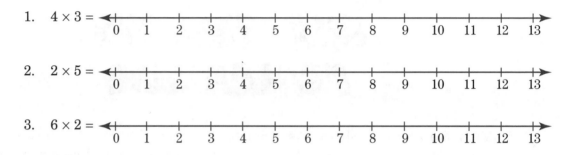

1. $4 \times 3 =$

2. $2 \times 5 =$

3. $6 \times 2 =$

Does this mean that you can't multiply more than two numbers together? No, it doesn't mean that. When you added more than two numbers, you added two together and then added another to that sum. You do the same thing when multiplying.

For example, what is the product of $2 \times (3 \times 4)$?

First, find the product of the 3×4 inside the parentheses and then multiply that product by 2. Keep in mind that multiplication is a *binary operation*—an operation performed on two numbers at a time. For instance, in the above example multiply two of the numbers first, get a product, multiply that product by the next number and get a final product. This could go on indefinitely but only two numbers are multiplied together at the same time.

Let's find the result of $2 \times (3 \times 4)$, using the number line. First, 3×4:

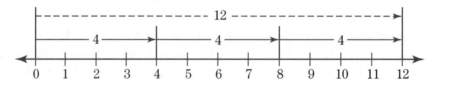

Since $3 \times 4 = 12$, find 2×12:

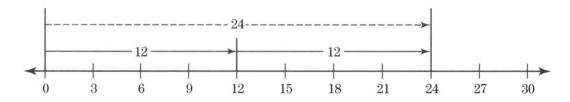

Thus you see that $3 \times 4 = 12$ and $2 \times 12 = 24$. Therefore,

$$2 \times (3 \times 4) = 24$$

Will the order of the multiplications matter? Suppose you were seeking an answer to $(2 \times 3) \times 4$. Here you start with the product of 2×3 inside the parentheses and then multiply that product by 4. First, 2×3:

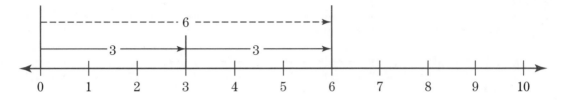

Since $2 \times 3 = 6$, find 6×4:

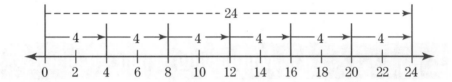

Thus you see that $2 \times 3 = 6$ and $6 \times 4 = 24$. Therefore,

$$(2 \times 3) \times 4 = 24$$

If three numbers are multiplied together, the product will be the same, regardless of the grouping order. That is,

$$(2 \times 3) \times 4 = 2 \times (3 \times 4)$$

PRACTICE EXERCISE 40

Illustrate the truth of each of the following problems on the number line and find the product.

1. $3 \times (1 \times 4) = (3 \times 1) \times 4$

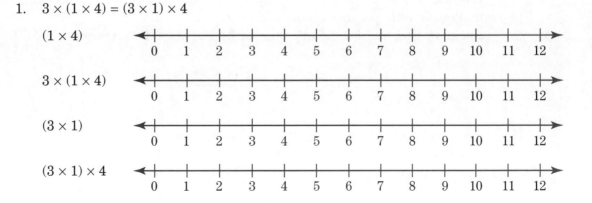

2. $(2 \times 2) \times 3 = 2 \times (2 \times 3)$

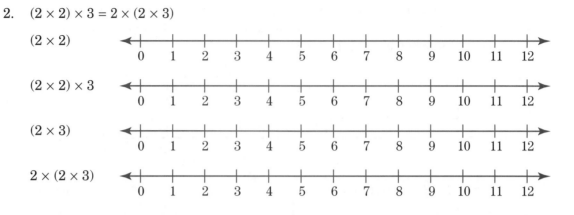

4.3 Multiplying Whole Numbers

EXAMPLE

Marilyn's son David receives an allowance of $5 a week, walks a dog for $10 a week, and baby-sits two evenings a week for which he earns $19 a week. The other day, David decided to figure out his yearly earnings. Can you help him? (Remember, there are 52 weeks in a year.)

SOLUTION

David knew that he could find the answer in a simple way. He could write $5 + 5 + 5 \ldots$ in column form 52 times, and $10 + 10 + 10 \ldots 52$ times and finally $19 + 19 + 19 \ldots 52$ times. But he knew that repeated additions, such as adding $5 a total of 52 times, can be more easily done by multiplication. He set up these three examples:

$$\$\ 5 \times 52 = ?$$
$$\$10 \times 52 = ?$$
$$\$19 \times 52 = ?$$

However, before you start to multiply large numbers it is best to review the simple multiplication facts you are already aware of.

You know that multiplying any number by 1 gives you a product equal to the number.

$$\begin{array}{r} 1 \\ \times 3 \\ \hline \end{array} = \overbrace{1 + 1 + 1}^{3 \text{ terms}} = 3$$

or

$$\begin{array}{r} 4 \\ \times 1 \\ \hline \end{array} = \overbrace{4}^{1 \text{ term}} = 4$$

In the same manner, multiplying by 0 is also a special case, since any number times 0 is always 0.

$$\begin{array}{r} 0 \\ \times 6 \\ \hline \end{array} = \overbrace{0 + 0 + 0 + 0 + 0 + 0}^{6 \text{ terms}} = 0$$

or

$$\begin{array}{r} 4 \\ \times 0 \\ \hline \end{array} = \overbrace{0}^{0 \text{ terms}} = 0$$

This practice exercise contains only the special multipliers of 1 and 0. Even though these examples are simple, do them carefully.

PRACTICE EXERCISE 41

Place the product under each example. Try for a perfect score.

1.	2.	3.	4.	5.	6.
1 ×0	0 ×7	6 ×0	0 ×2	5 ×0	0 ×6

7.	8.	9.	10.	11.	12.
0 ×3	4 ×0	0 ×9	3 ×0	2 ×1	1 ×1

13.	14.	15.	16.	17.	18.
0 ×1	4 ×1	9 ×1	1 ×5	8 ×0	0 ×8

19.	20.	21.	22.	23.	24.
0 ×0	9 ×0	0 ×5	7 ×0	3 ×1	1 ×8

25. 1	26. 1	27. 0	28. 1	29. 5	30. 2
×2	×9	×4	×3	×1	×0

31. 1	32. 1	33. 8	34. 6	35. 1	36. 7
×6	×4	×1	×1	×7	×1

4.4 Those Old Combinations Again

Skill and speed in multiplication depend very much on knowing how numbers combine. There are basic number combinations using all the symbols from 0 to 9. These are well worth practicing to gain skill because they save you so much work when you know them. They are sometimes called the multiplication tables.

There are 10 combinations for each number, or 100 altogether. Thirty-six of them were done by you in the last practice exercise. With the 64 more in this next exercise, you will have learned the hundred combinations in multiplication.

Once again we suggest that you do not write the answers in the book. Write your answers on a separate piece of paper so you can use this exercise for practice as many times as you wish. Don't write the problem. Place your paper under the first row and write each answer. Then fold back the paper and place it under the next row. Write the answers, fold back the paper, and so forth, until you have all eleven rows. The idea is to see how you can improve by recognizing the answers more and more quickly. Keep your own record of this.

PRACTICE EXERCISE 42

Find the product for each of the following problems. Remember not to write in the book, so that you can practice again and again.

1. 6	2. 2	3. 7	4. 5	5. 7	6. 6
×5	×9	×2	×9	×8	×4

7. 6	8. 6	9. 8	10. 7	11. 7	12. 9
×7	×9	×6	×7	×9	×7

13. 8
    ×5
    —

14. 4
    ×4
    —

15. 5
    ×6
    —

16. 3
    ×7
    —

17. 4
    ×9
    —

18. 5
    ×4
    —

19. 8
    ×3
    —

20. 7
    ×6
    —

21. 7
    ×3
    —

22. 4
    ×7
    —

23. 9
    ×9
    —

24. 8
    ×8
    —

25. 9
    ×3
    —

26. 5
    ×7
    —

27. 5
    ×8
    —

28. 4
    ×6
    —

29. 9
    ×4
    —

30. 3
    ×9
    —

31. 2
    ×8
    —

32. 4
    ×8
    —

33. 6
    ×6
    —

34. 2
    ×5
    —

35. 3
    ×5
    —

36. 8
    ×7
    —

37. 7
    ×4
    —

38. 9
    ×6
    —

39. 5
    ×5
    —

40. 9
    ×2
    —

41. 9
    ×8
    —

42. 5
    ×3
    —

43. 8
    ×9
    —

44. 2
    ×7
    —

45. 6
    ×8
    —

46. 2
    ×2
    —

47. 8
    ×4
    —

48. 3
    ×2
    —

49. 2
    ×3
    —

50. 9
    ×5
    —

51. 2
    ×4
    —

52. 3
    ×3
    —

53. 4
    ×5
    —

54. 5
    ×2
    —

55. 3
    ×6
    —

56. 3
    ×8
    —

57. 8
    ×2
    —

58. 7
    ×5
    —

59. 6
    ×2
    —

60. 3
    ×4
    —

61. 2
    ×6
    —

62. 4
    ×3
    —

63. 6
    ×3
    —

64. 4
    ×2
    —

4.5 Multiplying by a One-Digit Number

EXAMPLE

At the year-end sale, Mrs. De Vogt bought 3 needlepoint designs for $12 each. How much did they cost her?

SOLUTION

The problem is simply the repeated addition of $12 + $12 + $12, or $36. You know that a problem of this type can be done more easily and more quickly by multiplication. How do you multiply 3 × $12?

The number 12 can be expressed as 10 + 2; and 3 × 12 can be written as 3 × (10 + 2). The product is 36. If you look at this problem from a different angle, you see that 3 × 10 = 30 and 3 × 2 = 6; the sum of these two products 30 + 6 = 36.

To multiply, you find the product of 3 and each number in the multiplicand, 12.

Problem	Step 1	Step 2	Step 3
12 × 3	1 2 × 3 ――― 6	1 2 × 3 ――― 6 3 0 ―――	1 2 × 3 ――― 6 +3 0 ――― 3 6
	Multiply: 3 × 2 = 6 This is called a *partial* product.	Multiply: 3 × 1 (Recall that a 1 in the *tens* place is a 10.) This is the other *partial* product. 3 × 10 = 30	Add the partial product together: 30 + 6 = 36

Thus: 12 × 3 = 36.

Now that you know the *why* for multiplying numbers, let's look for a quicker method of multiplying.

Problem	Step 1	Step 2
12 × 3	1 2 × 3 —— 6	1 2 × 3 —— 3 6
	Multiply: 3 × 2 = 6 Place the product in the *ones* column.	Multiply by 3 the 1 in the *tens* column: 3 × 1 = 3 Place your answer in the *tens* column.

PRACTICE EXERCISE 43

Find each product. Work slowly and carefully, using the sample method just demonstrated.

1.	43 × 2	2.	12 × 4	3.	20 × 3	4.	31 × 3	5.	24 × 2

6.	11 × 8	7.	33 × 2	8.	78 × 1	9.	10 × 7	10.	65 × 1

EXAMPLE

Billy earned $13 a day at an after-school job. After seven work days, how much had he earned?

SOLUTION

The problem is to multiply 13×7.

Problem	Step 1	Step 2	Step 3
13 × 7	1 3 × 7 —— 2 1	1 3 × 7 —— 2 1 7 0 ——	1 3 × 7 —— 2 1 +7 0 —— 9 1
	Multiply: $7 \times 3 = 21$ This is the first *partial* product.	Multiply: $7 \times 10 = 70$ This is the second *partial* product.	Add the partial products: $70 + 21 = 91$

Thus: $13 \times 7 = 91$.

Again we look for an easier method of multiplying. Recall that when we found a sum greater than 10 in any column we *carried* to the next column. This same method can be applied to multiplication also, as shown in the following example:

Problem	Step 1	Step 2
13 × 7	2 1 3 × 7 —— 1	2 1 3 × 7 —— 9 1
	Multiply the 7 times the ones number 3: $7 \times 3 = 21$ Write the 1 in the *ones* column and *carry* the 2 to the *tens* column.	Multiply the 7 times the tens number 1: $7 \times 1 = 7$ Add on the carried number 2; $7 + 2 = 9$ Write the 9 in the *tens* column.

Thus: $13 \times 7 = 91$.

The same method can be used whenever you are multiplying by a single-digit number. It doesn't matter how large or small your multiplicand may be. Just remember that you multiply first and then add on, if necessary, the carried number.

Look over the next two examples before you attempt the practice exercise on page 104.

Problem	Step 1	Step 2	Step 3
125 × 6	3 125 × 6 ——— 0	1 3 125 × 6 ——— 5 0	1 3 125 × 6 ——— 7 5 0
	Multiply: $6 \times 5 = 30$ Place the 0 in the *ones* column and carry the 3 to the *tens* place.	Multiply: $6 \times 2 = 12$ Add the 3 you carried: $12 + 3 = 15$ Write the 5 in the *tens* column and and carry the 1 to the *hundreds* column.	Multiply: $6 \times 1 = 6$ Add the 1 you carried: $6 + 1 = 7$ Write the 7 in the *hundreds* column.

Thus: $125 \times 6 = 750$.

Problem	Step 1	Step 2	Step 3	Step 4
1,306 × 7	4 1306 × 7 ——— 2	4 1306 × 7 ——— 4 2	2 4 1306 × 7 ——— 1 4 2	2 4 1306 × 7 ——— 9 1 4 2
	Multiply: $7 \times 6 = 42$ Write the 2 in the *ones* column and carry the 4.	Multiply: $7 \times 0 = 0$ Add the 4 you carried: $0 + 4 = 4$ Write the 4 in the *tens* column.	Multiply: $7 \times 3 = 21$ Write the 1 in the *hundreds* column and carry the 2.	Multiply: $7 \times 1 = 7$ Add the 2 you carried: $7 + 2 = 9$ Write the 9 in the *thousands* column.

Thus: $1,306 \times 7 = 9,142$.

EXAMPLE

Let's return to the example in section 4.3. You are now able to help David figure his yearly earnings:

SOLUTION

$$
\begin{array}{r}
52 \\
\times\ 5 \\
\hline
260
\end{array}
\qquad\qquad
\begin{array}{r}
52 \\
\times\ 10 \\
\hline
520
\end{array}
\qquad\qquad
\begin{array}{r}
\scriptstyle 1 \\
52 \\
\times\ 19 \\
\hline
468 \\
52 \\
\hline
988
\end{array}
$$

$$
\begin{array}{r}
\$260 \\
520 \\
+\ 988 \\
\hline
\end{array}
$$

$1,768 his total earnings for the year

Note: The method for finding the last product will be covered in Section 4.7.

PRACTICE EXERCISE 44

Find the product for each of the following examples. The first five examples have been partially done for you.

1.
$$
\begin{array}{r}
\scriptstyle 4 \\
99 \\
\times\ 5 \\
\hline
5
\end{array}
$$
2.
$$
\begin{array}{r}
\scriptstyle 2 \\
77 \\
\times\ 4 \\
\hline
8
\end{array}
$$
3.
$$
\begin{array}{r}
\scriptstyle 4 \\
38 \\
\times\ 6 \\
\hline
8
\end{array}
$$
4.
$$
\begin{array}{r}
\scriptstyle 2 \\
58 \\
\times\ 3 \\
\hline
4
\end{array}
$$
5.
$$
\begin{array}{r}
\scriptstyle 1 \\
42 \\
\times\ 9 \\
\hline
8
\end{array}
$$

6.
$$
\begin{array}{r}
282 \\
\times\ 3 \\
\hline
\end{array}
$$
7.
$$
\begin{array}{r}
129 \\
\times\ 3 \\
\hline
\end{array}
$$
8.
$$
\begin{array}{r}
245 \\
\times\ 7 \\
\hline
\end{array}
$$
9.
$$
\begin{array}{r}
307 \\
\times\ 8 \\
\hline
\end{array}
$$
10.
$$
\begin{array}{r}
773 \\
\times\ 7 \\
\hline
\end{array}
$$

11.
$$
\begin{array}{r}
1,038 \\
\times\ 5 \\
\hline
\end{array}
$$
12.
$$
\begin{array}{r}
976 \\
\times\ 6 \\
\hline
\end{array}
$$
13.
$$
\begin{array}{r}
540 \\
\times\ 7 \\
\hline
\end{array}
$$
14.
$$
\begin{array}{r}
165 \\
\times\ 5 \\
\hline
\end{array}
$$
15.
$$
\begin{array}{r}
2,389 \\
\times\ 8 \\
\hline
\end{array}
$$

REVIEW OF IMPORTANT IDEAS

Some important ideas covered in Chapter 4 so far are:

 Multiplication is "repeated addition."

 The symbol × means to multiply.

 Numbers are multiplied by pairs only.

 The order of the multiplicand and multiplier doesn't affect the product.

 For a product greater than 10, carry the tens number to the next column and add it to the next product.

4.6 Word Problems

Review section 2.7 in Chapter 2 for general instructions needed to solve any word problem. The following are specific instructions for solving a multiplication example.

KEY WORDS OR PHRASES FOR PROBLEMS SOLVED BY MULTIPLYING

There are certain key words or phrases that indicate the problem will be solved by multiplication. Recall that multiplication means repeated *addition*, so all those key words can be used again. You must recognize that the problem contains *repeated* additions and that you will obtain a faster result by multiplication.

- *Product*—"Find the *product* of 46 times 9."
- *Times*—"Three *times* as many adults attended as children."
- *How many* and *how much*—"*How many* apartments are there in the building?" or "*How much* did it cost?"

PRACTICE EXERCISE 45

Search out the key words or phrases and then do each problem.

1. For the past 6 days in a row, I played tennis. If the cost is $25 a day, how much did it cost?

2. There are 6 floors in my apartment building and 8 apartments on each floor. How many apartments are there in my building?

3. A family of 4 went to a movie. What was the cost of admission if it was $8 for each person?

4. A car travels 18 mi. on a gallon of gasoline. How many miles can it travel on 8 gal. of gasoline?

5. Find the product of 46 times 9.

6. If a certain fabric costs $9 a yard, how much would 35 yards cost?

7. If 1,063 people visited the shrine each day, how many people visited the shrine in 7 days?

8. A restaurant has 246 tables. If each table seats 4 people and the restaurant is filled at 7:00 P.M., how many people are seated at the tables?

9. Last week 163 children went to the circus on Thursday night. Three times as many children went on Saturday afternoon. How many children were there on Saturday afternoon?

10. A concert sells out at 4,873 tickets. How many tickets would be sold if 4 performances sold out?

4.7 Multiplying by a Number Having Two or More Digits

Multiplication of this type is more involved but not more difficult. You will continue using all the laws you have learned concerning the multiplication of whole numbers.

For example, to find the product of 31 times 18, you could add 18 thirty-one times or add 31 eighteen times. You know that multiplication is an easier way of doing the example, so let's multiply 31 × 18.

Problem	Step 1	Step 2	Step 3
31 ×18	31 × 18 ——— 248	31 ×18 ——— 248 310	31 × 18 ——— 248 + 310 ——— 558
	Multiply: 8 × 31 = 248	Multiply: 10 × 31 = 310 (Remember the 1 in the *tens* place is 10.)	Add the partial products: 248 + 310 = 558

Thus: 31 × 18 = 558.

Notice that 31 × 10 = 310. An earlier example on page 106 showed that 7 × 10 = 70. Do you know the product of 15 × 10? It is 150. Multiplying by 10 can be accomplished by simply adding a *zero* to the right of the multiplicand. Look at these examples:

$$
\begin{array}{r} 36 \\ \times 10 \\ \hline 360 \end{array}
\qquad
\begin{array}{r} 493 \\ \times\ 10 \\ \hline 4{,}930 \end{array}
\qquad
\begin{array}{r} 1{,}063 \\ \times\ \ 10 \\ \hline 10{,}630 \end{array}
$$

To multiply by 100, multiply the same way as you did by 10, except you must add *two zeros* to the right of the multiplicand.

$$
\begin{array}{r} 36 \\ \times 100 \\ \hline 3600 \end{array}
\qquad
\begin{array}{r} 493 \\ \times\ 100 \\ \hline 49{,}300 \end{array}
\qquad
\begin{array}{r} 1{,}063 \\ \times\ \ 100 \\ \hline 106{,}300 \end{array}
$$

The same method is used for any multiplication of 10; 100; 1,000; 10,000; etc. For a multiplier of 20 you proceed as if you multiplied by a 2 and then a 10. For example:

Problem	Step 1	Step 2
36 ×20	36 ×20 — 72	36 × 20 — 720
	Multiply: $36 \times 2 = 72$	Multiplying by 10 results in adding a *zero* to the 72.

Thus: $36 \times 20 = 720$.

Let's try this short practice exercise first before continuing. Do it slowly and carefully.

PRACTICE EXERCISE 46

Multiply. The first three examples have been partially completed.

1. $14 \times 10 = 14$_____

2. $56 \times 100 = 56$_____

3. $23 \times 20 = 46$_____

4. $124 \times 100 =$

5. $36 \times 30 =$

6. $51 \times 70 =$

7. $306 \times 90 =$

8. $300 \times 20 =$

9. $123 \times 10 =$

10. $500 \times 200 =$

Let's turn to another example.

Problem	Step 1	Step 2	Step 3
47 ×39	4 7 × 3 9 ——— 4 2 3	4 7 × 3 9 ——— 4 2 3 1 4 1 0	4 7 × 3 9 ——— 4 2 3 +1 4 1 0 ——— 1,8 3 3
	Multiply: 9 × 47 = 423	Multiply: 30 × 47 = 1,410	Add the partial products: 423 + 1,410 = 1,833

Problem	Step 1	Step 2	Step 3
146 × 73	1 4 6 × 7 3 ——— 4 3 8	1 4 6 × 7 3 ——— 4 3 8 1 0 2 2 0	1 4 6 × 7 3 ——— 4 3 8 +1 0 2 2 0 ——— 1 0 6 5 8
	Multiply: 3 × 146 = 438	Multiply: 70 × 146 = 10,220	Add the partial products: 438 + 10,220 = 10,658

Our goal is to do a multiplication example as simply and as speedily as possible. Look at the same problem again, but this time let's do it in our heads.

It would look like this

```
      146
    × 73
    ─────
      438
  10 22
  ───────
  10,658
```

It would sound like this

Step 1: 3 × 6 = 18. Write the 8 and carry the 1.

Step 2: 3 × 4 = 12. Add the 1 you carried: 12 + 1 = 13. Write the 3 and carry the 1.

Step 3: 3 × 1 = 3. Add the 1 you carried: 3 + 1 = 4. Write the 4.

Step 4: 7 × 6 = 42. Write the 2 directly under the number you multiplied by (7) and carry the 4.

Step 5: 7 × 4 = 28. Add the 4 you carried: 28 + 4 = 32. Write the 2 and carry the 3.

Step 6: 7 × 1 = 7. Add the 3 you carried: 7 + 3 = 10. Write the 10.

Step 7: Add to find the final answer.

Practice will help you achieve better results. All of the above will help you to understand the process involved in the multiplication of large numbers. If you find you are running into difficulties doing the next exercise, go back to the illustrated examples in this section, reread them, and then try the practice exercise again.

PRACTICE EXERCISE 47

Multiply.

1. 58
    ×43

2. 39
    ×28

3. 56
    ×35

4. 92
    ×87

5. 61
    ×47

6. 830
    × 34

7. 605
    × 12

8. 391
    ×109

9. 172
    × 65

10. 989
    × 81

11. 1,006
    ×   39

12. 650
    × 45

13. 348
    ×290

14. 6,204
    × 109

15. 9,992
    ×   88

4.8 Multiplication Applied to Measuring

AREA

EXAMPLE

Reed family has an L-shaped living room with dimensions as illustrated in the diagram below. They decided to cover the floor with wall-to-wall carpeting. They found that carpeting cost $14 a square yard in a special half-price sale. How much did it cost to have the room carpeted?

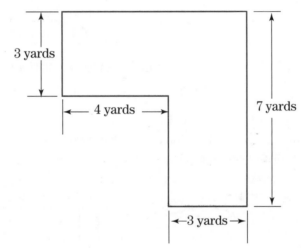

SOLUTION

If you wish to find the measurement of the interior of a figure, then you must find its *area*. To find the area, you must know the number of square units required to cover it or the number of square units the region contains. In our problem, you would use a square yard like this to find the area.

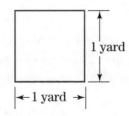

The L-shaped living room can be subdivided into two rectangles, as in the following diagram:

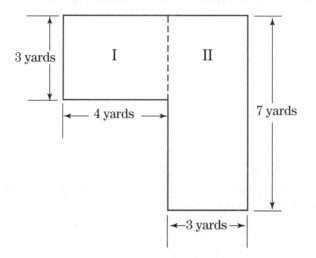

The problem is then simplified to finding the area of the two rectangles marked I and II above.

Thus, the area of Rectangle I is 12 square yards (yd.²), since this is what is needed to cover the interior of the rectangle. This could also be done by finding the product of the length and the width of the rectangle.

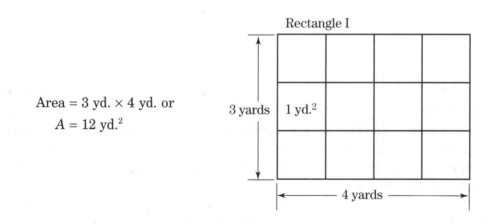

Area = 3 yd. × 4 yd. or

$A = 12$ yd.²

Recall that earlier in the book a statement of this type was called an *equation*. Let's express this in a more general manner: *The area of a rectangle is equal to the product of the length and width*, with both dimensions expressed in the same unit of measurement, or

Area = length × width
or in simplified form
$A = L \times W$

This equation is stated in letters rather than numbers and is called a *formula*. Let's use this formula to find the area of rectangle II.

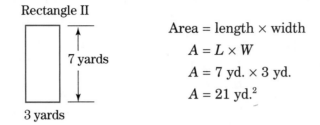

Rectangle II

7 yards

3 yards

Area = length × width

$A = L \times W$

$A = 7 \text{ yd.} \times 3 \text{ yd.}$

$A = 21 \text{ yd.}^2$

The area of the L-shaped living room is the sum of the areas of the two rectangles.

Area = rectangle I + rectangle II

$A = 12 \text{ yd.}^2 + 21 \text{ yd.}^2$

$A = 33 \text{ yd.}^2$

Since carpeting costs $14 a square yard, you must find the product of 14 × 33 to know the total cost.

$$\begin{array}{r} 33 \\ \times 14 \\ \hline \end{array}$$

$462 to have the room carpeted

PRACTICE EXERCISE 48

1. Find the area of a rectangle whose dimensions are 9 ft. by 36 ft.

2. What is the area of a room whose dimensions are 5 yd. by 6 yd.?

3. A sheet of drawing paper measures 14 in. by 18 in. What is its area?

4. What is the cost of sanding and one coat of varnish on an 8 ft. × 15 ft. floor if it costs $3 a square foot?

PERIMETER

The *perimeter* of a geometric figure, as mentioned in section 2.14 of Chapter 2, is the sum of the lengths of its sides. The perimeter of a *rectangle* 8 ft. by 15 ft. is shown below.

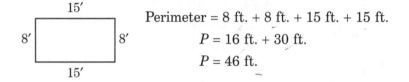

Perimeter = 8 ft. + 8 ft. + 15 ft. + 15 ft.

P = 16 ft. + 30 ft.

P = 46 ft.

This statement is another example of an *equation*. Suppose you had a rectangle which did not have any particular dimensions. How can you express its perimeter?

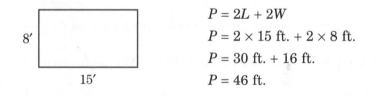

Perimeter = $L + L + W + W$ where

L = length and W = width.

This is a formula for finding the perimeter of the rectangle since it is a rule expressed in terms of letters rather than numbers. The repeated addition of $L + L$ can be written as $2 \times L$ and $W + W$ as $2 \times W$. In mathematics you can drop the *times* sign between the letter and number, and write $2L$ and $2W$. Even though the multiplication sign is not written it is understood to be there.

Thus, $P = L + L + W + W$ becomes $P = 2L + 2W$. This is the formula we can use to find the perimeter of any rectangle.

EXAMPLE

Find the perimeter of a rectangle whose dimensions are 8 ft. by 15 ft.

SOLUTION

$P = 2L + 2W$

$P = 2 \times 15$ ft. + 2×8 ft.

$P = 30$ ft. + 16 ft.

$P = 46$ ft.

How can we develop a formula for the perimeter of an equilateral triangle? An equilateral triangle is a triangle which has three equal sides. Let's represent the length of each side by the letter s.

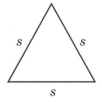

Then, $P = s + s + s$

$P = 3 \times s$ (repeated addition)

$P = 3s$ (dropping the *times* sign)

EXAMPLE

Find the perimeter of a square whose side measures of 6 in.

SOLUTION

The formula for the perimeter of a square is

$$P = 4s$$
$$P = 4 \times 6 \text{ in.}$$
$$P = 24 \text{ in.}$$

This step is called *substitution*. It means we replaced the letter s by the numerical value 6 in.

EXAMPLE

Each side of a pentagon is 1,500 ft. What is its perimeter?

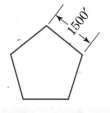

SOLUTION

A pentagon is a five-sided closed plane geometric figure as shown above. Each side measures 1,500 ft. Thus, the formula for the perimeter is

$$P = s + s + s + s + s$$
$$P = 5s$$
$$P = 5 \times s$$
$$P = 5 \times 1,500 \text{ ft.}$$
$$P = 7,500 \text{ ft.}$$

PRACTICE EXERCISE 49

Using formulas, solve the following problems:

1. Using the formula $P = 3s$, find the perimeter if the value of s is 34 ft.

2. How many feet of split fencing are needed to enclose a rectangular-shaped garden with dimensions of 15 ft. by 12 ft.?

3. Find the perimeter of a square if each side measures 16 ft.

4. Using the formula $P = 2L + 2W$, what is the value of P if $L = 32$ ft. and $W = 10$ ft.?

5. Using the formula $P = 12x$, what is the value of P if $x = 17$ in.?

6. Write a formula for the perimeter of this triangle. (Remember: The perimeter is the sum of all the sides of the figure, so $P = a + a + b$.) Complete the formula, writing it in its simplest form as was shown earlier.

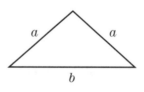

7. Mr. Cassidy has 62 ft. of wire and wants to string it in the shape of a triangle. Two sides of the triangle must measure 23 ft. each. What should be the length of the third side?

TERMS YOU SHOULD REMEMBER

Area   The measure of space within an enclosed region.
Perimeter   The sum of the lengths of the sides of a closed geometric figure.
Square inch   A square whose sides are 1 in.
Square foot   A square whose sides are 1 ft.
Square yard   A square whose sides are 1 yd.

4.9 More Difficult Word Problems Requiring Mu

EXAMPLE

A shopkeeper bought 76 dolls. Each doll cost $13. How much did he pay for all of them?

SOLUTION

This is a problem in repeated addition or multiplication. You want to add $13 a total of 76 times or 76 × $13. Since 76 × 13 is the same as 13 × 76 choose the number you want as your multiplier. Choosing 13 as the multiplier makes the example look like this:

$$\begin{array}{r} 76 \\ \times\ 13 \\ \hline 228 \\ 76 \\ \hline 988 \end{array}$$

Step 1: 3 × 6 = 18. Write the 8 and carry the 1.

Step 2: 3 × 7 = 21. Add the 1 you carried; 21 + 1 = 22.

Step 3: 1 × 6 = 6. Write the 6 in the column exactly under the 1.

Step 4: 1 × 7 = 7. Write the 7 to the left of the 6. Then add to find the final answer.

Before beginning this practice exercise, reread section 4.6 to find the key words that indicate multiplication example. There are 10 problems in this exercise. Try for a perfect score.

PRACTICE EXERCISE 50

1. A sewing machine operator makes 125 articles daily. At the end of a five-day work week, how many articles has she completed?

2. A candy store owner sold 652 pieces of candy. Each piece of candy costs 45¢. How many cents does he collect if he sells all his candy?

3. A shipping clerk mails 15 cartons to each of 740 customers. What is the total number of cartons she mails?

4. How much does a man earn in 14 weeks if his weekly earnings are $535?

5. A car-leasing company rents 1,435 cars monthly. How many cars does it rent by the end of one year? (Remember, 12 months = 1 year.)

s 2,432 members. If each member pays $420 dues a year, how much
union collect?

a Parlor, each slice of pizza costs 99¢. If each pie is cut into 8 slices,
ey (in cents) does the restaurant get for two pies?

WOW! PIZZA!
LET ME AT IT!

8. On each boat trip, a "dragger" returns with 3,800 lb. of fish. How many pounds of
fish will she bring in after 61 trips?

9. Sam purchases 200 leather wallets at $11 each. If he sells all of them, collecting
$3,200, how much profit does he make?

10. The rent for a 3-room apartment is $725 a month. How many dollars do you pay for
rent in 3 years?

REVIEW OF IMPORTANT IDEAS

 An equation is a sentence of equality.

 A formula is a rule stated in letters rather than numbers.

 The area of a rectangle = length × width.

 The perimeter of a geometric figure is equal to the sum of its sides.
All dimensions must be in the same unit of measurement.

 Replacing a letter with a number in a formula is called *substitution*.

LET'S SEE HOW YOUR SKILL IN MULTIPLICATION HAS IMPROVED

This chapter covered multiplication of whole numbers. Many different types of problems were included and your skill in computation has probably improved since the pretest. Here is a way to find out, since the chapter test will let you check on your own understanding.

A Score of	Means That You
19–22	Did very well. You can move to Chapter 5.
17–18	Know this material except for a few points. Reread the sections about the ones you missed.
14–16	Need to check carefully on the sections you missed.
0–13	Need to review the chapter again to refresh your memory and improve your skills.

Questions	Are Found in Section
14	4.2
1	4.4
2–6	4.5
16, 17	4.6
7–13	4.7
20, 22	4.8
15, 18, 19, 21	4.9

Chapter Test 4

Write your answers in the spaces provided. Use a calculator to check your answers after you have completed the entire chapter test.

1. $\begin{array}{r} 8 \\ \times 6 \\ \hline \end{array}$ 2. $\begin{array}{r} 43 \\ \times 2 \\ \hline \end{array}$ 3. $\begin{array}{r} 313 \\ \times 3 \\ \hline \end{array}$ 4. $\begin{array}{r} 64 \\ \times 2 \\ \hline \end{array}$ 5. $\begin{array}{r} 304 \\ \times 5 \\ \hline \end{array}$

6. $\begin{array}{r} 440 \\ \times 2 \\ \hline \end{array}$ 7. $\begin{array}{r} 24 \\ \times 13 \\ \hline \end{array}$ 8. $\begin{array}{r} 63 \\ \times 26 \\ \hline \end{array}$ 9. $\begin{array}{r} 87 \\ \times 56 \\ \hline \end{array}$ 10. $\begin{array}{r} 402 \\ \times 86 \\ \hline \end{array}$

11. $\begin{array}{r} 610 \\ \times 80 \\ \hline \end{array}$ 12. $\begin{array}{r} 600 \\ \times 300 \\ \hline \end{array}$ 13. $\begin{array}{r} 487 \\ \times 89 \\ \hline \end{array}$

14.   Find the product of 2 × 4 on the number line.

$$\xleftarrow{\quad}\underset{0\ \ 1\ \ 2\ \ 3\ \ 4\ \ 5\ \ 6\ \ 7\ \ 8\ \ 9\ \ 10}{\mid\ \mid\ \mid\ \mid\ \mid\ \mid\ \mid\ \mid\ \mid\ \mid\ \mid}\xrightarrow{\quad}$$

15.   How many dollars do you save in a Christmas Club savings account if you save $25 a week for 50 weeks?

16.   A local business needs to purchase 6 modems at $168 apiece. How much will the business spend all together?

17.   Mr. Raynor wishes to landscape his front yard with four hemlock bushes and two birch trees. A hemlock bush costs $99 and a birch tree costs $169. How much will Mr. Raynor spend on all six plants?

18.   Your car travels 12 mi. for each gallon of gasoline. How many miles can you travel on 12 gal.?

19.   If a stadium seats 54,260 persons, how many people would have attended 23 events if it was filled each time?

20.   A rectangular-shaped garden has dimensions of 26 ft. by 45 ft. What is the area of the garden?

21.   Find the product of 1,723 and 391.

22.   Find the value of P in the formula $P = 6s$, if $s = 39$ ft.

PUZZLE TIME

Ready to do another cross-number puzzle? This one will help you to review multiplication of whole numbers. The solution appears on page 435.

ACROSS

2. $67 \times 9 =$
3. $38 \times 9 =$
6. $342 \times 5 =$
9. $479 \times 6 =$
11. $4 \times 7 =$
12. $6 \times 9 =$
14. $195 \times 5 =$
17. $28 \times 3 =$
19. $9 \times 7 =$
20. $163 \times 5 =$
22. $5 \times 5 =$
24. $4 \times 3 =$
26. $78 \times 54 =$
30. $147 \times 15 =$
31. $48 \times 18 =$
32. $14 \times 27 =$

DOWN

1. $12 \times 11 =$
2. $4 \times 15 =$
4. $8 \times 6 =$
5. $9 \times 3 =$
7. $24 \times 3 =$
8. $7 \times 27 =$
10. $9 \times 5 =$
13. $27 \times 18 =$
15. $54 \times 14 =$
16. $29 \times 19 =$
18. $24 \times 18 =$
21. $32 \times 16 =$
23. $6 \times 9 =$
25. $4 \times 5 =$
27. $7 \times 4 =$
28. $8 \times 2 =$
29. $12 \times 20 =$
30. $14 \times 2 =$

HOLD IT!

You have done well. With the knowledge you have gathered in these chapters, you will find that working with numbers helps you in everyday situations. Chapter 5 covers division and will complete the four basic operations with numbers.

Practice your skills as often as you can. They are only tools that develop further learning.

Before you start Chapter 5, think about this:

Reading in mathematics improves as you know more words or terms.

Reading of mathematics requires close attention to relationships.
1. Which ideas are connected or related?
2. How does one fact lead to or cause another?
3. Which idea follows next in order of sequence?

This is a general review of some of the skills you have learned up to this point. You are making good progress if you get all the answers to these problems correct.

1. 68	2. 4,352	3. 56	4. 932	5. 5,000
137	−1,527	× 7	473	−2,733
+ 97			+948	

6. 402	7. 7	8. 973	9. 38
× 37	6	−691	×19
	5		
	8		
	+9		

10. At a major automobile factory there are 26 departments with 24 workers each, 18 departments with 28 workers each, and 25 departments with 31 workers each. What is the total working force?

ANSWERS
Try checking your answers on a calculator first.

1. 302	2. 2,825	3. 392	4. 2,353	5. 2,267
6. 14,874	7. 35	8. 282	9. 722	

10. 26	28	31	624
× 24	× 18	× 25	504
624	504	775	+ 775
			1,903

Let's Divide It Up!

I f six cans of X-Cola cost $3.60, what is the cost of one can? If a 12-foot board of lumber costs $15.00, what is the cost of one foot? These are two of many instances where *division* is needed to find the answer. *Division* is our fourth and final operation associated with whole numbers.

How well you can do on this topic can be checked with the pretest which follows.

Pretest 5—See What You Know and Remember

Work these exercises carefully, doing as many problems as you can. Write each answer in the space provided.

1. $6\overline{)42}$

2. $69 \div 3 =$

3. $2\overline{)426}$

4. $5\overline{)505}$

5. $4\overline{)208}$

6. $8\overline{)880}$

7. $\dfrac{65}{9} =$

8. $513 \div 3 =$

9. $4\overline{)824}$

10. $7\overline{)896}$

11. What would be the share for each person if we divided $63 among 9 people?

12.   Illustrate $6 \div 2$ on the number line below.

13.   Seven people buy an equal share in the purchase of a lottery ticket. If they win $56,049, how much does each one receive?

14. $85 \div 17 =$ 15. $24\overline{)504}$ 16. $32\overline{)1,664}$ 17. $42\overline{)6,066}$

18.   Find the mean, median, and mode of the heights of five women whose individual heights are: 62 in., 65 in., 58 in., 62 in., and 68 in.

19.   The area of a rectangle contains 132 ft.^{2}. If the length is 22 ft., find the width.

$$?' \qquad \boxed{\text{Area} = 132 \text{ ft.}^2}$$

22'

20.   Mr. Smith showed a yearly income of $26,250 on his federal income tax return last year. This represented his earnings for 35 weeks during the year. How much did he earn each week?

Now turn to the back of the book to check your answers. Add up all that you had correct. Count by the number of separate answers, not by the numbers of questions. In the pretest, there were 20 questions, but 22 separate answers.

A Score of   *Means That You*

20–22   Did very well. You can move to Chapter 6.
17–19   Know this material except for a few points. Read the sections about the ones you missed.
14–16   Need to check carefully on the sections you missed.
0–13   Need to work with the chapter to refresh your memory and improve your skills.

5.1 What Is Division?

If you recall, it was stated that subtraction is the inverse of addition. *Division* is the inverse of multiplication. Consider the following problem:

$$3\overline{)12}$$

The problem asks you to divide 12 into 3 equal parts or $3 \times ? = 12$. Since $3 \times 4 = 12$, then

$$3\overline{)12}^{\,4}$$

We picture $3\overline{)12}$ as 12 items divided into 3 groups.

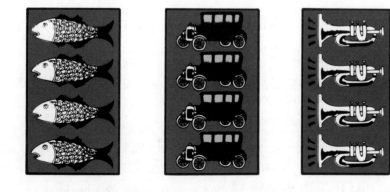

There are 4 items in each group. Thus:

$$3\overline{)12}^{\,4}$$

Multiplication was an operation of repeated additions. Division is an operation of repeated subtractions. Using the same example, $3\overline{)12}$, you subtract 3 from 12 continually until you can't subtract any more. The number of times you were able to subtract 3 is the answer to the division example.

$$
\begin{array}{rl}
12 & \\
-\ 3 & \quad (1) \\
\hline
9 & \\
-\ 3 & \quad (2) \\
\hline
6 & \\
-\ 3 & \quad (3) \\
\hline
3 & \\
-\ 3 & \quad (4) \\
\hline
0 &
\end{array}
$$

Since 3 was subtracted 4 times, you say that

$$3\overline{)12}^{\,4}$$

Division can be considered as (1) the inverse of multiplication or (2) repeated subtractions.

5.2 The Language of Division

EXAMPLE

What would be the share for each person if we divided $12 among 3 people?

SOLUTION
To find out what each person would receive, you must divide $12 by 3. This could be written in any of the following ways:

$$3\overline{)12} \quad \text{or} \quad 12 \div 3 \quad \text{or} \quad \frac{12}{3} \quad \text{or} \quad 12/3$$

You see that there is more than one way to indicate division.
The various parts of the division example are:

$$
\begin{array}{l}
4 \longleftarrow \text{quotient} \\
3\overline{)12} \longleftarrow \text{dividend} \\
\uparrow \\
 \longleftarrow \text{divisor}
\end{array}
$$

We say that 3 divides *into* 12 or 12 is divided *by* 3. In either case, the *quotient* or answer is always 4.

You notice that division has a language too. There are terms you must know to understand division.

TERMS YOU SHOULD REMEMBER

Divide   To separate into groups or shares.
Dividend   A number to be divided.
Divisor   The number by which the dividend is to be divided.
Quotient   The number resulting from the division of one number by another.

5.3 You Still Need Those Combinations

In each of the other three operations—addition, subtraction, multiplication—there was a total of 100 possible combinations. In division there are only 90. Do you know why?

It was stated earlier that division is the inverse operation of multiplication. This means that you find the quotient by multiplying the divisor by the quotient, which gives you a product equal to the dividend.

$$\text{Look at the example } \frac{6}{0} = ?, \text{ which is the same as } 0\overline{)6}.$$

This is a problem of finding the answer to the example $0 \times ? = 6$. Obviously there is *no* value which will make this a true statement. *Division by zero is meaningless, because there is no possible quotient.* Therefore, the divisor can be any number at all except 0.

This limits you to only 9 single-digit divisors and each one can be divided into 10 dividends. Thus, there are only 90 possible division combinations if we *limit* possible quotients to 1-digit numbers.

Dividing by 1 also has an interesting result. Look at these division examples.

$$1\overset{?}{\overline{)5}} \quad 3 \div 1 = ? \quad \frac{9}{1} = ?$$

Applying the inverse operation of multiplication for the first example, you find the quotient by multiplying the divisor (1) by the quotient (?) which gives you a product equal to the dividend (5), or $1 \times ? = 5$. The quotient is 5; $1 \times 5 = 5$. Similarly,

$$1 \times ? = 3 \quad 1 \times ? = 9$$
$$? = 3 \qquad ? = 9$$

In general, dividing by 1 results in a quotient equal to the dividend.

What happens when you divide 0 by any number except 0?

$$5\overset{?}{\overline{)0}} \qquad 0 \div 9 = ? \qquad \frac{0}{1} = ?$$

Using the same procedure as before, you have

$$5 \times ? = 0 \qquad 9 \times ? = 0 \qquad 1 \times ? = 0$$

Each quotient is 0. In general, dividing 0 by any number except 0 results in a quotient of 0.

In the practice exercise which follows, all of the 90 possible division combinations are given. Review each of them carefully so you can gain speed and skill.

Again we suggest that you do *not* write the answers in the book. Write your answers on a separate piece of paper so you can use this exercise for practice as many times as you wish. The idea is to see how you can improve by recognizing the answers more and more quickly.

Remember that division relies on your power to multiply. Although you just completed the chapter on multiplication, it might be wise to return to Chapter 4, practice exercises 41 and 42, and review the 100 multiplication combinations.

PRACTICE EXERCISE 51

Divide.

1. $1\overline{)1}$	2. $6\overline{)48}$	3. $3\overline{)9}$	4. $2\overline{)6}$	5. $5\overline{)20}$
6. $8\overline{)16}$	7. $4\overline{)4}$	8. $6\overline{)42}$	9. $5\overline{)0}$	10. $9\overline{)9}$
11. $2\overline{)8}$	12. $7\overline{)35}$	13. $1\overline{)2}$	14. $9\overline{)45}$	15. $4\overline{)8}$
16. $4\overline{)0}$	17. $7\overline{)42}$	18. $3\overline{)27}$	19. $9\overline{)54}$	20. $5\overline{)10}$
21. $3\overline{)6}$	22. $4\overline{)24}$	23. $6\overline{)36}$	24. $1\overline{)3}$	25. $8\overline{)24}$

26. $2\overline{)14}$ 27. $7\overline{)7}$ 28. $9\overline{)72}$ 29. $5\overline{)40}$ 30. $3\overline{)0}$

31. $4\overline{)28}$ 32. $1\overline{)7}$ 33. $7\overline{)49}$ 34. $9\overline{)36}$ 35. $6\overline{)30}$

36. $9\overline{)0}$ 37. $5\overline{)35}$ 38. $3\overline{)3}$ 39. $9\overline{)63}$ 40. $8\overline{)32}$

41. $5\overline{)15}$ 42. $9\overline{)81}$ 43. $6\overline{)0}$ 44. $7\overline{)14}$ 45. $3\overline{)24}$

46. $8 \div 1 =$ 47. $32 \div 4 =$ 48. $48 \div 8 =$ 49. $24 \div 6 =$ 50. $10 \div 2 =$

51. $5\overline{)30}$ 52. $2\overline{)4}$ 53. $9\overline{)27}$ 54. $4\overline{)16}$ 55. $8\overline{)0}$

56. $3\overline{)21}$ 57. $6\overline{)18}$ 58. $7\overline{)56}$ 59. $1\overline{)5}$ 60. $5\overline{)5}$

61. $6\overline{)54}$ 62. $1\overline{)0}$ 63. $8\overline{)40}$ 64. $3\overline{)18}$ 65. $2\overline{)18}$

66. $\dfrac{12}{2} =$ 67. $\dfrac{9}{1} =$ 68. $\dfrac{12}{4} =$ 69. $\dfrac{18}{9} =$ 70. $\dfrac{25}{5} =$

71. $1\overline{)4}$ 72. $7\overline{)28}$ 73. $5\overline{)45}$ 74. $6\overline{)12}$ 75. $4\overline{)36}$

76. $7\overline{)0}$ 77. $3\overline{)12}$ 78. $8\overline{)72}$ 79. $7\overline{)63}$ 80. $2\overline{)2}$

81. $8\overline{)56}$ 82. $6\overline{)6}$ 83. $8\overline{)8}$ 84. $1\overline{)6}$ 85. $7\overline{)21}$

86. $2\overline{)16}$ 87. $4\overline{)20}$ 88. $8\overline{)64}$ 89. $2\overline{)0}$ 90. $3\overline{)15}$

5.4 Help from the Number Line

You represent division on the number line as a series of subtractions. Repeating the problem 12 ÷ 3, you locate 12 on the number line and subtract groups of 3 units starting from the right.

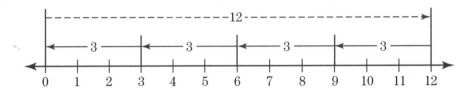

You subtract 4 equal groups of 3 units each, so

$$\frac{12}{3} = 4$$

Suppose you divide 13 by 5. How do you represent that division on the number line? Begin in the same manner as in the previous example. Subtract groups of 5 beginning from the right.

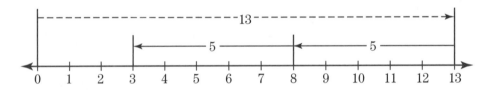

You see that you can only subtract *two* groups of 5. After subtracting two groups of 5, you end on 3 on the number line. You say that *two* groups of 5 may be subtracted and 3 units still remain; therefore 13 ÷ 5 = 2 with a remainder of 3. We write it as 13 ÷ 5 = 2 R 3 (R = remainder).

PRACTICE EXERCISE 52

Illustrate each of the following division problems on a number line. Find each quotient.

1. 8 ÷ 2 =   1 2 3 4 5 6 7 8

2. 3)‾10‾   1 2 3 4 5 6 7 8 9 10

3. $16 \div 4 =$

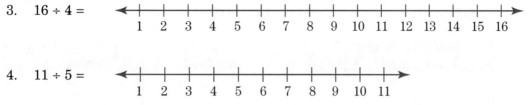

4. $11 \div 5 =$

5.5 Dividing by a One-Digit Number

EXAMPLE

A work crew of 48 men must be assigned to 4 different jobs. If each job requires an equal number of men, how many men will be assigned to each job?

SOLUTION
To find out the size of each work crew, you must divide 48 by 4. It looks like this:

$$4\overline{)48}$$

Rewrite the dividend as 4 tens and 8 ones.

$$4\overline{)4 \text{ tens and } 8 \text{ ones}}$$

Dividing each part of the dividend by 4 yields a quotient of

$$\overline{4)4 \text{ tens and } 8 \text{ ones}}^{\,1 \text{ ten and } 2 \text{ ones}} \quad \text{or} \quad 12$$

Thus, you see that you can divide the divisor into the dividend in the following way.

Problem	Step 1	Step 2	Check
$4\overline{)48}$	$4\overline{)48}^{\,1}$	$4\overline{)48}^{\,12}$	$\begin{array}{r} 12 \\ \times\ 4 \\ \hline 48 \end{array}$
	Divide the divisor (4) into the *tens* place number (4): $4 \div 4 = 1$ Place the 1 in the quotient over the 4 in the dividend.	Divide the divisor (4) into the *ones* place number (8): $8 \div 4 = 2$ Place the 2 in the quotient over the 8 in the dividend.	

Thus: $48 \div 4 = 12$.

Let's look at another example before starting the practice exercise.

Problem	Step 1	Step 2	Check
3)63̄	2 3)63̄	21 3)63̄	21 × 3 — 63
	Divide the divisor (3) into the *tens* number (6): 6 ÷ 3 = 2 Place the 2 in the quotient over the *tens* number (6).	Divide the divisor (3) into the *ones* number (3): 3 ÷ 3 = 1 Place the 1 in the quotient over the *ones* number (3).	

Thus: 63 ÷ 3 = 21.

This same method is used regardless of the size of the number in the dividend. You begin by dividing into the number on the left. Then divide the next number and the next number until no more numbers remain.

PRACTICE EXERCISE 53

Divide. Check by multiplying.

1. 3)93̄   2. 4)84̄   3. 5)55̄   4. 2)86̄   5. 3)69̄

6. 4)48̄   7. 8)88̄   8. 2)62̄   9. 6)66̄   10. 3)36̄

EXAMPLE

There are 488 employees in the A & R Auto Company. How can they be divided so that an equal number will be sent to 4 different departments?

SOLUTION

Divide as shown below.

Problem	Step 1	Step 2	Step 3	Check
$4\overline{)488}$	$\dfrac{1}{4\overline{)488}}$	$\dfrac{12}{4\overline{)488}}$	$\dfrac{122}{4\overline{)488}}$	$\begin{array}{r} 122 \\ \times\ 3 \\ \hline 488 \end{array}$
	Divide: $4 \div 4 = 1$ Place the 1 in the quotient over the *hundreds* number (4).	Divide: $8 \div 4 = 2$ Place the 2 in the quotient over the *tens* number (8).	Divide: $8 \div 4 = 2$ Place the 2 in the quotient over the *ones* number (8).	

Thus: $488 \div 4 = 122$.

EXAMPLE

Three partners earn a total of $96,096 yearly. How much does each one earn?

SOLUTION

Look at this problem:

Problem	Step 1	Step 2	Step 3
$3\overline{)\$96,096}$	$\dfrac{3}{3\overline{)\$96,096}}$	$\dfrac{32}{3\overline{)\$96,096}}$	$\dfrac{32,0}{3\overline{)\$96,096}}$
	Divide: $9 \div 3 - 3$	Divide: $6 \div 3 = 2$	Divide: $0 \div 3 = 3$

Step 4	Step 5	Check
$\dfrac{32,30}{3\overline{)\$96,096}}$	$\dfrac{\$32,032}{3\overline{)\$96,096}}$	$\begin{array}{r} \$32,032 \\ \times\ \ \ \ \ 3 \\ \hline \$96,096 \end{array}$
Divide: $9 \div 3 = 3$	Divide: $6 \div 3 = 2$	

Thus: $\$96,096 \div 3 = \$32,032$.

PRACTICE EXERCISE 54

Divide. Check each answer by multiplying.

1. $3\overline{)669}$ 2. $4\overline{)804}$ 3. $2\overline{)8,642}$ 4. $5\overline{)505}$ 5. $3\overline{)9,336}$

6. $6\overline{)606}$ 7. $2\overline{)842}$ 8. $4\overline{)8,044}$ 9. $3\overline{)39,303}$ 10. $8\overline{)8,888}$

11. $2\overline{)20,024}$ 12. $6\overline{)666}$ 13. $3\overline{)3,036}$ 14. $3\overline{)966}$ 15. $2\overline{)868}$

5.6 When the Quotient Contains Fewer Places Than the Dividend

Have you noticed in each example you have done thus far that the quotient contains as many places as the dividend? That happens when the divisor divides into the first number on the left. What happens if it does not?

EXAMPLE

Five cartons weigh 455 lb. Each carton weighs the same. How many pounds does each carton weigh?

SOLUTION
Dividing 455 by 5 will give the answer to the problem.

Problem	Step 1	Step 2	Check
5)‾455‾	$\overset{9}{5)\overline{455}}$	$\overset{91}{5)\overline{455}}$	$\begin{array}{r} 91 \\ \times\ 5 \\ \hline 455 \end{array}$
	Divide the 5 into the *hundreds* number (4): $4 \div 5 = ?$ *or* $5 \times ? = 4$ There is no whole number which fits, so divide the 5 into the first two numbers (45): $45 \div 5 = 9$ Place the 9 in the quotient over the *tens* number (5).	Divide the 5 into the *ones* number (5): $5 \div 5 = 1$ Place the 1 in the quotient over the *ones* number (5).	

Thus: $455 \div 5 = 91$.

PRACTICE EXERCISE 55

Divide. Check by multiplying.

1. 9)‾549‾   2. 8)‾248‾   3. 3)‾216‾   4. 7)‾357‾   5. 9)‾729‾

6. 4)‾324‾   7. 5)‾155‾   8. 7)‾217‾   9. 2)‾128‾   10. 6)‾486‾

5.7 Zero as a Place Holder

EXAMPLE

Alba earns $912 for 3 weeks' work. How much does she earn each week?

SOLUTION

The quotient of $912 \div 3$ will tell us what Alba earns weekly. Sometimes you will find that it is not the first number on the left in the dividend which is too small to divide into but another number in the dividend. In the problem, it is the 1. Look at the following illustration:

Problem	Step 1	Step 2	Step 3	Check
$3\overline{)\$912}$	$3\overline{)\$912}^{\ 3}$	$3\overline{)\$912}^{\ 30}$	$3\overline{)\$912}^{\ \$304}$	$\begin{array}{r}\$304\\ \times\ \ \ 3\\ \hline \$912\end{array}$
	Divide: $9 \div 3 = 3$ Place the 3 in the quotient above the 9.	Divide: $1 \div 3 = ?$ *or* $3 \times ? = 1$ This is not possible, so *place a zero in the quotient above the tens number (1).*	Divide 3 into the *tens* and *ones* numbers (12): $12 \div 3 = 4$ Place the 4 above the *ones* number in the quotient.	

Thus: $912 \div 3 = 304$.

EXAMPLE

A CD player costs $220. You put $60 down and agree to pay the remainder in 8 equal monthly installments. How much must you pay each month?

SOLUTION

Since the CD player costs $220 and you put $60 down, the balance you owe is only $220 − $60 = $160. If you are going to pay $160 in 8 months, you must divide to find your monthly payments. Let's look at this problem:

Problem	Step 1	Step 2	Check
$8\overline{)\$160}$	$8\overline{)\$160}^{\,2}$	$8\overline{)\,\$160}^{\,\$20^*}$	$\begin{array}{r} \$\,20 \\ \times\quad 8 \\ \hline \$160 \end{array}$
	Divide: $1 \div 8 = ?$ *or* $8 \times ? = 1$ This is not possible, so divide the first two number (16) by 8: $16 \div 8 = 2$ Place the 2 in the quotient above the *tens* number (6).	Divide: $0 \div 8 = 0$ Place the answer (0) in the quotient over the *ones* number (0).	

*Don't forget to include the *zero* in the quotient. Omitting the zero gives a quotient of 2, which does not check when multiplying.

Thus: $\$160 \div 8 = \$20$.

PRACTICE EXERCISE 56

Divide. Check each answer by multiplying. Don't forget to include the zero when it is needed.

1. $6\overline{)624}$ 2. $7\overline{)280}$ 3. $5\overline{)525}$ 4. $3\overline{)318}$

5. $7\overline{)735}$ 6. $8\overline{)720}$ 7. $9\overline{)810}$ 8. $8\overline{)816}$

9. $2\overline{)2,610}$ 10. $4\overline{)4,124}$ 11. $3\overline{)3,660}$ 12. $2\overline{)2,120}$

5.8 Dividing and Leaving a Remainder

In section 5.4, "Help From the Number Line," an example

$$5\overline{)13}$$

was illustrated. It was found that two groups of 5 could be subtracted and there would be 3 units remaining. Therefore, we say that $13 \div 5 = 2$ with a *remainder* of 3. A short way of writing the remainder is R, so the problem would look like this:

Problem	Step 1	Step 2	Check	
$5\overline{)13}$	$\begin{array}{r} 2 \\ 5\overline{)13} \end{array}$	$\begin{array}{r} 2\ R\ 3 \\ 5\overline{)13} \end{array}$	$\begin{array}{r} 5 \\ \times\ 2 \\ \hline 10 \end{array}$	$\begin{array}{r} 10 \\ +\ 3 \\ \hline 13 \end{array}$
	Since 5 does not divide into 1 you divide it into the two numbers (13): $13 \div 5 = ?$ *or* $5 \times ? = 13$. *No number divides this evenly*, for 2 is too small and 3 is too large: $5 \times 2 = 10$ and $5 \times 3 = 15$. The smaller one (2) is our answer so place it in the quotient above the *ones* number (3).	$5 \times 2 = 10$ This leaves a remainder of 3.		

Thus: $\begin{array}{r} 2\ R\ 3 \\ 5\overline{)13} \end{array}$

In the check you multiply the quotient (2) by the divisor (5) and add the remainder (3) to get the dividend (13).

$$2 \times 5 = 10$$
$$\underline{+\ 3}$$
$$13$$

PRACTICE EXERCISE 57

Divide. Check each answer by multiplying.

1. $2\overline{)17}$ 2. $5\overline{)39}$ 3. $7\overline{)50}$ 4. $9\overline{)46}$ 5. $4\overline{)19}$

6. $3\overline{)25}$ 7. $8\overline{)35}$ 8. $3\overline{)20}$ 9. $6\overline{)47}$ 10. $9\overline{)83}$

11. $6\overline{)45}$ 12. $4\overline{)37}$ 13. $8\overline{)67}$ 14. $5\overline{)17}$ 15. $7\overline{)26}$

5.9 Carrying Remainders

EXAMPLE

Sixty-five people volunteer for 5 different jobs. If an equal number are assigned to each job, how many people will be on each job?

SOLUTION

Once again the problem calls for dividing one number by another. The problem is:

$$5\overline{)65}$$

Rewriting the example, as done previously, looks like this:

$$5\overline{)6 \text{ tens and } 5 \text{ ones}}$$

Dividing 6 tens and 5 ones into *five* groups is difficult but since 5 divides evenly into 5 tens, let's rewrite the dividend as 5 tens + 1 ten + 5 ones, or 5 tens and 15 ones (recall that 1 ten = 10 ones).

$$5\overline{)5 \text{ tens and } 15 \text{ ones}}$$

Dividing each part of the dividend by 5 yields a quotient of

$$\frac{1 \text{ ten and } 3 \text{ ones}}{5\overline{)5 \text{ tens and } 15 \text{ ones}}} \quad \text{or} \quad 13$$

Following the same procedure in the original example would look like this:

Problem	Step 1	Step 2	Check
$5\overline{)65}$	$\begin{array}{r} 1 \\ 5\overline{)65} \\ -5 \\ \hline 15 \end{array}$	$\begin{array}{r} 13 \\ 5\overline{)65} \\ -\ 5 \\ \hline 15 \\ -\ 15 \end{array}$	$\begin{array}{r} 13 \\ \times\ 5 \\ \hline 65 \end{array}$
	Divide the divisor (5) into the *tens* number (6): $6 \div 5 = ?$ cannot be divided evenly. The answer is 1 with a remainder of 1. The answer (1) is placed in the quotient above the *tens* number (6). The remainder (1) or one ten is carried to the *ones* number (5), making it 15.	Divide the 5 into the new *ones* number (15): $15 \div 5 = 3$. Place the 3 in the quotient above the *ones* number (5).	

Thus: $65 \div 5 = 13$.

This same method can be applied to any example where the divisor doesn't divide the first number or any other number evenly. Look at the next two examples before attempting the practice exercise that follows.

EXAMPLE

The sum of $519 is to be divided evenly among 3 people. How much money does each one receive?

SOLUTION

Problem	Step 1	Step 2	Step 3	Check
3)$519	$$\begin{array}{r} 1 \\ 3\overline{)\$519} \\ -3 \\ \hline 21 \end{array}$$	$$\begin{array}{r} 17 \\ 3\overline{)\$519} \\ -3 \\ \hline 21 \\ -21 \\ \hline 09 \end{array}$$	$$\begin{array}{r} \$173 \\ 3\overline{)\$519} \\ -3 \\ \hline 21 \\ -21 \\ \hline 09 \\ -9 \\ \hline \end{array}$$	$$\begin{array}{r} \$173 \\ \times\ \ \ 3 \\ \hline \$519 \end{array}$$
	Divide: $5 \div 3 = 1\ R\ 2$ Place the 1 in the quotient above the 5 and carry the remainder 2 to the *tens* number. The *tens* number is now 21.	Divide: $21 \div 3 = 7$ Place the 7 above the *tens* number (1).	Divide: $9 \div 3 = 3$ Place the 3 above the *ones* number (9).	

Thus: $519 \div 3 = $173.

EXAMPLE

Six clerks handle accounts for 768 customers. If each clerk handles an equal share, how many customers does each one have?

SOLUTION

Problem	Step 1	Step 2	Step 3	Check
$6\overline{)768}$	$\overset{1}{6\overline{)768}}$	$\overset{12}{6\overline{)768}}$	$\overset{128}{6\overline{)768}}$	$\begin{array}{r} 128 \\ \times\ 6 \\ \hline 768 \end{array}$
	Divide: 7 ÷ 6 = 1 R 1 Place the 1 in the quotient and carry 1, making the *tens* place 16. *Remember—* 1 hundred = 10 tens.	Divide: 16 ÷ 6 = 2 R 4 Place the 2 in the quotient and carry the 4, making the *ones* place 48. *Remember—* 4 tens = 40 ones.	Divide: 48 ÷ 6 = 8 Place the 8 in the quotient.	

Thus: 768 ÷ 6 = 128.

PRACTICE EXERCISE 58

Divide. Check each answer by multiplying. Be sure to indicate the remainder if there is one.

1. $6\overline{)78}$ 2. $3\overline{)249}$ 3. $4\overline{)76}$ 4. $7\overline{)497}$

5. $3\overline{)231}$ 6. $5\overline{)146}$ 7. $2\overline{)1,302}$ 8. $6\overline{)4,746}$

9. $2\overline{)1,359}$ 10. $8\overline{)6,504}$ 11. $9\overline{)7,002}$ 12. $7\overline{)23,817}$

5.10 Word Problems Requiring Division

Reading word problems and understanding their solution require skills that have been presented already. The need is for more practice so that these habits come easily. Key words associated with solving problems by division have been enumerated in the illustrated examples already. The key word most commonly found in each problem is *each*. "How much does *each* one receive?"

Now try these problems.

PRACTICE EXERCISE 59

1. Your daughter brings home 5 friends for cookies and milk. If there are 12 cookies to be divided equally among the 6 children, how many cookies does *each* one receive?

2. Tomato juice is on sale at 2 cans for 86¢. How many cents does *one* can cost?

3. A boat makes 7 trips and returns with 350 lb. of scallops. It brings back an equal amount for each trip. How many pounds does it bring back *each* time?

4. Mr. Gandolfo made 144 gal. of wine and must place it in barrels that *each* hold 9 gal. How many barrels does he need?

5. Mike makes 4 telephone calls and is charged $12. How much does *each* call cost?

6. The annual prize money of $210 is to be divided equally among the first 3 winners. How much money will *each* of the winners receive?

7. You put $400 down on a pre-owned automobile which costs $12,400 and agree to pay the rest in 8 equal installments. How much is *each* installment?

8. A truck is carrying 6 equal bales of cotton whose total weight is 2,070 lb. What is the weight of *each* bale?

9. A salesman drives on the average 424 mi. each week. If his car uses 1 gal. of gasoline for each 8 mi. he drives, how many gallons of gasoline does he use *each* week?

10. Bill finds that he has spent $312 in the past 6 months for the purchase of CDs. How many dollars has he spent *each* month if he spends an equal amount each month?

5.11 Dividing by More Than a One-Digit Number

When dividing by more than a one-digit number, the process of estimation is best used. The method is identical to the one described earlier. Since you will be multiplying the divisor by numbers like 10, 20, 50, 100, etc., it is advisable for you to review Practice Exercise 46 in Chapter 4, before you continue. Be sure you remember the short-cut method for multiplying before you look at the example.

EXAMPLE

If 672 parking tickets were issued in a 32-day period, approximately how many were issued each day?

SOLUTION

Dividing 672 by 32 will give the answer to this problem. Use the procedure that follows:

Problem	Step 1	Step 2	Check
$32\overline{)672}$	$\begin{array}{r} 2 \\ 32\overline{)672} \\ 64\downarrow \\ \hline 32 \end{array}$	$\begin{array}{r} 21 \\ 32\overline{)672} \\ 64\downarrow \\ \hline 32 \\ 32 \\ \hline 0 \end{array}$	$\begin{array}{r} 32 \\ \times\ 21 \\ \hline 32 \\ 64 \\ \hline 672 \end{array}$
	Since 32 does not divide into 6, you divide 32 into the first two numbers (67). $67 \div 32 = 2$ Multiply 2×32 and place the product 64 under the 67 in the dividend. Subtract and bring down the 2 making the next dividend 32.	Divide: $32 \div 32 = 1$ Place the 1 in the quotient, multiply $1 \times 32 = 32$ and subtract, leaving a remainder of 0.	

Thus: $672 \div 32 = 21$.

You see that division by any number of digits is treated the same as before. To be sure you understand more difficult examples, two more illustrations will follow.

EXAMPLE

Find the quotient of 5,544 divided by 42.

SOLUTION

Problem	Step 1	Step 2	Step 3	Check
$42\overline{)5{,}544}$	$\begin{array}{r}1\\42\overline{)5{,}544}\\42\!\downarrow\\\hline134\end{array}$	$\begin{array}{r}13\\42\overline{)5{,}544}\\42\!\downarrow\\\hline134\\126\!\downarrow\\\hline84\end{array}$	$\begin{array}{r}132\\42\overline{)5{,}544}\\42\!\downarrow\\\hline134\\126\!\downarrow\\\hline84\\84\\\hline\end{array}$	$\begin{array}{r}132\\\times\ \ 42\\\hline264\\528\\\hline5{,}544\end{array}$
	Divide 42 into the first two numbers (55). $55 \div 42 = 1$ Place the 1 in the quotient above the 5. Multiply: $1 \times 42 = 42$ and place the product under the 55 in the dividend. Subtract and bring down the 4 in the tens column making the next dividend 134.	Divide: $134 \div 42 = 3$ Place the 3 in the quotient above the 4. Multiply $3 \times 42 = 126$ and place the product under the 134. Subtract: $134 - 126 = 8$. Bring down the 4 in the ones column making the next dividend 84.	Divide: $84 \div 42 = 2$ Place the 2 in the quotient above the 4. Multiply: $2 \times 42 = 84$ and place the product under the 84. Subtract, leaving a remainder of 0.	

Thus: $5{,}544 \div 42 = 132$.

You see that the problem 5,544 ÷ 42 = ? can be completed as follows:

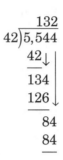

$$
\begin{array}{r}
132 \\
42\overline{)5{,}544} \\
42\downarrow \\
\overline{134} \\
126\downarrow \\
\overline{84} \\
84 \\
\overline{}
\end{array}
$$

Divisions may also have remainders as illustrated in this example.

EXAMPLE

What is the quotient and remainder when 6,578 is divided by 365?

SOLUTION

Problem	Step 1	Step 2	Check
365)6,578	$\begin{array}{r}1\\365\overline{)6{,}578}\\365\downarrow\\\overline{2928}\end{array}$	$\begin{array}{r}18\text{R}8\\365\overline{)6{,}578}\\365\downarrow\\\overline{2928}\\2920\\\overline{8}\end{array}$	$\begin{array}{r}365\\\times\ \ 18\\\overline{2920}\\365\\\overline{6{,}570}\\+\ \ \ 8\\\overline{6{,}578}\end{array}$
	Divide: 657 ÷ 365 = 1 Place the 1 in the quotient above the 7. Multiply: 1 × 365 = 365 Place the product 365 below 657 in the dividend. Subtract: 657 − 365 = 292. Bring down the 8 making the next dividend 2928.	Divide: 2928 ÷ 365 = 8. Multiply: 8 × 365 = 2920 Place the 8 in the quotient above the 8. Subtract: 2928 − 2920 = 8. The remainder is 8.	

Thus: 6,578 ÷ 365 = 18 R 8.

A remainder must always be less than the divisor. It should be divided again if it is not.

PRACTICE EXERCISE 60

Divide. Check each answer by multiplying. Some of the exercises have remainders. Try using the short-form method described earlier.

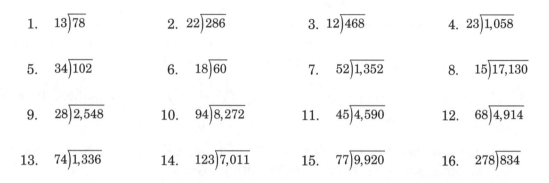

1. $13\overline{)78}$ 2. $22\overline{)286}$ 3. $12\overline{)468}$ 4. $23\overline{)1,058}$

5. $34\overline{)102}$ 6. $18\overline{)60}$ 7. $52\overline{)1,352}$ 8. $15\overline{)17,130}$

9. $28\overline{)2,548}$ 10. $94\overline{)8,272}$ 11. $45\overline{)4,590}$ 12. $68\overline{)4,914}$

13. $74\overline{)1,336}$ 14. $123\overline{)7,011}$ 15. $77\overline{)9,920}$ 16. $278\overline{)834}$

TERMS YOU SHOULD REMEMBER

Remainder   The dividend minus the product of the divisor and quotient.
Partial quotient   An incomplete quantity resulting from dividing part of the quotient by the divisor.

REVIEW OF IMPORTANT IDEAS

Some of the most important ideas thus far in Chapter 5 are:

Division is repeated subtraction.

Division is the inverse operation of multiplication.

Division depends on your ability to multiply.

Remainders may occur when dividing. All problems do not divide evenly.

5.12 More Difficult Word Problems Requiring Division

Read section 5.10 to recall the key words associated with problems solved by dividing before doing this next exercise.

PRACTICE EXERCISE 61

1. There are 216 adults attending 8 classes of English in the evening adult high school. On the average, how many were enrolled in each class?

2. You earn $37,776 in a year. How many dollars do you earn monthly? (12 months = 1 year)

3. You purchase 2 snowmobiles for $11,280 and agree to pay for them in 24 equal monthly payments. What is your monthly payment?

4. A truck is carrying a load of 108 barrels. The total weight of the barrels is 6,804 lb. If each barrel is the same weight, what does one barrel weigh?

5. A man purchases a piece of lumber 192 in. long. How many pieces 16 in. long can be cut from it? (Disregard any waste occurring from cutting.)

6. Mr. Jones, at the end of 47 weeks of work, had earned $21,855. How many dollars did he earn weekly?

7. A car rental agency rents 16,464 cars each year. On the average, how many cars do they rent monthly? (12 months = 1 year)

8. A real estate salesman travels 420 mi. each week. If his car uses 1 gal. of gasoline for each 12 mi. he drives, how many gal. of gasoline does he use in each week?

9. June requires 54 in. of material to make a skirt. If she orders 432 in. of cloth, how many skirts can she make?

10. A truck carries bales of cotton whose total weight is 13,875 lb. If each bale weighs 375 lb., how many bales are on the truck?

5.13 Mean, Median, and Mode

The mean, median, and mode are referred to as measures of central tendency. This means that each of them is a value near the middle or center of all the other numbers when they are arranged in order of size. The method for obtaining each one suggests its definition.

MEAN, MEDIAN, AND MODE

- *Mean*—Find the mean by adding the given values and then dividing the sum by the number of addends.
- *Median*—Arrange the values in ascending or descending order and find that value which is in the center of all the numbers.
- *Mode*—Find the value among all the numbers which occurs most frequently. There may or may not be a mode.

In order for you to get a better idea of how these *three* measures of central tendency behave, look at the following example:

EXAMPLE

If the heights of the first seven persons who enter a store are 65 in., 72 in., 75 in., 67 in., 60 in., 65 in., and 65 in., find, if possible, the mean, median, and mode.

SOLUTION
The listed heights of the seven persons are called the *data*. From the data we determine the mean, median, and mode.

1. *Mean*—65 in. + 72 in. + 75 in. + 67 in. + 60 in. + 65 in. + 65 in. = 469 in. There are 7 addends, so 469 ÷ 7 = 67 in.
2. *Median*—60 in., 65 in., 65 in., (65 in.), 67 in., 72 in., 75 in. The center value is the fourth one, 65 in.
3. *Mode*—The value of 65 in. occurs three times while all the others occur once. Thus, 65 in. is the mode.

Thus: The mean height is 67 in.
The median height is 65 in.
The mode height is 65 in.

The mean, median, and mode may be the same or may be different, as just shown. You do see, through, that they all tend toward the center of the data.

PRACTICE EXERCISE 62

---

1. The four children in the Russo family weigh 132 lb., 117 lb., 94 lb., and 73 lb. What is their mean weight?

2. Mr. Charles received the following grades on his math tests: 85%, 94%, 64%, 88%, and 89%.

   a. What was the mean grade? _____

   b. What was the median grade? _____

   c. Is there a mode? _____ If so, what is it? _____

3. In one week of driving her car, Ms. Betancourt used 13, 11, 5, 11, 10, 9, and 4 gal. of gasoline each day.

   a. What was the mean number of gallons used per day? _____

   b. What was the median number of gallons used per day? _____

   c. Is there a mode? _____ If so, what is it? _____

4. If the total weight of the 11 members of a football team was 2,552 lb., then what is the average weight (mean) for each member of the team?

5. If the Cartwright family of 5 took a weekend trip and spent a total of $365, then what was the average (mean) expense for each member of the family?

6. Mike, an office worker, timed himself at various times of the day to determine the average (mean) number of words typed in one minute. He found that as the day progressed he typed with less speed. These were his findings: 63, 61, 57, 54, 43, and 40 words per minute. What was his average number of words per minute?

5.14 Division Applied to Area Measurement

EXAMPLE

Mr. Sims had a gallon of paint which could cover 520 ft.^{2} of rectangular wall surface. If the height of the wall is 8 ft., then what is the length, in feet, of the wall surface that can be painted with his gallon of paint?

SOLUTION

Since the walls are rectangular in shape, the problem becomes simply an example in which you know the area of the rectangle and its width. You are looking for its length.

$$L \times W = A$$
$$8 \text{ ft.} \times ? = 520 \text{ ft.}^2$$
$$? = 65 \text{ ft.}$$

Area = 520 ft.^{2} 8′

$x′$

Thus, Mr. Sims can cover 65 ft. of wall with his gallon of paint.

PRACTICE EXERCISE 63

1. A rectangular garden contains 1,462 ft.^{2} of surface that can be planted. If the length of the rectangle is 43 ft., find its width.

2. If the area of a rectangular plot of ground contains 612 ft.^{2} and its length is 34 ft., what is its width?

3. The printed material on a rectangular sheet of paper covered 108 in.^{2} If its length is 12 in., find its width.

CHECK WHAT YOU HAVE LEARNED

The following test lets you see how well you have learned the ideas in Chapter 5. In counting up your answers, remember that there were 22 separate answers.

A Score of	Means That You
20–22	Did very well. You can move on to Chapter 6.
17–19	Know the material except for a few points. Reread the sections about the ones you missed.
14–16	Need to check carefully the sections you missed.
0–13	Need to review the chapter again to refresh your memory and improve your skills.

Questions	Are Covered in Section
1	5.3
12	5.4
2–4	5.5
5	5.6
6, 9	5.7
7	5.8
8, 10	5.9
11, 13	5.10
14–17	5.11
20	5.12
18	5.13
19	5.14

Chapter Test 5

Write each answer in the space provided. Wherever possible, use a calculator to check your answers after you have completed the entire chapter test.

1. $7\overline{)35}$ 2. $84 \div 4 =$ 3. $2\overline{)628}$ 4. $6\overline{)606}$ 5. $3\overline{)186}$

6. $9\overline{)990}$ 7. $\dfrac{51}{9} =$ 8. $348 \div 2 =$ 9. $3\overline{)618}$ 10. $4\overline{)512}$

11. What would be the share for each person if we divide $16 among 8 people?

12. Illustrate $9 \div 3$ on the number line below.

0   1   2   3   4   5   6   7   8   9   10

13. Six people buy an equal share in the purchase of a lottery ticket. If they win $54,048, how much does each one receive?

14. $76 \div 19 =$ 15. $26\overline{)546}$ 16. $42\overline{)2,184}$ 17. $32\overline{)4,626}$

18. Find the mean, median, and mode for the following data: 32 ft., 30 ft., 41 ft., 38 ft., 36 ft., 30 ft., 32 ft., 32 ft., 35 ft.

19. The area of a rectangle contains 138 yd.^{2} If the length is 23 yd., find the width.

$A = 138$ yd.^{2} ? yd.

23 yd.

20. At a recent fund-raising dinner, $78,500 was collected from the sale of $25 tickets. How many people attended the affair?

PUZZLE TIME

Division of Whole Numbers

ACROSS

1. $6\overline{)72}$

3. $3\overline{)48}$

4. $7\overline{)1,981}$

5. $5\overline{)125}$

6. $6\overline{)360}$

7. $12\overline{)192}$

10. $5\overline{)31,125}$

12. $24\overline{)89,304}$

15. $73\overline{)730}$

16. $324 \div 18 =$

18. $216 \div 6 =$

19. $4\overline{)3,276}$

20. $11\overline{)781}$

21. $1,040 \div 13 =$

DOWN

1. $6\overline{)1,086}$

2. $4\overline{)92}$

3. $17\overline{)255}$

5. $6\overline{)1,236}$

7. $23\overline{)276}$

8. $650 \div 10 =$

9. $2,736 \div 38 =$

11. $9\overline{)261}$

12. $17\overline{)527}$

13. $910 \div 13 =$

14. $11\overline{)1,276}$

17. $15\overline{)6,150}$

18. $186 \div 6 =$

19. $528 \div 6$

You will find the solution on page 440.

PUZZLE TIME

All Operations

ACROSS

1. $86 - 49$
3. $8 \times 7 =$
4. $686 \div 7 =$
6. $9 \times 7 =$
8. $7 \times 2 =$
9. $10 \times 9 + 2 =$
11. $16 \times 6 =$
12. $2 \times 7 + 1 =$
13. $12 \times 11 =$
14. $2{,}744 \div 2 =$
16. $11 \times 11 =$
18. $11 \times 8 - 10 =$
19. $6 \times 97 - 564 =$
21. $93 \times 3 + 5 =$
23. $111 \times 7 =$
26. $2{,}852 \div 4 =$
27. $30 \times 7 =$
29. $7 \times 7 - 7 =$
31. $10 \times 5 =$
33. $1{,}776 \div 3 =$
35. $8{,}060 - 939 =$
38. $444 - 321 =$
41. $9 \times 8 - 23 =$
43. $4 \times 8 =$
44. $7 \times 6 + 1 =$
45. $14 \times 4 =$
47. $12 \times 5 =$
48. $10 \times 9 =$
49. $13 \times 3 =$
50. $12 \times 8 - 11 =$

DOWN

1.	$3 \times 12 - 5 =$	22.	$9 \times 9 =$
2.	$9{,}583 - 2{,}100=$	24.	$14 \times 4 + 15 =$
3.	$181 \times 3 =$	25.	$10 \times 7 =$
5.	$4{,}060 \div 5=$	28.	$7 \times 5 =$
6.	$2{,}304 \times 3 =$	30.	$9 \times 3 =$
7.	$6 \times 6 =$	31.	$204 \div 4 =$
9.	$2{,}739 \div 3 =$	32.	$7 \times 9 =$
10.	$512 - 255 =$	34.	$12{,}893 - 3{,}573 =$
13.	$7 \times 3 - 2 =$	36.	$12 \times 12 =$
14.	$9 \times 2 =$	37.	$63 \times 4 + 41 =$
15.	$7 \times 3 =$	39.	$7{,}374 \div 3 =$
17.	$156 \div 13 =$	40.	$13 \times 13 =$
18.	$44{,}604 \div 6 =$	42.	$146 \times 3 + 1 =$
20.	$1{,}744 \times 5 =$	43.	$6 \times 6 =$
21.	$9 \times 3 =$	46.	$14 \times 5 - 5 =$

The solution to this puzzle is on page 440.

With the completion of this puzzle you will have finished the *four* fundamental operations for whole numbers. You will be ready for Chapter 6, which introduces the fraction. Before you begin though, there are a few points that you should be aware of.

It takes time and a lot of your energy to succeed.

You will make errors as you study and do mathematics.

The number of your errors will decrease as you continue studying.

The idea that you must be perfect or that you must score a perfect paper can act like a brake to slow you down in getting ready to take or in taking tests.

Although each pretest and chapter test does not require a perfect paper, do try to achieve at least the satisfactory grade each time.

HOLD IT!

Are you developing your ability in learning to compute? Try this review exercise to see your progress in the four fundamental operations. Check your answers on a calculator after you have completed 1–10.

1. $\begin{array}{r} 103 \\ \times\ 35 \\ \hline \end{array}$ 2. $3{,}220 + 758 + 2{,}608 =$ 3. $\begin{array}{r} 7{,}003 \\ -\ 2{,}746 \\ \hline \end{array}$ 4. $28\overline{)392}$ 5. $106\overline{)25{,}122}$

6. In a recent survey conducted by the *Dispatch*, it was discovered that 5 small cities had a total budget of $14,316,210. What was the average budget for each city?

7. A zoo plans to temporarily house a dolphin for 16 days. It has been told the dolphin requires 65 pounds of fish a day. How many pounds of fish are needed for the 16 days?

8. A line trawler makes 12 trips to sea and returns with a total poundage of 13,404. On the average, how many pounds does the ship catch per trip?

9. A house which cost $137,500 is now selling for $129,995. How much less is its new price?

10. Mr. Ramirez pays child support of $375 a month to his ex-wife. How much does this amount to over a 3-year period?

ANSWERS

1. 3,605
2. 6,586
3. 4,257
4. 14
5. 237
6. $2,863,242
7. 1,040 lb.
8. 1,117 lb.
9. $7,505
10. $13,500

The Important Parts— Fractions

Everyone has the need to know and to use fractions. You sometimes buy portions of things rather than the whole things. You may buy part of a pound rather than the whole pound. You may measure part of a cup rather than the whole cup. You may purchase part of a yard of material, not the whole yard. These are only a few instances in which you need to use fractions. You can probably think of many more.

The pretest which follows can show how well you understand fractions.

Pretest 6—See What You Know and Remember

Try your past skills again on this test. Do as many problems as you can. Write each answer in the space provided. Simplify the answer if possible.

1. Using this diagram, write, as a fraction, the portion of the rectangle that has been shaded. _____

2. If I have 8 items and eat 2 of them, represent, as a fraction, the portion I have eaten. _____

3. A fraction consists of two parts, a *numerator* and a *denominator*. In the fraction $\frac{3}{16}$, name the numerator and the denominator.

    3 is the   _____

    16 is the   _____

4. Identify each fraction as a *proper* or *improper* fraction.

    a. $\frac{3}{2}$ _____

    b. $\frac{2}{5}$ _____

    c. $\frac{4}{4}$ _____

5. A line is divided into 12 equal segments. What is the name given to one of these parts?   _____

6. A man has 5 items. If he sells 2 of them, what fractional part remains?   _____

7. In a classroom of 27 adults, 17 are men.

    a. What fraction of the class is men?   _____

    b. What fraction of the class is women?   _____

8. Write the fraction $\frac{12}{32}$ in its simplest form.   _____

9. To change $\frac{2}{7}$ to an equivalent fraction with a new denominator of 28 you must multiply the numerator and the denominator by a certain number. What number is it?   _____

10. Change the improper fraction $\frac{14}{5}$ to a mixed number..   _____

11. Change the mixed number $3\frac{1}{3}$ to an improper fraction   _____

12. There are approximately 10 hours of daylight each day during the month of October. There are 24 hours in a day.

    a. What fraction of the day is daylight?   _____

    b. What fraction of the day is night?   _____

13. Find the lowest common denominator for these fractions:

$$\frac{1}{6}, \quad \frac{2}{3}, \quad \frac{3}{4}$$ _____

14. Which is the larger of the two fractions:

$$\frac{7}{8} \quad \text{or} \quad \frac{11}{12}$$ _____

15. Arrange these three fractions in order of size, starting with the smallest one first.

$$\frac{2}{3}, \quad \frac{5}{12}, \quad \frac{1}{2}$$ _____

Now turn to the back of the book to check your answers. Add up all that you had correct. Count by the number of separate answers, not by the number of questions. In this pretest there were 15 questions but 20 separate answers.

A Score of	Means That You
18–20	Did very well. You can move to Chapter 7.
16–17	Know this material except for a few points. Read the sections about the ones you missed.
13–15	Need to check carefully on the sections you missed.
0–12	Need to work with this chapter to refresh your memory and improve your skills.

Questions	Are Covered in Section
1–2, 5–7, 12	6.1
3	6.2
4	6.3
8	6.4
9	6.5
13–15	6.6
10	6.7
11	6.8

6.1 What Is a Fraction?

Mrs. White baked a custard pie and cut it into 8 equal slices. The relationship of *one* of these slices to the whole pie can be represented by a *unit fraction*. The unit fraction in this case is:

$$\frac{1}{8} = \frac{\text{one of the slices}}{\text{the number of slices in the whole pie}}$$

and is illustrated as:

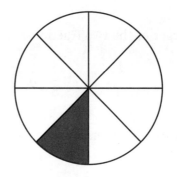

 A fraction can be written using a horizontal line or a slanted line to separate the numerator from the denominator: $\frac{5}{8}$, is the same thing as 5/8.

 A fraction can represent part of a larger quantity or group. What fractional part is 3 men out of a group of 8 men?

$\frac{3}{8}$ is the required fraction.

$$\frac{3}{8} = \frac{\text{three of the men}}{\text{all the men}}$$

EXAMPLE

If Mrs. White has a family of 5 and they each eat one of the 8 equal pieces of the custard pie, what fraction indicates the portion of the pie that was eaten? What fraction represents the remaining portion?

SOLUTION

Since the pie was divided into 8 pieces and 5 of them were eaten, then $\frac{5}{8}$ is the

representing the eaten part; $\frac{3}{8}$ is the part that still remains. This is pictured below:

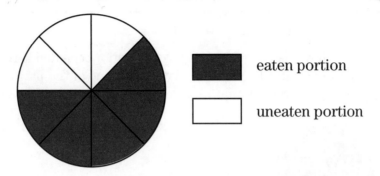

eaten portion

uneaten portion

Thus, a fraction can represent whole parts of a group, 3 men being $\frac{3}{8}$ of the group of

8 men.

 A third meaning of a fraction is revealed by its symbol. Thus, in $\frac{3}{4}$, the

line between 3 and 4 means that the 3 is being divided by the 4:

$$3 \div 4 \quad \text{or} \quad 4\overline{)3}$$

This is illustrated by dividing 3 units into 4 equal parts:

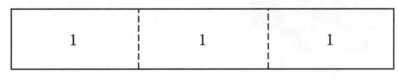

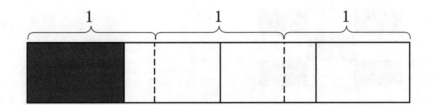

The shaded portion represents $\frac{1}{4}$ of 3 or $\frac{3}{4}$.

 The last meaning of a fraction is a comparison between two numbers.

Comparing 3 to 4 is written as $\frac{3}{4}$ or $3:4$. This is called a *ratio*.

PRACTICE EXERCISE 64

For each diagram in the first five problems, answer the following questions:

a. Into how many equal parts has the figure been divided?
b. Write the unit fraction associated with one of the parts.
c. Write the fraction that represents the shaded portion of each figure.

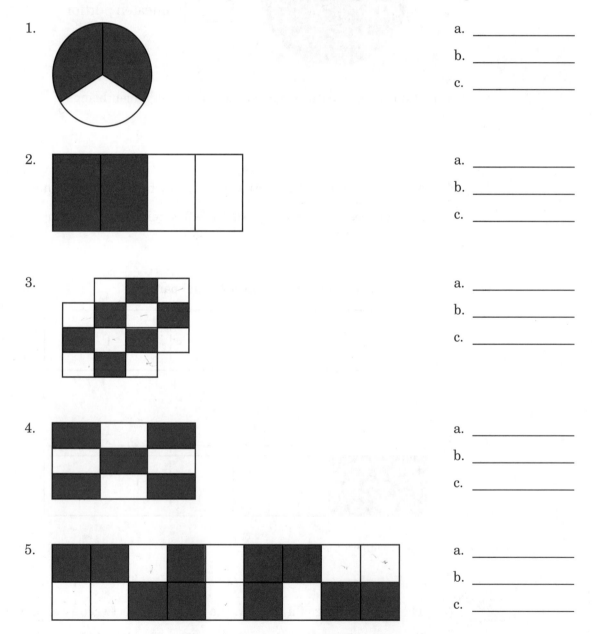

1.   a. _____

                                                        b. _____

                                                        c. _____

2.   a. _____

                                                        b. _____

                                                        c. _____

3.   a. _____

                                                        b. _____

                                                        c. _____

4.   a. _____

                                                        b. _____

                                                        c. _____

5.   a. _____

                                                        b. _____

                                                        c. _____

6. Which is the *unit fraction*?

    a. $\dfrac{3}{8}$ b. $\dfrac{2}{5}$ c. $\dfrac{1}{3}$ d. $\dfrac{3}{1}$ _____

7. Mrs. White used 5 eggs to make a cake. Write the fraction that
    represents the part of a dozen eggs used. (12 eggs = 1 dozen)   _____

8. Joe took 7 foul shots in a basketball game. Write the fraction
    of the total number of shots that were successful if he made 4
    baskets.   _____

9. The fraction $\dfrac{5}{8}$ means

    a. $5\overline{)8}$ b. $8\overline{)5}$ _____

10. The ratio 6 ft. to 9 ft. is written as the fraction

    a. $\dfrac{9}{6}$ b. $\dfrac{6}{9}$ c. Neither of these   _____

6.2 The Language of a Fraction

EXAMPLE

Three of the four cupcakes  were eaten at dinner. Write the fraction
which represents the portion of the cupcakes which were eaten.

SOLUTION

Since there were originally 4 cupcakes and 3 were eaten, then $\dfrac{3}{4}$ is the fraction representing the portion that was eaten.

$$\dfrac{3}{4} = \dfrac{\text{the number of eaten cupcakes}}{\text{the total number of cupcakes}} = \dfrac{\text{numerator}}{\text{denominator}}$$

The 4 tells us the total number of parts into which the whole has been divided. This part
of a fraction is called the *denominator*. The 3 is called the *numerator*; it tells us the number
of parts we are talking about. Both the numerator and the denominator of a fraction are
called *terms* of the fraction. The unit fraction must have a numerator equal to 1. A fraction may have any value for the denominator except 0. Remember, in the last chapter in

division, it was shown that you cannot divide by 0; therefore no meaningful fraction can have a denominator of 0.

$$\frac{3}{4} = \frac{\text{numerator}}{\text{denominator}}$$

You must know the following terms to understand fractions.

TERMS YOU SHOULD REMEMBER

Fraction   A part of something or a quotient of two quantities.
Denominator   The quantity written below the line indicating division in a fraction.
Numerator   The quantity written above the line in a fraction.
Terms of a fraction   The numerator and denominator of a fraction.

6.3 What Kinds of Fractions Are There?

There are two kinds of fractions. The fraction that is most common to us is the *proper fraction*. These are all proper fractions: $\frac{3}{4}, \frac{5}{10}, \frac{7}{8}$. *Proper fractions have a numerator which is less than the denominator.* Thus, the value of any proper fraction must be less than 1.

The second type of fraction has a numerator equal to or greater than the denominator. Examples of *improper fractions* are: $\frac{6}{6}, \frac{9}{4}, \frac{32}{5}$. *Improper fractions have a value equal to or greater than 1.*

Proper fraction	*Illustration (shaded part)*	

Value less than 1

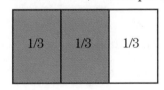

two-thirds
$= \frac{2}{3}$

Improper fraction

a.   Value equal to 1

three-thirds
$= \frac{3}{3}$

b. Value greater
than 1

| 1/3 | 1/3 | 1/3 |

| 1/3 | 1/3 | 1/3 |

four-thirds
$= \dfrac{4}{3}$

PRACTICE EXERCISE 65

1. Identify each of these as a *proper* or an *improper* fraction:

a. $\dfrac{5}{2}$ _____

b. $\dfrac{4}{5}$ _____

c. $\dfrac{9}{10}$ _____

d. $\dfrac{6}{6}$ _____

e. $\dfrac{7}{4}$ _____

In problems 2 through 4 assume that each separate rectangle or triangle represents a unit, 1. The combined shaded pieces in each diagram then represent a certain fractional part of a unit.

a. Write each fraction.

b. Identify the kind of fraction it is.

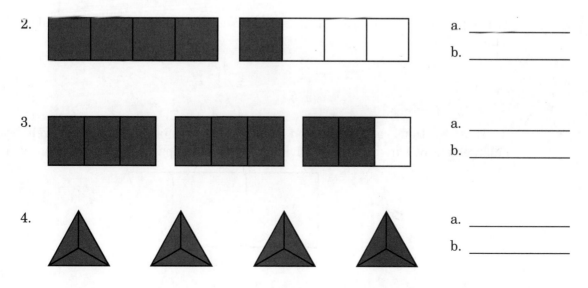

2. a. _____
    b. _____

3. a. _____
    b. _____

4. a. _____
    b. _____

5. a. An entire pie contains $\dfrac{?}{8}$.   a. _____

    b. What kind of fraction is this?   b. _____

6. A carpenter marked off 7 of the equal divisions on a 12 in.
    ruler.

    a. What fractional part of the ruler did he mark off?   a. _____

    b. What kind of fraction do we call this?   b. _____

6.4 Simplifying Fractions

Which fraction is larger: $\dfrac{4}{8}$ or $\dfrac{1}{2}$?

$\dfrac{4}{8}$ is illustrated as:

$\dfrac{4}{8} = \dfrac{\text{the number of shaded parts}}{\text{the total number of parts}}$

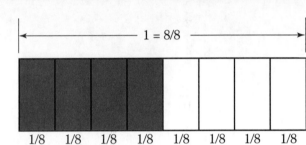

If we assume the whole is the same for both fractions, then

$\dfrac{1}{2}$ is illustrated as:

$\dfrac{1}{2} = \dfrac{\text{the number of shaded parts}}{\text{the total number of parts}}$

Which fraction is the larger? Obviously, neither one since they are both equal to each other.
Thus:

$$\frac{4}{8} = \frac{1}{2}$$

These fractions can be found on a ruler along with other fractions as well. Look at an
enlargement of 1 in. on a ruler. Divide it into 2 equal parts. Each part is marked.

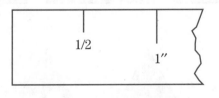

Divide it into 4 equal parts. Each part is $\frac{1}{4}$ of an inch and marked.

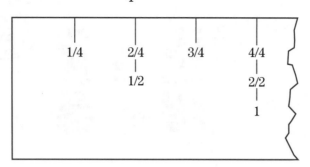

Obviously,

$$\frac{2}{4} = \frac{1}{2} \quad \text{and} \quad \frac{4}{4} = \frac{2}{2} = 1$$

If we continue the division into 8 equal parts, each part is $\frac{1}{8}$ of an inch and is marked accordingly.

Again,

$$\frac{2}{8} = \frac{1}{4}$$

$$\frac{4}{8} = \frac{2}{4} = \frac{1}{2}$$

$$\frac{6}{8} = \frac{3}{4}$$

$$\frac{8}{8} = \frac{4}{4} = \frac{2}{2} = 1$$

Note that in each of the four cases, a fraction with larger terms can be changed to one with smaller terms to which it is equivalent by dividing both terms of the original fraction by the same number, thus reducing its size. The fraction $\frac{1}{2}$ as compared to $\frac{2}{4}$ or $\frac{4}{8}$ is written in its *simplest form*. Some people call it a *reduced fraction*.

The fraction $\dfrac{2}{4}$ is not a fraction in its simplest form, for both the numerator and denominator can be divided by 2. To learn how to simplify a fraction look at this problem:

Problem	Step 1	Step 2
$\dfrac{2}{4} =$	$\dfrac{2 \div 2}{4 \div 2} =$	$\dfrac{1}{2}$
	Find the *largest* number which divides evenly into both the numerator and denominator. In this case, the number is 2.	Divide each term of the fraction by 2. The fraction is equal to $\dfrac{1}{2}$.

Thus: $\dfrac{2}{4} = \dfrac{1}{2}$.

No number other than 1 divides evenly into both terms of the fraction $\dfrac{1}{2}$; therefore, $\dfrac{1}{2}$ cannot be simplified further. Another fraction written in simplest form is $\dfrac{8}{15}$.

EXAMPLE

Simplify the fraction $\dfrac{8}{12}$.

SOLUTION

Problem	Step 1	Step 2
$\dfrac{8}{12} =$	$\dfrac{8 \div 4}{12 \div 4} =$	$\dfrac{2}{3}$
	4 is the *largest* number which divides evenly (without leaving a remainder) into both the numerator (8) and the denominator (12)	Divide each term by 4: $\dfrac{8}{12} = \dfrac{2}{3}$

Thus: $\dfrac{8}{12} = \dfrac{2}{3}$.

PRACTICE EXERCISE 66

Write each fraction in its simplest (or reduced) form. Some fractions cannot be simplified.

1. $\dfrac{2}{8} =$ 2. $\dfrac{5}{10} =$ 3. $\dfrac{4}{20} =$ 4. $\dfrac{3}{12} =$ 5. $\dfrac{6}{12} =$

6. $\dfrac{3}{13} =$ 7. $\dfrac{8}{16} =$ 8. $\dfrac{8}{18} =$ 9. $\dfrac{12}{17} =$ 10. $\dfrac{28}{32} =$

11. $\dfrac{6}{15} =$ 12. $\dfrac{9}{12} =$ 13. $\dfrac{14}{28} =$ 14. $\dfrac{10}{15} =$ 15. $\dfrac{21}{24} =$

EXAMPLE

Simplify the fraction $\dfrac{26}{39}$.

SOLUTION

Is there a number which will divide evenly into both 26 and 39? If there is, it is fairly difficult to find. Remember, if there is no number other than 1 which divides into both terms of the fraction, then the fraction cannot be simplified. If there is such a number, then what is it and how do you find it?

The numerator 26 has a number of divisors. Divisors are numbers which will divide evenly into a number. The divisors of 26 are 1, 2, 13, and 26. Each divisor is called a *factor* of the number. Thus, 1, 2, 13, and 26 are all *factors* of 26.

The denominator 39 has factors of 1, 3, 13, and 39. Thus the numbers 26 and 39 have a *common factor* or common divisor of 13. The number 13 can divide evenly into both terms of the fraction, permitting you to simplify the fraction like this:

Problem	Step 1	Step 2
$\dfrac{26}{39} =$	$\dfrac{26 \div 13}{39 \div 13} =$	$\dfrac{2}{3}$
	Each term can be divided evenly by 13.	Dividing each term results in the fraction $\dfrac{2}{3}$.

Thus: $\dfrac{26}{39} = \dfrac{2}{3}$.

Every number has a common factor or divisor of 1. Since division by 1 results in the same quotient as the original number, there is no point in dividing by 1.

EXAMPLE

Simplify the fraction $\dfrac{33}{49}$.

SOLUTION

The numerator 33 has factors of 1, 3, 11, and 33. The denominator 49 has factors of 1, 7, and 49. Since they have no common factors or divisors other than 1, the fraction is already in its simplest form.

PRACTICE EXERCISE 67

Reduce, if possible, each of the following fractions to its simplest form (lowest term).

1. $\dfrac{25}{50} =$ 2. $\dfrac{44}{77} =$ 3. $\dfrac{35}{47} =$ 4. $\dfrac{45}{63} =$ 5. $\dfrac{35}{49} =$

6. $\dfrac{34}{68} =$ 7. $\dfrac{25}{95} =$ 8. $\dfrac{38}{68} =$ 9. $\dfrac{21}{35} =$ 10. $\dfrac{42}{48} =$

6.5 Equivalent Fractions

When studying the ruler in section 6.4, you learned that $\dfrac{4}{8} = \dfrac{2}{4} = \dfrac{1}{2}$. Each fraction is written differently but the value of all three fractions is the same. You learned that $\dfrac{4}{8}$ and $\dfrac{2}{4}$ can be simplified to $\dfrac{1}{2}$ by dividing both terms of the fraction by the largest common divisor. Reversing this procedure will allow you to find other fractions with larger terms which are equivalent to a given fraction. This is illustrated below.

EXAMPLE

Find two fractions equivalent to $\dfrac{1}{3}$.

SOLUTION

Reversing the procedure means that you must multiply, rather than divide, each term of the fraction by a common number. This common number is called a *common multiplier*. Suppose your common multiplier is 2. This is illustrated as

$$\frac{1 \times 2}{3 \times 2} = \frac{2}{6}$$

Another multiplier can be 4. Thus:

$$\frac{1 \times 4}{3 \times 4} = \frac{4}{12}$$

Is $\dfrac{1}{3} = \dfrac{2}{6} = \dfrac{4}{12}$?

Are these three fractions equivalent fractions? Picture each of these fractions as parts of equal sized rectangles.

1/3   2/6   4/12

The shaded portions of each rectangle are equal despite the fact that the fractions contain different terms. The form of the fraction changes but the value remains the same.

Problem	Step 1	Step 2
$\dfrac{1}{3} = \dfrac{?}{6}$	$\dfrac{1 \times 2}{3 \times 2} = \dfrac{?}{6}$	$\dfrac{2}{6}$
	Determine the value of the common multiplier: $3 \times ? = 6; ? = 2$	Multiply both terms of the fraction by 2. Thus: $\dfrac{1}{3} = \dfrac{2}{6}$

Thus: $\dfrac{1}{3} = \dfrac{2}{6}$.

Problem	Step 1	Step 2
$\dfrac{3}{4} = \dfrac{?}{32}$	$\dfrac{3 \times 8}{4 \times 8} = \dfrac{?}{32}$	$\dfrac{24}{32}$
	Determine the value of the common multiplier: $4 \times ? = 32; ? = 8$	Multiply both terms of the fraction by 8.

Thus: $\dfrac{3}{4} = \dfrac{24}{32}$.

PRACTICE EXERCISE 68

Change each of the following fractions to the new denominator. (Change each fraction to higher terms).

1. $\dfrac{2}{3} = \dfrac{}{6}$

2. $\dfrac{2}{5} = \dfrac{}{35}$

3. $\dfrac{3}{5} = \dfrac{}{25}$

4. $\dfrac{3}{4} = \dfrac{}{8}$

5. $\dfrac{1}{8} = \dfrac{}{24}$

6. $\dfrac{1}{2} = \dfrac{}{8}$

7. $\dfrac{2}{9} = \dfrac{}{36}$

8. $\dfrac{3}{8} = \dfrac{}{40}$

9. $\dfrac{9}{10} = \dfrac{}{40}$

10. $\dfrac{5}{6} = \dfrac{}{18}$

11. $\dfrac{2}{5} = \dfrac{}{15}$

12. $\dfrac{2}{3} = \dfrac{}{12}$

13. $\dfrac{1}{6} = \dfrac{}{24}$

14. $\dfrac{7}{8} = \dfrac{}{16}$

15. $\dfrac{3}{4} = \dfrac{}{16}$

TERMS YOU SHOULD REMEMBER

Divisor   The quantity by which the dividend is to be divided.

Factor   One of two or more quantities having a designated product: 2 and 3 are factors of 6.

Simplest form   The way of writing a fraction that uses the smallest numbers (lowest terms).

Equivalent   Equal in value.

Lowest terms   The parts of a fraction whose terms cannot be reduced because they have no common factor.

Higher terms   The parts of a fraction whose terms can be reduced because they have a common factor other than 1.

6.6 Which Fraction Is Larger or Smaller?

Which would you rather have, $\frac{1}{4}$ of a dollar of $\frac{5}{12}$ of a dollar?

Certainly, you would choose the larger fractional part but which fraction is the larger?

If the denominators of both fractions are the same, it is an easy task to find the larger or smaller fraction. Look at the example below:

$$\frac{5}{12} = \frac{5 \text{ of those parts}}{\text{number the whole is divided into}}$$

$$\frac{3}{12} = \frac{3 \text{ of those parts}}{\text{number the whole is divided into}}$$

If the whole in both fractions is the same, then obviously $\frac{5}{12}$ is greater than $\frac{3}{12}$. Recall the inequality symbol:

$$\frac{5}{12} > \frac{3}{12}$$

What happens if the denominators are not alike? Which fraction is larger, $\frac{1}{4}$ or $\frac{5}{12}$?

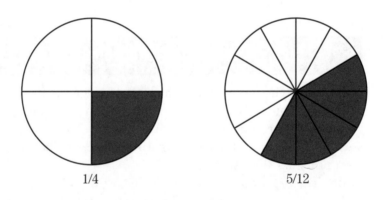

1/4   5/12

You can see that

$$\frac{5}{12} > \frac{1}{4}$$

This picture or illustration method is not satisfactory when the two fractions are very close in value.

Another method you can use to compare fractions is:

 Change each fraction to the same denominator.

 Choose as the larger fraction the one which has the larger numerator.

Sometimes when changing each fraction to the same denominator you change the denominator with the smaller number to the same number as the one with the larger denominator. Look at the next example:

Problem	Step 1	Step 2	Step 3
Which is larger? $\dfrac{5}{12}$ or $\dfrac{1}{4}$	$\dfrac{1}{4} = \dfrac{?}{12}$	$\dfrac{1 \times 3}{4 \times 3} = \dfrac{3}{12}$	$\dfrac{5}{12} > \dfrac{3}{12}$ or $\dfrac{5}{12} > \dfrac{1}{4}$
	Change the fraction with the smaller denominator, $\dfrac{1}{4}$, to an equivalent fraction whose denominator is 12ths.	Multiply both terms of the fraction by 3. Thus: $\dfrac{1}{4} = \dfrac{3}{12}$	Compare the fractions: $\dfrac{5}{12} > \dfrac{3}{12}$ or $\dfrac{5}{12} > \dfrac{1}{4}$

Thus: $\dfrac{5}{12} > \dfrac{1}{4}$.

PRACTICE EXERCISE 69

Which is the larger fraction?

1. $\dfrac{7}{12}$ or $\dfrac{9}{12}$ _____

2. $\dfrac{5}{6}$ or $\dfrac{4}{6}$ _____

3. $\dfrac{2}{3}$ or $\dfrac{5}{12}$ _____

4. $\dfrac{3}{5}$ or $\dfrac{11}{20}$ _____

Which is the smaller fraction?

5. $\dfrac{9}{10}$ or $\dfrac{8}{10}$ _____

6. $\dfrac{13}{16}$ or $\dfrac{11}{16}$ _____

7. $\dfrac{2}{3}$ or $\dfrac{5}{24}$ _____

8. $\dfrac{5}{6}$ or $\dfrac{7}{18}$ _____

EXAMPLE

Compare the fractions $\dfrac{1}{2}$ and $\dfrac{1}{3}$.

SOLUTION

Sometimes it is very difficult to change the denominator of the fraction with the smaller number to the same number as the one with the larger denominator. In the case of $\dfrac{1}{2} = \dfrac{?}{3}$, or $2 \times ? = 3$, there is no whole number which can be used, so you must find another way of doing this example. You must find a *common denominator* for both fractions. A *common denominator* is a number that is divisible by both denominators. In this example, 6 is divisible by both 2 and 3. Other numbers are also divisible by 2 and 3, but 6 is the *lowest (or least) common denominator* (abbreviated as L.C.D.).

Problem	Step 1	Step 2	Step 3
Compare the fractions:	$\dfrac{1}{2} = \dfrac{?}{6}$	$\dfrac{1 \times 3}{2 \times 3} = \dfrac{3}{6}$	$\dfrac{3}{6} > \dfrac{2}{6}$ or
$\dfrac{1}{2}$ and $\dfrac{1}{3}$	$\dfrac{1}{3} = \dfrac{?}{6}$	$\dfrac{1 \times 2}{3 \times 2} = \dfrac{2}{6}$	$\dfrac{1}{2} > \dfrac{1}{3}$

Thus: $\dfrac{1}{2} > \dfrac{1}{3}$.

EXAMPLE

Arrange the fraction $\dfrac{1}{2}, \dfrac{5}{8}, \dfrac{2}{3}$ in order of size, the largest first.

SOLUTION

First find the *lowest common denominator* for the three fractions. What number will 2, 8, and 3 divide into evenly?

The fraction $\dfrac{1}{2}$ can be changed to an equivalent fraction whose denominator is 2, 4, 6, 8, 10, 12, 14, 16, 18, 20, 22, (24), 26, etc.

The fraction $\dfrac{5}{8}$ can be changed to an equivalent fraction whose denominator is 8, 16, (24), 32, 40, etc.

The fraction $\dfrac{2}{3}$ can be changed to an equivalent fraction whose denominator is 3, 6, 9, 12, 15, 18, 21, (24), 27, 30, etc.

The numbers 2, 8, and 3 will all divide evenly into 24. Thus, 24 is the *lowest common denominator*.

Problem	Step 1	Step 2	Step 3
Compare: $\dfrac{1}{2}, \dfrac{5}{8}, \dfrac{2}{3}$	$\dfrac{1}{2} = \dfrac{?}{24}$ $\dfrac{5}{8} = \dfrac{?}{24}$ $\dfrac{2}{3} = \dfrac{?}{24}$	$\dfrac{1 \times 12}{2 \times 12} = \dfrac{12}{24}$ $\dfrac{5 \times 3}{8 \times 3} = \dfrac{15}{24}$ $\dfrac{2 \times 8}{3 \times 8} = \dfrac{16}{24}$	$\dfrac{16}{24} > \dfrac{15}{24} > \dfrac{12}{24}$ or $\dfrac{2}{3} > \dfrac{5}{8} > \dfrac{1}{2}$

Thus: $\dfrac{2}{3} > \dfrac{5}{8} > \dfrac{1}{2}$.

PRACTICE EXERCISE 70

Find the lowest common denominator for each group of examples.

1. $\dfrac{2}{3}$ and $\dfrac{5}{9}$ _____

2. $\dfrac{1}{4}$ and $\dfrac{2}{3}$ _____

3. $\dfrac{1}{2}$ and $\dfrac{4}{5}$ _____

4. $\dfrac{4}{5}$ and $\dfrac{2}{3}$ _____

5. $\dfrac{3}{10}$ and $\dfrac{7}{30}$ _____

6. $\dfrac{4}{5}$ and $\dfrac{4}{15}$ _____

Arrange the fractions in order of size, the smallest one first.

7. $\dfrac{1}{2}, \dfrac{5}{6}, \dfrac{3}{5}$ _____

8. $\dfrac{4}{5}, \dfrac{5}{12}, \dfrac{5}{6}$ _____

Arrange the fractions in order of size, the largest one first.

9. $\dfrac{1}{2}, \dfrac{4}{5}, \dfrac{2}{3}$ _____

10. $\dfrac{4}{15}, \dfrac{3}{10}, \dfrac{7}{30}$ _____

6.7 Changing Improper Fractions to Mixed Numbers

You learned earlier that the *value of an improper fraction must be equal to or greater than 1*. Fractions such as $\frac{4}{4}, \frac{5}{4}, \frac{6}{4}$, etc., are *improper fractions*.

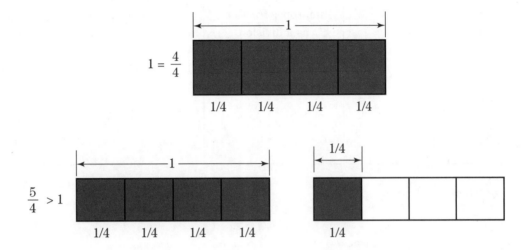

$$1 = \frac{4}{4}$$

$$\frac{5}{4} > 1$$

$\frac{5}{4} = 1$ and $\frac{1}{4}$, which is written as $1\frac{1}{4}$. This is called a *mixed number*. It is made up of two parts:

$$1 = \text{whole number}$$

$$\frac{1}{4} = \text{proper fraction}$$

An improper fraction can be changed into a *mixed number* either by using the method just shown or by dividing. Recall that the line separating the 5 and 4 means that the numerator (5) is being divided by the denominator (4):

$$5 \div 4 \quad \text{or} \quad 4\overline{)5}$$

Problem	Step 1	Step 2	Step 3
$\dfrac{5}{4} = ?$	$4\overline{)5}$	$4\overset{1\ \text{R}\,1}{\overline{)5}}$	$4\overline{)\ 5}\ \ 1\dfrac{1}{4}$
	Write the improper fraction as a division example: $4\overline{)5}$	Divide the denominator into the numerator: $4\overset{1\ \text{R}\,1}{\overline{)5}}$	Write the remainder as the numerator and the divisor as the denominator of a fraction $\dfrac{1}{4}$ *Write it in its simplest form.*

Thus: $\dfrac{5}{4} = 1\dfrac{1}{4}$.

Problem	Step 1	Step 2	Step 3
$\dfrac{6}{4} = ?$	$4\overline{)6}$	$4\overset{1\ \text{R}\,2}{\overline{)6}}$	$4\overline{)\ 6}\ \ 1\dfrac{2}{4} = 1\dfrac{1}{2}$
	Divide the denominator (4) into the numerator (6): $4\overline{)6}$	Divide: 1 R 2	Write the remainder as the numerator and the divisor as the denominator of a fraction *in its simplest form:* $\dfrac{2}{4} = \dfrac{1}{2}$

Thus: $\dfrac{6}{4} = 1\dfrac{1}{2}$.

PRACTICE EXERCISE 71

Write as a simple fraction, the shaded portion of the following, assuming that each circle, rectangle, or triangle represents a whole unit.

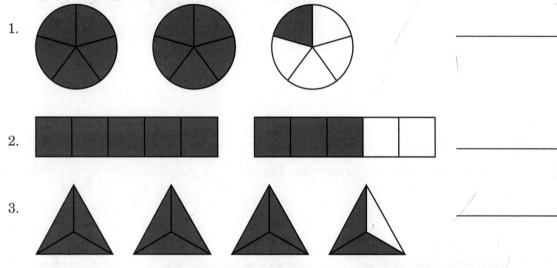

1. _____

2. _____

3. _____

Write each of these mixed numbers in its simplest form.

4. $1\dfrac{6}{8} =$ 5. $2\dfrac{5}{15} =$ 6. $5\dfrac{2}{6} =$

Write each of these improper fractions as a whole or mixed number. Each fraction should be written in simplest form.

7. $\dfrac{13}{5} =$ 8. $\dfrac{2}{2} =$ 9. $\dfrac{13}{2} =$ 10. $\dfrac{8}{6} =$

11. $\dfrac{41}{8} =$ 12. $\dfrac{32}{3} =$ 13. $\dfrac{19}{4} =$ 14. $\dfrac{5}{3} =$

15. $\dfrac{17}{8} =$ 16. $\dfrac{9}{3} =$ 17. $\dfrac{80}{9} =$ 18. $\dfrac{37}{5} =$

19. $\dfrac{43}{2} =$ 20. $\dfrac{32}{5} =$ 21. $\dfrac{18}{4} =$ 22. $\dfrac{5}{5} =$

23. $\dfrac{19}{3} =$ 24. $\dfrac{12}{4} =$ 25. $\dfrac{22}{6} =$

6.8 Changing Mixed Numbers to Improper Fractions

A MIXED NUMBER CONSISTS OF A WHOLE NUMBER AND A PROPER FRACTION!

A mixed number consists of a whole number and a proper fraction. A mixed number can be changed into an improper fraction only. The mixed number $2\frac{1}{2}$ is pictured as:

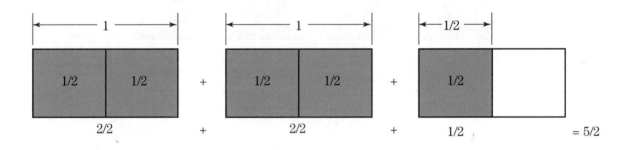

Therefore: $2\frac{1}{2} = \frac{5}{2}$.

EXAMPLE

Change $2\frac{1}{2}$ into an improper fraction.

SOLUTION

Problem	Step 1	Step 2	Step 3
$2\dfrac{1}{2} = \dfrac{?}{2}$	$\dfrac{2 \times 2 + 1}{2} = \dfrac{4+1}{2}$	$\dfrac{4+1}{2} = \dfrac{5}{2}$	$2\dfrac{1}{2} = \dfrac{5}{2}$
	Multiply the whole number (2) by the denominator (2). Add the numerator (1) to the product. Keep the same denominator.	Add the numerators: $\dfrac{4+1}{2} = \dfrac{5}{2}$	

Thus: $2\dfrac{1}{2} = \dfrac{5}{2}$.

EXAMPLE

Change $4\dfrac{2}{3}$ into an improper fraction.

SOLUTION

Problem	Step 1	Step 2	Step 3
$4\dfrac{2}{3} = \dfrac{?}{3}$	$\dfrac{4 \times 3 + 2}{3} = \dfrac{12+2}{3}$	$\dfrac{12+2}{3} = \dfrac{14}{3}$	$4\dfrac{2}{3} = \dfrac{14}{3}$
	Multiply the whole number (4) by the denominator (3). Add the numerator (2) to the product. Keep the same denominator.	Add the numerators. $\dfrac{12+2}{3} = \dfrac{14}{3}$	

Thus: $4\dfrac{2}{3} = \dfrac{14}{3}$.

PRACTICE EXERCISE 72

Change each of the following whole or mixed numbers into improper fractions.

1. $4\dfrac{1}{9} = \dfrac{}{9}$

2. $3\dfrac{1}{2} = \dfrac{}{2}$

3. $6\dfrac{3}{5} = \dfrac{}{5}$

4. $6 = \dfrac{}{5}$

5. $8\dfrac{3}{4} = \dfrac{}{4}$

6. $10\dfrac{3}{7} = \dfrac{}{7}$

7. $7\dfrac{1}{10} = \dfrac{}{10}$

8. $11\dfrac{3}{5} = \dfrac{}{5}$

9. $12\dfrac{5}{8} = \dfrac{}{8}$

10. $4\dfrac{5}{7} = \dfrac{}{7}$

11. $6\dfrac{1}{6} = \dfrac{}{6}$

12. $5 = \dfrac{}{4}$

13. $8 = \dfrac{}{3}$

14. $14\dfrac{2}{3} = \dfrac{}{3}$

15. $23\dfrac{13}{16} = \dfrac{}{16}$

TERMS YOU SHOULD REMEMBER

Proper fraction   A numerical fraction in which the numerator is less than the denominator.

Improper fraction   A numerical fraction in which the numerator is equal to or greater than the denominator.

Reduced fraction   A fraction of the same value but with smaller numerical terms.

Mixed number   A number consisting of a whole number joined to a fraction.

Common denominator   A quantity into which all the denominators of a set of fractions can be evenly divided.

REVIEW OF IMPORTANT IDEAS

Some of the most important ideas in Chapter 6 were:

 A fraction is a quotient of two quantities.

 There are proper and improper fractions.

Fractions can be *reduced*, or written in simplest form.

 Equivalent fractions are fractions whose values are the same but whose respective numerators and denominators may be different.

CHECK WHAT YOU HAVE LEARNED

The following test lets you see how well you have learned the ideas in Chapter 6. Before you begin, though, there is one hint you should be aware of.

You see how the method of doing an example depends on the basic *how and why* rather than on memorization. Nevertheless, there are times when memorizing is important. Memorizing the basic number facts of addition, subtraction, multiplication, and division is a necessity. If you attempt to memorize methods for solving each example or problem, you are doomed to fail. You will always be dependent upon the memorized fact and, if you forget that, you will be unable to solve the problem. Don't let this happen to you. Know the *how and why*, and then the solution to each problem becomes meaningful and understandable.

In counting up your answers, remember that there were 20 separate answers.

A Score of	Means That You
18–20	Did very well. You can move to Chapter 7.
16–17	Know this material except for a few points. Reread the sections about the ones you missed.
13–15	Need to check carefully on the sections you missed.
0–12	Need to review the chapter to refresh your memory and improve your skills.

Questions	Are Covered in Section
1–2, 5–7, 12	6.1
3	6.2
4	6.3
8	6.4
9	6.5
13–15	6.6
10	6.7
11	6.8

Chapter Test 6

Have you worked out your difficulties with the simple operations of fractions? Try to get a satisfactory grade on this chapter test.

Reduce all fractions to their simplest form.

1. Using the diagram shown, write as a fraction the portion of the rectangle that has been shaded.   _____

2. A shopper buys a package of 8 apples and eats 6. What fractional part has been eaten?   _____

3. In the fraction $\dfrac{15}{16}$, name the numerator and the denominator.

    15 is the   _____

    16 is the   _____

4. Identify each fraction below as a *proper* or an *improper* fraction.

    a. $\dfrac{2}{3}$ a. _____

    b. $\dfrac{5}{5}$ b. _____

    c. $\dfrac{6}{5}$ c. _____

5. A line is divided into 3 equal segments. Using a fraction, tell what name is given to one of these parts.   _____

6. If there are 60 minutes in each hour, what fractional part of an hour is 40 minutes?   _____

7. At a party of 25 adults, 15 are women.

    a. What fraction of the party is women?   a. _____

    b. What fraction of the party is men?   b. _____

8. Write the fraction $\dfrac{24}{32}$ in its simplest form.   _____

9. To change $\dfrac{3}{5}$ to an equivalent fraction with a new denominator of 25, you must multiply the numerator and denominator by a certain number. What is that number?   _____

10. Change the improper fraction $\dfrac{23}{8}$ to a mixed number.   _____

11. Change the mixed number $4\dfrac{2}{7}$ to an improper fraction.   _____

12. In an adult education course there are 2 women with blonde hair out of a total of 8 women in the class.

    a. What fraction of the women has blonde hair?   a. _____

    b. What fraction of the women does not have blonde hair?   b. _____

13. Find the lowest common denominator for the fractions:

    $$\dfrac{3}{8}, \dfrac{1}{3}, \text{ and } \dfrac{5}{12}$$

    _____

14. Which is the larger of the two fractions:

    $$\dfrac{3}{4} \text{ or } \dfrac{19}{24}?$$

    _____

15. Arrange these three fractions in order of size, starting with the smallest one:

    $$\dfrac{7}{8}, \dfrac{13}{16}, \dfrac{3}{4}$$

    _____

YOU ARE ON YOUR WAY

You have completed the entire work on whole numbers and you have just entered into the world of fractions. You know that in your everyday life, fractions appear just as frequently as whole numbers. The four operations of addition, subtraction, multiplication, and division of fractions will be studied in the next three chapters.

Piece by Piece– Adding Fractions

In the last chapter you learned about fractions and their relationship to whole and mixed numbers. This chapter discusses how to *use* fractions or how to *operate* with them.

The topic is *addition* of fractions, the first operation dealing with fractions. You know you can figure the total of two or more fractions by adding them together. No matter what fractions or how many fractions you are asked to add together, it can be done by following certain basic steps. Check your skill on this topic in the pretest which follows.

Pretest 7–See What You Know and Remember

Work these exercises carefully, doing as many of these problems as you can. Write your answers in simplified form below each question.

1. $\dfrac{3}{8}$
    $+\dfrac{4}{8}$

2. $\dfrac{1}{6}$
    $+\dfrac{1}{6}$

3. $\dfrac{2}{5}$
    $+\dfrac{3}{5}$

4. $\dfrac{7}{9}$
    $+\dfrac{4}{9}$

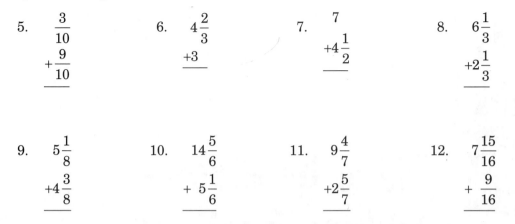

5. $\frac{3}{10}$
$+\frac{9}{10}$
___

6. $4\frac{2}{3}$
$+3$
___

7. 7
$+4\frac{1}{2}$
___

8. $6\frac{1}{3}$
$+2\frac{1}{3}$
___

9. $5\frac{1}{8}$
$+4\frac{3}{8}$
___

10. $14\frac{5}{6}$
$+5\frac{1}{6}$
___

11. $9\frac{4}{7}$
$+2\frac{5}{7}$
___

12. $7\frac{15}{16}$
$+\frac{9}{16}$
___

13. A recipe calls for adding $\frac{3}{8}$ of a cup of water now and $\frac{1}{8}$ of a cup later. How much water is used in this recipe?

14. Mrs. Pink shops for food and returns with $3\frac{1}{4}$ lb. of potatoes, 1 gal. of milk which weighs $3\frac{3}{4}$ lb., and a six-pack of soda which weighs $5\frac{1}{4}$ lbs. What was the total number of pounds Mrs. Pink carried home?

15. Find the least common denominator for these *three* fractions: $\frac{1}{3}, \frac{3}{4}, \frac{5}{6}$

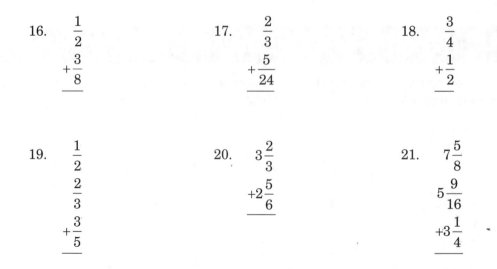

16. $\frac{1}{2}$
$+\frac{3}{8}$
___

17. $\frac{2}{3}$
$+\frac{5}{24}$
___

18. $\frac{3}{4}$
$+\frac{1}{2}$
___

19. $\frac{1}{2}$
$\frac{2}{3}$
$+\frac{3}{5}$
___

20. $3\frac{2}{3}$
$+2\frac{5}{6}$
___

21. $7\frac{5}{8}$
$5\frac{9}{16}$
$+3\frac{1}{4}$
___

22.   How many feet of fencing are required to enclose the vegetable garden pictured below?

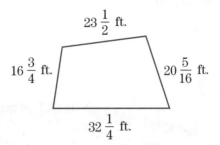

23.   How many yards of rope were removed from a spool if $4\frac{5}{6}$ yd. were used on Monday, $10\frac{1}{2}$ yd. on Tuesday, and $7\frac{1}{12}$ yd. on Wednesday?

Check your answers by turning to the back of the book. Add up all that you had correct.

A Score of	Means That You
21–23	Did very well. You can move to Chapter 8.
18–20	Know the material except for a few points. Read the sections about the ones you missed.
15–17	Need to check carefully on the sections you missed.
0–14	Need to work with the chapter to refresh your memory and improve your skills.

Questions	Are Covered in Section
1	7.1
2	7.2
3	7.3
4–5	7.4
6–8	7.5
9	7.6
10	7.7
11–12	7.8
13–14	7.9
15–19	7.10
20–21	7.11
22–23	7.12

7.1 Fractions Can Be Added

EXAMPLE

Mrs. Calucci baked a dish of lasagna for her family and cut it into 6 equal portions. If Mrs. Calucci serves 3 portions, or $\frac{3}{6}$, for lunch and 2 portions, or $\frac{2}{6}$, for dinner, what part of the lasagna has she served?

SOLUTION

To find out how much was served, you must find the *sum* of $\frac{3}{6}$ and $\frac{2}{6}$. This means you must *add* the two fractions together. The diagram below will help you to discover how fractions are added.

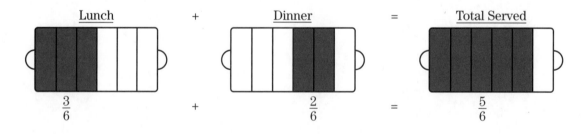

Mrs. Calucci served $\frac{5}{6}$ of her dish of lasagna. These fractions are called *like fractions* since they have the same number in the denominator, namely, sixths or 6ths.

To add like fractions, *add* only the numerators; the denominator remains the same. Look at this example:

Problem	Step 1	Step 2
$\dfrac{3}{6} + \dfrac{2}{6} =$	$\dfrac{3+2}{6}$	$\dfrac{5}{6}$
	Add the numerators: $3 + 2 = 5$	Place this sum (5) over the same denominator (6): $\dfrac{5}{6}$

Thus: $\dfrac{3}{6} + \dfrac{2}{6} = \dfrac{5}{6}$.

TERMS YOU SHOULD REMEMBER

Add   To join to another.
Sum   The result of adding two or more numbers.
Like fractions   Fractions whose denominators are the same number.
Numerator   The number in the top portion of a fraction (dividend).
Denominator   The number in the bottom portion of a fraction (divisor).

PRACTICE EXERCISE 73

Find the *sum* of each of the following problems. Follow the steps shown in the previous example.

1. $\dfrac{1}{5} + \dfrac{2}{5} = \dfrac{}{5}$

2. $\dfrac{1}{3} + \dfrac{1}{3} = \dfrac{}{3}$

3. $\dfrac{2}{4} + \dfrac{1}{4} = \dfrac{}{4}$

4. $\dfrac{4}{7} + \dfrac{1}{7} = \dfrac{}{7}$

5. $\dfrac{2}{9} + \dfrac{4}{9} + \dfrac{1}{9} =$

6. $\dfrac{6}{11} + \dfrac{3}{11} =$

7. $\dfrac{1}{6} + \dfrac{1}{6} + \dfrac{3}{6} =$

8. $\dfrac{1}{8} + \dfrac{2}{8} + \dfrac{4}{8} =$

9. $\dfrac{5}{12} + \dfrac{3}{12} + \dfrac{3}{12} =$

10. $\dfrac{2}{10} + \dfrac{4}{10} + \dfrac{1}{10} =$

7.2 After Adding, What Next?

EXAMPLE

A recipe calls for $\frac{1}{4}$ cup of water now and another $\frac{1}{4}$ cup of water later. How much water is needed in the recipe?

SOLUTION

You must add the fractions $\frac{1}{4} + \frac{1}{4}$ to find how much water is needed. Look at the following example:

Problem	Step 1	Step 2	Step 3
$\frac{1}{4} + \frac{1}{4}$	$\frac{1+1}{4}$	$\frac{2}{4}$	$\frac{2 \div 2}{4 \div 2} = \frac{1}{2}$
	Add the numerators together: $1 + 1 = 2$	Place this sum over the same denominator (4): $\frac{2}{4}$	Write the fraction $\frac{2}{4}$ in its simplest form. Divide the numerator and denominator by 2: $\frac{2}{4} = \frac{1}{2}$

Thus: $\frac{1}{4} + \frac{1}{4} = \frac{1}{2}$.

PRACTICE EXERCISE 74

Add the fractions and write each sum in its simplest form.

1. $\frac{1}{9} + \frac{2}{9} =$

2. $\frac{1}{6} + \frac{3}{6} =$

3. $\frac{3}{8} + \frac{1}{8} =$

4. $\frac{3}{10} + \frac{2}{10} =$

5. $\dfrac{1}{12}+\dfrac{3}{12}=$ 6. $\dfrac{1}{6}+\dfrac{1}{6}+\dfrac{1}{6}=$

7. $\dfrac{1}{10}+\dfrac{2}{10}+\dfrac{5}{10}=$ 8. $\dfrac{7}{12}+\dfrac{1}{12}+\dfrac{1}{12}=$

9. $\dfrac{2}{8}+\dfrac{3}{8}+\dfrac{1}{8}=$ 10. $\dfrac{2}{9}+\dfrac{2}{9}+\dfrac{2}{9}=$

7.3 Fractions Whose Sum Is One

EXAMPLE

David jogged $\dfrac{3}{4}$ of a mile before he got tired and had to rest. He then ran another $\dfrac{1}{4}$ of a mile. How many miles did he run altogether?

SOLUTION

You must find the sum of the two fractions $\dfrac{3}{4}+\dfrac{1}{4}$ to determine the number of miles he ran altogether.

Problem	Step 1	Step 2	Step 3
$\dfrac{3}{4}+\dfrac{1}{4}$	$\dfrac{3+1}{4}$	$\dfrac{4}{4}$	$\dfrac{4}{4}=1$
	Add the numerators: $3+1=4$	Place the sum (4) over the same denominator (4): $\dfrac{4}{4}$	Simplify the fraction: $\dfrac{4}{4}=1$ (Recall that the value of a fraction is 1 if it has the same numerator and denominator. $\dfrac{0}{0}$ is the exception to this rule.)

Thus: $\dfrac{3}{4}+\dfrac{1}{4}=1$.

PRACTICE EXERCISE 75

Add each of the following fractions. Simplify each sum when it is possible.

1. $\dfrac{2}{5} + \dfrac{3}{5} =$

2. $\dfrac{1}{2} + \dfrac{1}{2} =$

3. $\dfrac{3}{8} + \dfrac{4}{8} =$

4. $\dfrac{1}{3} + \dfrac{2}{3} =$

5. $\dfrac{1}{6} + \dfrac{5}{6} =$

6. $\dfrac{1}{7} + \dfrac{4}{7} + \dfrac{2}{7} =$

7. $\dfrac{4}{8} + \dfrac{3}{8} + \dfrac{1}{8} =$

8. $\dfrac{4}{10} + \dfrac{3}{10} + \dfrac{2}{10} =$

9. $\dfrac{4}{12} + \dfrac{4}{12} + \dfrac{4}{12} =$

10. $\dfrac{3}{6} + \dfrac{1}{6} + \dfrac{2}{6} =$

7.4 Fractions Whose Sum Is Greater Than One

EXAMPLE

Mr. Garcia is a carpenter who made two measurements ($\dfrac{7}{8}$ in. and $\dfrac{5}{8}$ in.), as shown below on a ruler. What was the total number of inches he measured?

*Note: The fraction $\dfrac{7}{8}$ can

also be written as 7/8.

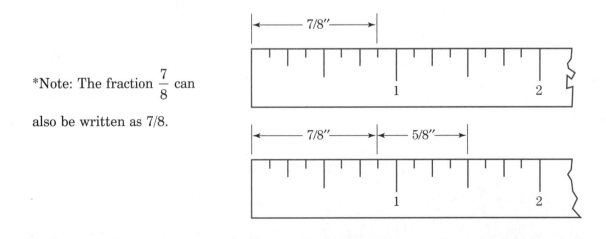

SOLUTION

You must add the fractions $\frac{7}{8}$ and $\frac{5}{8}$ to find the total number of inches.

Problem	Step 1	Step 2	Step 3
$\frac{7}{8}+\frac{5}{8}$	$\frac{7+5}{8}$	$\frac{12}{8}$	$8\overline{)12}\;\;^{1\;R\;4}$ or $1\frac{4}{8}=1\frac{1}{2}$
	Add the numerators: $7+5=12$	Place this sum (12) over the same denominator (8): $\frac{12}{8}$	Change the improper fraction into a mixed number and reduce to simplest form: $\frac{12}{8}=1\frac{4}{8}=1\frac{1}{2}$

Thus: $\frac{7}{8}+\frac{5}{8}=1\frac{1}{2}$.

PRACTICE EXERCISE 76

Add each of the following problems. Write each sum in its simplest form.

1. $\frac{3}{4}+\frac{2}{4}=$

2. $\frac{2}{3}+\frac{2}{3}=$

3. $\frac{2}{5}+\frac{4}{5}=$

4. $\frac{4}{9}+\frac{7}{9}=$

5. $\frac{7}{10}+\frac{5}{10}=$

6. $\frac{3}{8}+\frac{3}{8}+\frac{6}{8}=$

7. $\frac{1}{12}+\frac{7}{12}+\frac{6}{12}=$

8. $\frac{5}{6}+\frac{5}{6}+\frac{5}{6}=$

9. $\frac{2}{7}+\frac{3}{7}+\frac{4}{7}=$

10. $\frac{2}{3}+\frac{2}{3}+\frac{2}{3}=$

7.5 Mixed Numbers Can Also Be Added

If you can add whole numbers and you can add fractions, then you can add mixed numbers.

EXAMPLE

A school banner requires $6\frac{1}{3}$ yards of royal blue fabric and $2\frac{1}{3}$ yards of gold fabric. How many yards of fabric are needed in total?

SOLUTION

You must add $6\frac{1}{3} + 2\frac{1}{3}$ to find the answer to your problem. Add the fractions first and then the whole numbers. Look at the following example.

Problem	Step 1	Step 2
$6\frac{1}{3}$ $+2\frac{1}{3}$	$6\frac{1}{3}$ $+2\frac{1}{3}$ $\frac{2}{3}$	$6\frac{1}{3}$ $+2\frac{1}{3}$ $8\frac{2}{3}$
	Add the fractions: $\frac{1}{3} + \frac{1}{3} = \frac{2}{3}$	Add the whole numbers $6 + 2 = 8$

Thus: $6\frac{1}{3} + 2\frac{1}{3} = 8\frac{2}{3}$.

PRACTICE EXERCISE 77

Add these examples.

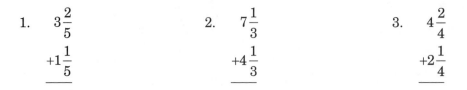

1. $3\frac{2}{5}$
    $+1\frac{1}{5}$

2. $7\frac{1}{3}$
    $+4\frac{1}{3}$

3. $4\frac{2}{4}$
    $+2\frac{1}{4}$

4. $\dfrac{4}{7}$

 $+7\dfrac{1}{7}$
 ___

5. $4\dfrac{2}{9}$

 $3\dfrac{4}{9}$

 $+1\dfrac{1}{9}$
 ___

6. $13\dfrac{6}{11}$

 $+\quad\dfrac{3}{11}$
 ___

7. $4\dfrac{2}{3}$

 $\underline{+3}\left(\text{Hint: }\dfrac{2}{3}+0=\dfrac{2}{3}\right)$

8. 5

 $+2\dfrac{1}{2}$
 ___

7.6 Mixed Numbers Whose Sum May Be Simplified

If the sum of the mixed numbers is also a mixed number, sometimes the fractional part can be written in simplified form. Let's see what happens when this occurs.

Problem	Step 1	Step 2	Step 3
$4\dfrac{2}{9}$ $+2\dfrac{1}{9}$ ___	$4\dfrac{2}{9}$ $+2\dfrac{1}{9}$ ___ $\dfrac{3}{9}$	$4\dfrac{2}{9}$ $+2\dfrac{1}{9}$ ___ $6\dfrac{3}{9}$	$6\dfrac{3}{9}=6\dfrac{1}{3}$
	Add the fractions: $\dfrac{2}{9}+\dfrac{1}{9}=\dfrac{3}{9}$	Add the whole numbers: $4+2=6$	Simplify the fraction (if possible): $\dfrac{3}{9}=\dfrac{1}{3}$

Thus: $4\dfrac{2}{9}+2\dfrac{1}{9}=6\dfrac{1}{3}$.

PRACTICE EXERCISE 78

Add the following mixed numbers, reducing all fractions to the simplest terms.

1. $5\dfrac{3}{8}$

    $+1\dfrac{1}{8}$
    _____

2. $4\dfrac{1}{6}$

    $+3\dfrac{3}{6}$
    _____

3. $23\dfrac{1}{12}$

    $+\dfrac{3}{12}$
    _____

4. $13\dfrac{3}{10}$

    $+9\dfrac{2}{10}$
    _____

5. $10\dfrac{1}{10}$

    $9\dfrac{2}{10}$

    $+7\dfrac{5}{10}$
    _____

6. $9\dfrac{7}{12}$

    $\dfrac{1}{12}$

    $+3\dfrac{1}{12}$
    _____

7. $7\dfrac{2}{8}$

    $4\dfrac{3}{8}$

    $+\dfrac{1}{8}$
    _____

8. $\dfrac{2}{9}$

    $4\dfrac{2}{9}$

    $+3\dfrac{2}{9}$
    _____

7.7 Mixed Numbers Whose Fractional Parts Add Up to One

If the sum of the fractions is one, you add that to the whole number sum. Look at this problem:

Problem	Step 1	Step 2	Step 3
$4\dfrac{2}{5}$ $+3\dfrac{3}{5}$	$4\dfrac{2}{5}$ $+3\dfrac{3}{5}$ $\overline{\dfrac{5}{5}}$	$4\dfrac{2}{5}$ $+3\dfrac{3}{5}$ $\overline{7\dfrac{5}{5}}$	$\dfrac{5}{5}=1$ $7+1=8$
	Add the fractions: $\dfrac{2}{5}+\dfrac{3}{5}=\dfrac{5}{5}$	Add the whole numbers: $4+3=7$	Change $\dfrac{5}{5}$ to 1. Add 1 to the whole number sum: $7+1=8$

Thus: $4\dfrac{2}{5}+3\dfrac{3}{5}=8$.

PRACTICE EXERCISE 79

Add these examples.

1. $5\dfrac{1}{2}$
    $+2\dfrac{1}{2}$

2. $4\dfrac{3}{8}$
    $+2\dfrac{5}{8}$

3. $10\dfrac{2}{3}$
    $+8\dfrac{1}{3}$

4. $22\dfrac{5}{6}$
    $+19\dfrac{1}{6}$

5. $3\dfrac{1}{7}$
    $4\dfrac{4}{7}$
    $+2\dfrac{2}{7}$

6. $\dfrac{4}{8}$
    $1\dfrac{3}{8}$
    $+6\dfrac{1}{8}$

7. $4\dfrac{4}{12}$
    $4\dfrac{4}{12}$
    $+4\dfrac{4}{12}$

8. $6\dfrac{3}{6}$
    $5\dfrac{1}{6}$
    $+\dfrac{2}{6}$

7.8 Mixed Numbers with Fractional Parts Whose Sum Is Greater Than One

EXAMPLE

Mr. Lopez is an ironworker. He wants to cut three pieces of iron of the following lengths: $3\frac{1}{4}$ ft., $2\frac{1}{4}$ ft., and $\frac{3}{4}$ ft. How long a piece must he start with if waste from cutting is not considered?

SOLUTION

The problem is to find the sum of $3\frac{1}{4} + 2\frac{1}{4} + \frac{3}{4}$. Here is how to solve the example.

Problem	Step 1	Step 2	Step 3
$3\frac{1}{4}$ $2\frac{1}{4}$ $+\frac{3}{4}$	$3\frac{1}{4}$ $2\frac{1}{4}$ $+\frac{3}{4}$ $\frac{5}{4}$	$3\frac{1}{4}$ $2\frac{1}{4}$ $+\frac{3}{4}$ $5\frac{5}{4}$	$\frac{5}{4} = 1\frac{1}{4}$, so $5 + 1\frac{1}{4} = 6\frac{1}{4}$
	Add the fractions: $\frac{1}{4} + \frac{1}{4} + \frac{3}{4} = \frac{5}{4}$	Add the whole numbers: $3 + 2 = 5$	Since $\frac{5}{4} = 1\frac{1}{4}$, add that to the whole number 5: $\frac{5}{4} + 1\frac{1}{4} = 6\frac{1}{4}$

Thus: $3\frac{1}{4}$ ft. $+ 2\frac{1}{4}$ ft. $+ \frac{3}{4}$ ft. $= 6\frac{1}{4}$ ft.

PRACTICE EXERCISE 80

Add these mixed number problems, reducing fractions to the lowest terms.

1. $5\dfrac{2}{3}$
    $+2\dfrac{2}{3}$
    _____

2. $6\dfrac{2}{5}$
    $+4\dfrac{4}{5}$
    _____

3. $4\dfrac{7}{9}$
    $+7\dfrac{4}{9}$
    _____

4. $8\dfrac{7}{10}$
    $+4\dfrac{5}{10}$
    _____

5. $4\dfrac{3}{8}$
    $\dfrac{3}{8}$
    $+7\dfrac{6}{8}$
    _____

6. $6\dfrac{1}{12}$
    $7\dfrac{7}{12}$
    $+1\dfrac{6}{12}$
    _____

7. $5\dfrac{2}{7}$
    $\dfrac{3}{7}$
    $+2\dfrac{4}{7}$
    _____

8. $\dfrac{2}{3}$
    $4\dfrac{2}{3}$
    $+5\dfrac{2}{3}$
    _____

7.9 Word Problems

Before you begin, go back to the general instructions in Chapter 2, section 2.7, to review the methods used to solve any word problem. The rules for solving a problem in addition also apply to problems in adding fractions.

PRACTICE EXERCISE 81

Solve these word problems. Reduce fractions to lowest terms.

1. A recipe calls for $\frac{1}{4}$ cup of sugar now and $\frac{3}{4}$ cup of sugar later. What is the total amount of sugar used?

2. If Juan paints $\frac{2}{7}$ of the room today and $\frac{3}{7}$ of it tomorrow, how much of the room will he have painted?

3. The Maple Grove drama club needs $4\frac{1}{8}$ yd. of fabric for one stage prop and $6\frac{5}{8}$ yd. for another prop. How many yards of fabric are needed for both?

4. Alice is a part-time hospital employee who works $6\frac{1}{4}$ h. on Monday and $3\frac{1}{4}$ h. on Wednesday. What is the total number of hours she works on these two days?

5. Three lengths of $4\frac{5}{6}$ yd., $10\frac{1}{6}$ yd., and 7 yd. are cut from a large ball of twine. How many yards have been removed?

6. Your employer withholds $\frac{6}{20}$ of your salary for Federal income taxes, $\frac{1}{20}$ for state taxes, and $\frac{1}{20}$ for city taxes. How much of your pay is taken out for taxes?

7. In September $2\frac{1}{16}$ in. of rain fell and in October $4\frac{5}{16}$ in. fell. How much rain fell in those two months?

8. Jack was on a diet and lost $6\frac{7}{8}$ lb. during a three-month period; he lost another $1\frac{3}{8}$ lb. during the summer. How much less does he weigh now since he began his diet?

9. Manuel needs three pieces of wood, one $16\frac{3}{16}$ in. long, a second one $5\frac{7}{16}$ in. long, and a third one $11\frac{7}{16}$ in. long. What is the total length of wood he needs?

10. In a March of Dimes Walkathon, Daisy walked $3\frac{1}{4}$ mi., Antoinette walked $5\frac{3}{4}$ mi., and Luz walked $1\frac{3}{4}$ mi. What is the total number of miles walked by the three girls?

7.10 What About Those Denominators?

All of the fractions you have added so far have had the *same* denominator. When they have the same denominator, you keep the same denominator and add only the numerators. See the following example:

$$\frac{1}{5}+\frac{2}{5}=\frac{1+2}{5}=\frac{3}{5}$$

How do you add two fractions which have *unlike* denominators?

$$\frac{2}{3}+\frac{1}{6}=?$$

You know how to add fractions that have like denominators so let's see how you can relate this problem to the method you already know.

Can you rewrite a fraction so that it has a different denominator? These are *equivalent fractions*, which you learned about in the last chapter. To change two or more fractions into equivalent fractions you must first find the *lowest common denominator*. Review section 6.6 in Chapter 6 if it is necessary.

Remember: A common denominator is a number into which all the other denominators will divide without leaving a remainder.

Common denominators for the problem above are 6, 12, 18, 24, etc., since 3 and 6 can divide evenly into each of these numbers. You choose 6 since it is the *lowest common denominator* (L.C.D.).

Change $\frac{2}{3}$ to *sixths* by *multiplying* both the numerator and denominator by 2:

$$\frac{2}{3}=\frac{2\times2}{3\times2}=\frac{4}{6}$$

Now you add:

$$\frac{4}{6}+\frac{1}{6}=\frac{5}{6}$$

Look at the example.

EXAMPLE

Mrs. Morse purchased $\frac{5}{8}$ yd. of fabric to make a tote bag. She found she needed $\frac{1}{4}$ yd. more fabric to make pockets. How many yards of fabric in all will Mrs. Morse need to purchase?

SOLUTION

Problem	Step 1	Step 2	Step 3
$\frac{5}{8}+\frac{1}{4}=$	L.C.D. = 8	$\frac{5}{8}=\frac{5}{8}$ $\frac{1}{4}=\frac{1\times2}{4\times2}=\frac{2}{8}$	$\frac{5}{8}+\frac{2}{8}=\frac{7}{8}$
	The L.C.D. for the fractions is 8, since 8 is divisible by both 4 and 8.	Change each fraction into an equivalent fraction whose denominator is 8: $\frac{5}{8}=\frac{5}{8}$ $\frac{1}{4}=\frac{2}{8}$	Add the fractions: $\frac{5}{8}+\frac{2}{8}=\frac{7}{8}$

Thus: $\frac{5}{8}$ yd. $+\frac{1}{4}$ yd. $=\frac{7}{8}$ yd.

EXAMPLE

Find the sum of $\frac{5}{6}$ and $\frac{2}{5}$.

SOLUTION

Problem	Step 1	Step 2	Step 3
$\dfrac{5}{6}+\dfrac{2}{5}=$	L.C.D. = 30	$\dfrac{5}{6}=\dfrac{5\times5}{6\times5}=\dfrac{25}{30}$ $\dfrac{2}{5}=\dfrac{2\times6}{5\times6}=\dfrac{12}{30}$	$\dfrac{25}{30}+\dfrac{12}{30}=\dfrac{37}{30}$ $=1\dfrac{7}{30}$
	The L.C.D. for the fractions is 30. Both 6 and 5 divide evenly into 30.	Change each fraction into an equivalent fraction whose denominator is 30: $\dfrac{5}{6}=\dfrac{25}{30}$ $\dfrac{2}{5}=\dfrac{12}{30}$	Add the fractions: $\dfrac{25}{30}+\dfrac{12}{30}=\dfrac{37}{30}$ Simplify the improper fraction: $\dfrac{37}{30}=1\dfrac{7}{30}$

Thus: $\dfrac{5}{6}+\dfrac{2}{5}=\dfrac{37}{30}=1\dfrac{7}{30}$.

PRACTICE EXERCISE 82

Add the following, and simplify, if possible.

1. $\dfrac{1}{8}+\dfrac{3}{4}=$

2. $\dfrac{1}{3}+\dfrac{1}{6}=$

3. $\dfrac{1}{7}+\dfrac{2}{3}=$

4. $\dfrac{2}{5}+\dfrac{5}{8}=$

5. $\dfrac{2}{3}+\dfrac{3}{5}=$

6. $\dfrac{3}{8}+\dfrac{3}{4}=$

7. $\dfrac{3}{4}+\dfrac{1}{3}=$

8. $\dfrac{1}{6}+\dfrac{3}{4}=$

9. $\dfrac{1}{4}+\dfrac{9}{10}=$

10. $\dfrac{3}{10}+\dfrac{1}{4}+\dfrac{2}{3}=$

11. $\dfrac{1}{2}+\dfrac{7}{8}+\dfrac{3}{4}=$

12. $\dfrac{1}{2}+\dfrac{2}{3}+\dfrac{3}{5}=$

7.11 Mixing Them All Up

Mixed numbers will not give you any trouble now that you know how to add fractions with unlike denominators. After adding the fractions, you add the whole numbers and combine the two. Look at the examples below.

EXAMPLE

Mr. Carrasco finds that he needs two pieces of $\frac{1}{4}$ in. steel rod. Their lengths are $8\frac{5}{8}$ in. and $6\frac{1}{4}$ in. How much steel must he purchase?

SOLUTION

Mr. Carrasco knew he had to find the sum of $8\frac{5}{8}$ and $6\frac{1}{4}$ to know the length of rod he needed.

Problem	Step 1	Step 2	Step 3
$8\frac{5}{8}+6\frac{1}{4}=$	L.C.D. = 8	$8\frac{5}{8}=8\frac{5}{8}$ $6\frac{1}{4}=6\frac{1\times2}{4\times2}=6\frac{2}{8}$	$8\frac{5}{8}+6\frac{2}{8}=14\frac{7}{8}$
	Find the L.C.D. for the fractions $\frac{5}{8}$ and $\frac{1}{4}$: L.C.D. = 8	Change each fraction to the L.C.D. (8): $\frac{5}{8}=\frac{5}{8}$ $\frac{1\times2}{4\times2}=\frac{2}{8}$	Add the fractions: $\frac{5}{8}+\frac{2}{8}=\frac{7}{8}$ Add the whole numbers: $8+6=14$ Combine into a mixed number: $14\frac{7}{8}$

Thus: $8\frac{5}{8}$ in. $=6\frac{1}{4}$ in. $=14\frac{7}{8}$ in.

EXAMPLE

Mrs. Clark needs $2\frac{5}{6}$ cups of milk to make a pudding and $1\frac{1}{3}$ cups of milk to make a cake. Find the total amount of milk required.

SOLUTION

Problem	Step 1	Step 2	Step 3
$2\frac{5}{6}+1\frac{1}{3}=$	L.C.D. = 6	$2\frac{5}{6}=2\frac{5}{6}$ $1\frac{1}{3}=1\frac{1\times2}{3\times2}=1\frac{2}{6}$	$2\frac{5}{6}+1\frac{2}{6}=3\frac{7}{6}$ $=3+1\frac{1}{6}$ $=4\frac{1}{6}$
	The L.C.D. for $\frac{5}{6}$ and $\frac{1}{3}$ is 6.	Change each fraction to the L.C.D. (6): $\frac{5}{6}=\frac{5}{6}$ $\frac{1}{3}=\frac{1\times2}{3\times2}=\frac{2}{6}$	Add: $2\frac{5}{6}+1\frac{2}{6}=3\frac{7}{6}$ Change the fraction $\frac{7}{6}$ to a mixed number: $\frac{7}{6}=1\frac{1}{6}$ Add: $3+1\frac{1}{6}=4\frac{1}{6}$

Thus: $2\frac{5}{6}+1\frac{1}{3}=4\frac{1}{6}$.

PRACTICE EXERCISE 83

Add.

1. $3\dfrac{1}{3}+2\dfrac{1}{2}=$

2. $7\dfrac{1}{2}+3\dfrac{3}{4}=$

3. $6\dfrac{3}{4}+4\dfrac{2}{5}=$

4. $3\dfrac{3}{4}+14\dfrac{7}{8}=$

5. $5\dfrac{3}{4}+3\dfrac{1}{7}=$

6. $7\dfrac{5}{8}+5\dfrac{9}{16}=$

7. $5\dfrac{3}{16}+\dfrac{5}{8}=$

8. $2\dfrac{1}{2}+1\dfrac{3}{5}=$

9. $8\dfrac{7}{8}+3\dfrac{5}{12}=$

10. $2\dfrac{1}{2}+\dfrac{2}{3}+3\dfrac{3}{5}=$

11. $3\dfrac{2}{3}+1\dfrac{1}{12}+10\dfrac{1}{4}=$

7.12 More Difficult Word Problems

Before beginning the practice exercise, review the simpler problems in section 7.9. Be sure you really understand them before proceeding to these more difficult problems.

PRACTICE EXERCISE 84

1. An automobile mechanic worked $3\dfrac{1}{3}$ h. on Monday and $4\dfrac{5}{6}$ h. on Tuesday repairing a car. How many hours had he worked on both days?

2. The distance from city A to city B is $13\dfrac{3}{10}$ miles. City C is $15\dfrac{1}{2}$ miles from city B. How far is it from city A to city C if you must pass through city B?

3. Daisy and Carmen went shopping together. Daisy bought $1\frac{1}{2}$ yd. of material for a skirt and Carmen bought $2\frac{5}{8}$ yd. of material for a dress. How many yards of material did they buy?

4. A bookcase has three shelves whose heights are $8\frac{7}{8}$ in., $6\frac{1}{4}$ in., and $7\frac{1}{2}$ in. What is the total height of the shelves?

5. In a department store $3\frac{2}{3}$, $7\frac{3}{4}$, 3, and $6\frac{1}{2}$ dozen pairs of pantyhose were sold over four days. How many dozen pairs were sold altogether?

6. To complete a job, a plumber finds that he needs pipe of various lengths: $8\frac{5}{8}$ in., $10\frac{1}{2}$ in., $7\frac{3}{16}$ in., and 9 in. What is the total length of pipe needed?

7. Find the total length of the bolt drawn below.

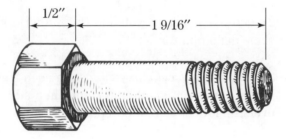

8. Find the length of the template drawn below.

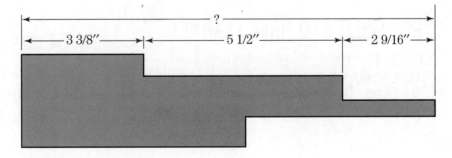

9. At the supermarket Rosa bought a $6\frac{1}{2}$ lb. pork tenderloin, an $8\frac{5}{8}$ lb. turkey, and 2 lb. of chopped meat. What was the total weight of her purchases?

10. A plot of ground is in the shape of a triangle. Find the amount of fencing needed to enclose it if the lengths of its sides are $73\frac{5}{6}$ ft., $63\frac{1}{2}$ ft., and $69\frac{1}{12}$ ft.

TERM YOU SHOULD REMEMBER

Unlike fractions   Fractions whose denominators are not the same number.

REVIEW OF IMPORTANT IDEAS

Some of the most important ideas in Chapter 7 were:

 To add *like fractions* you add the numerators and place the sum over the same denominator.

 To add *unlike fractions* you change each fraction to a fraction with the lowest common denominator and then add the numerators. Place the sum over the lowest common denominator.

 To add *mixed numbers* you add the fractions and the whole numbers separately. Combine the sums into a mixed number.

CHECK WHAT YOU HAVE LEARNED

This chapter covered addition of proper fractions and mixed numbers. Much of this material is probably not new to you but you might have had to review the method needed to add fractions. The chapter test will let you check on your own understanding. Work the exercises carefully and try to improve your score.

A Score of	Means That You
21–23	Did very well. You can move to Chapter 8.
18–20	Know this material except for a few points. Reread the sections about the ones you missed.
15–17	Need to check carefully on the sections you missed.
0–14	Need to review the chapter again to refresh your memory and improve your skills.

Questions	Are Covered in Section
1	7.1
2	7.2
3	7.3
4–5	7.4
6–8	7.5
9	7.6
10	7.7
11–12	7.8
13–14	7.9
15–19	7.10
20–21	7.11
22–23	7.12

Chapter Test 7

Add the following and write your answers in simplified form below each question.

1. $$\frac{3}{5}$$
$$+\frac{1}{5}$$

2. $$\frac{1}{8}$$
$$+\frac{1}{8}$$

3. $$\frac{4}{7}$$
$$+\frac{3}{7}$$

4. $$\frac{7}{11}$$
$$+\frac{5}{11}$$

5. $$\frac{3}{4}$$
$$+\frac{3}{4}$$

6. $$5\frac{1}{6}$$
$$+2$$

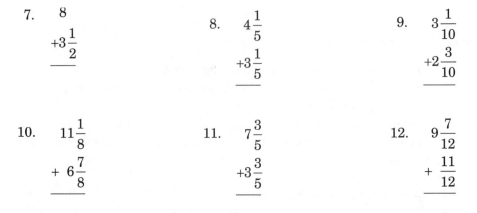

7. 8
    $+3\dfrac{1}{2}$

8. $4\dfrac{1}{5}$
    $+3\dfrac{1}{5}$

9. $3\dfrac{1}{10}$
    $+2\dfrac{3}{10}$

10. $11\dfrac{1}{8}$
    $+\ 6\dfrac{7}{8}$

11. $7\dfrac{3}{5}$
    $+3\dfrac{3}{5}$

12. $9\dfrac{7}{12}$
    $+\ \ \dfrac{11}{12}$

13. Two girls went swimming in a pool. One spent $\dfrac{5}{16}$ of an hour in the pool while the other was in the pool $\dfrac{3}{16}$ of an hour. How long were the girls in the pool?

14. Antonio was studying for his final exams. He read $14\dfrac{3}{4}$ pages of English, $9\dfrac{1}{4}$ pages of science, and $12\dfrac{3}{4}$ pages of math. How many pages did he study?

15. Find the least common denominator for these three fractions: $\dfrac{1}{4}, \dfrac{5}{8}, \dfrac{2}{3}$

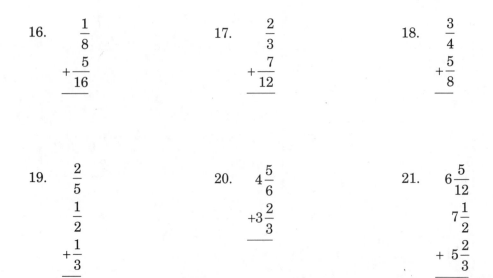

16. $\dfrac{1}{8}$
    $+\dfrac{5}{16}$

17. $\dfrac{2}{3}$
    $+\dfrac{7}{12}$

18. $\dfrac{3}{4}$
    $+\dfrac{5}{8}$

19. $\dfrac{2}{5}$
    $\dfrac{1}{2}$
    $+\dfrac{1}{3}$

20. $4\dfrac{5}{6}$
    $+3\dfrac{2}{3}$

21. $6\dfrac{5}{12}$
    $7\dfrac{1}{2}$
    $+\ 5\dfrac{2}{3}$

22.   How many yards of fencing are required to enclose the four-sided figure drawn below?

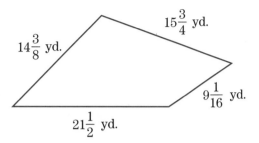

23.   How many feet of wood were removed if $3\frac{3}{4}$ ft., $5\frac{1}{2}$ ft., and $1\frac{5}{12}$ ft. were cut from a 16-ft. board?

HOW TO STUDY MATHEMATICS

Before you continue to the next chapter, here are a few suggestions for studying mathematics. There are certain ideas that must be remembered, such as the meaning of symbols or terms, the formulas used, or the laws or rules that govern which process to use.

If you have not been using mathematics frequently or have not been studying recently, you need time and help to get started again. These suggestions may help you rebuild your skills in mathematics.

HINTS FOR STUDYING MATHEMATICS

- Look through any new material (a chapter or a section) to get a general idea of the topic.
- Locate new terms or symbols and their explanations so that the use of these terms is not confusing.
- Read explanations before you try to work any problem.
- Learn the sequence or order of steps by working the problems as explained by the book.
- Stop and think whether you can explain the problem. Why did you add or multiply or use whatever method you did?
- Use practice problems as a way of checking how well you have learned the techniques. Keep trying practice problems until they become easy for you.
- Don't let your mistakes make you angry or frustrated. Go back over what you have done to see whether your mistake was the way you set up the problem (step by step) or an error in figuring.

Subtracting Fractions

Subtracting fractions is the second basic operation you need to know in working with fractions. Finding the difference between two amounts is a subtraction problem. You will see that subtracting fractions is not much different from adding fractions. In fact, subtracting fractions uses many of the same principles as adding fractions.

The examples which follow will test your knowledge in this type. Do them carefully.

Do as many of these problems as you can. Some are more difficult than others. Write each answer in simplified form in the space provided.

1. $\dfrac{2}{3}$
    $-\dfrac{1}{3}$

2. $\dfrac{3}{4}$
    $-\dfrac{1}{4}$

3. $\dfrac{9}{8}$
    $-\dfrac{1}{8}$

4. $\dfrac{23}{16}$
    $-\dfrac{3}{16}$

5. $2\dfrac{4}{5}$
    $-1\dfrac{2}{5}$

6. $4\dfrac{5}{8}$
    $-3\dfrac{3}{8}$

7. $8\dfrac{4}{7}$
    $-\dfrac{3}{7}$

8. 1
    $-\dfrac{2}{5}$

9. $5\dfrac{1}{5}$
    $-\dfrac{2}{5}$

10. $4\frac{1}{6}$
$-2\frac{5}{6}$
_____

11. $11\frac{1}{2}$
$-\ 2$
_____

12. 9
$-4\frac{1}{3}$
_____

13. Sonia bought $2\frac{1}{4}$ lb. of chopped meat. She used $1\frac{3}{4}$ lb. for spaghetti sauce. How many pounds of chopped meat does she have left?

14. An evening adult education class is $\frac{3}{4}$ of an hour long. Helen is $\frac{1}{4}$ of an hour late for class. How long does Helen spend in the classroom?

15. $\frac{2}{3}$
$-\frac{1}{6}$
_____

16. $\frac{1}{2}$
$-\frac{1}{5}$
_____

17. $3\frac{2}{3}$
$-2\frac{2}{5}$
_____

18. $4\frac{1}{4}$
$-1\frac{1}{2}$
_____

19. $13\frac{1}{8}$
$-\ 4\frac{3}{5}$
_____

20. Maria's mother, a part-time worker, worked $31\frac{1}{4}$ h. last week and $6\frac{2}{3}$ h. less this week. How many hours did she work this week?

21. From the sum of $2\frac{1}{2}$ and $3\frac{3}{5}$ subtract $1\frac{1}{4}$.

22. Antonia had two pieces of fabric, one measuring $2\frac{1}{2}$ yd. and a second measuring $6\frac{3}{4}$ yd. If she used $3\frac{5}{8}$ yd. of fabric for a pants suit, how much material still remains?

23. From a roll cloth $24\frac{7}{8}$ yd. long, a piece $12\frac{3}{4}$ yd. is cut. How many yards of material remain?

Check your answers by turning to the back of the book. Add the ones you had correct.

A Score of   Means That You

21–23   Did very well. You can move to Chapter 9.

18–20   Know the material except for a few points. Read the sections about the ones
               you missed.

15–17   Need to check carefully on the sections you missed.

  0–14   Need to work with the chapter to refresh your memory and improve your
               skills.

Questions	*Are Covered in Section*
1	8.1
2	8.2
3, 4	8.3
5–7, 11	8.4
8, 12	8.5
9, 10	8.6
13, 14	8.7
15, 16	8.8
17, 21	8.9
18, 19	8.10
20, 22, 23	8.11

8.1 Subtracting Fractions

EXAMPLE

Mrs. Jones baked a pie and divided it into 8 equal portions. Each portion was $\frac{1}{8}$ of the whole pie. She ate a piece for lunch, leaving $\frac{7}{8}$ of the pie. At dinner $\frac{4}{8}$ more of the pie was eaten. What part of the pie remains?

SOLUTION

To find out the part of the pie that still remains, you must subtract the fractions:

$$\frac{7}{8} - \frac{4}{8} = ?$$

The diagram below will help you to discover how fractions are subtracted.

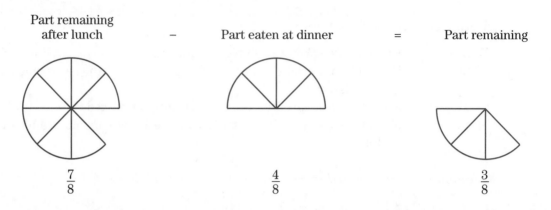

| Part remaining after lunch | − | Part eaten at dinner | = | Part remaining |

$$\frac{7}{8} \qquad \frac{4}{8} \qquad \frac{3}{8}$$

Thus: Mrs. Jones and her family ate $\frac{5}{8}$ of the pie, leaving $\frac{3}{8}$.

These fraction problems are handled exactly like the addition of fractions except that the operation is subtraction. Do you recall the names of the terms in a subtraction example?

$$\begin{array}{r} \frac{7}{8} \\ -\frac{4}{8} \\ \hline \frac{3}{8} \end{array} \text{ is the difference}$$

Fractions such as $\frac{7}{8}$ and $\frac{3}{8}$ are called *like fractions* since they have the same number in the denominator, namely 8. The fractions are *eighths* or 8ths.

In subtracting like fractions, you subtract only the numerators; the denominator remains the same. Look at this example:

Problem	Step 1	Step 2
$\frac{7}{8} - \frac{4}{8} =$	$\frac{7-4}{8}$	$\frac{3}{8}$
	Subtract the numerators: $7 - 4 = 3$	Place the difference (3) over the same denominator (8): $\frac{3}{8}$

Thus: $\frac{7}{8} - \frac{4}{8} = \frac{3}{8}$.

> **TERMS YOU SHOULD REMEMBER**
>
> *Subtract*   To take away.
> *Difference*   That which is left after a part has been taken away.
> *Like fractions*   Fractions whose denominators are the same number.

PRACTICE EXERCISE 85

Find the difference for each of these problems. Follow the steps shown in the previous example.

1. $\dfrac{3}{5} - \dfrac{2}{5} =$

2. $\dfrac{4}{7} - \dfrac{3}{7} =$

3. $\dfrac{6}{11} - \dfrac{3}{11} =$

4. $\dfrac{2}{3} - \dfrac{1}{3} =$

5. $\dfrac{5}{8} - \dfrac{2}{8} =$

6. $\dfrac{3}{4} - \dfrac{2}{4} =$

7. $\dfrac{11}{12} - \dfrac{4}{12} =$

8. $\dfrac{9}{10} - \dfrac{2}{10} =$

9. $\dfrac{6}{6} - \dfrac{1}{6} =$

10. $\dfrac{1}{9} - \dfrac{1}{9} =$

8.2 After Subtracting, What Next?

EXAMPLE

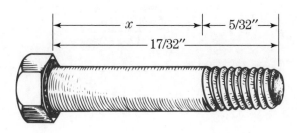

Find the length x in the diagram of the bolt drawn above.

SOLUTION

Subracting the fractions $\dfrac{17}{32} - \dfrac{5}{32}$ will give you the answer to the example. Here is the solution:

Problem	Step 1	Step 2	Step 3
$\dfrac{17}{32} - \dfrac{5}{32} =$	$\dfrac{17-5}{32}$	$\dfrac{12}{32}$	$\dfrac{12 \div 4}{32 \div 4} = \dfrac{3}{8}$
	Subtract the numerators: $17 - 5 = 12$	Place this difference (12) over the same denominator (32): $\dfrac{12}{32}$	Reduce the fraction by dividing each term by 4: $\dfrac{12 \div 4}{32 \div 4} = \dfrac{3}{8}$

Thus: $\dfrac{17}{32} - \dfrac{5}{32} = \dfrac{12}{32} = \dfrac{3}{8}$ in.

PRACTICE EXERCISE 86

Subtract the fractions and write each difference as a reduced fraction.

1. $\dfrac{5}{6} - \dfrac{1}{6} =$

2. $\dfrac{7}{8} - \dfrac{1}{8} =$

3. $\dfrac{11}{12} - \dfrac{7}{12} =$

4. $\dfrac{11}{20} - \dfrac{3}{20} =$

5. $\dfrac{3}{4} - \dfrac{1}{4} =$

6. $\dfrac{9}{10} - \dfrac{3}{10} =$

7. $\dfrac{11}{16} - \dfrac{5}{16} =$

8. $\dfrac{19}{32} - \dfrac{15}{32} =$

9. $\dfrac{13}{14} - \dfrac{11}{14} =$

8.3 Like Fractions Whose Difference Is One or Greater Than One

EXAMPLE

Two like fractions whose difference is *one*:

$$\frac{4}{3} - \frac{1}{3} = ?$$

SOLUTION

Problem	Step 1	Step 2	Step 3
$\dfrac{4}{3} - \dfrac{1}{3} =$	$\dfrac{4-1}{3}$	$\dfrac{3}{3}$	$\dfrac{3}{3} = 1$
	Subtract the numerators: $4 - 1 = 3$	Place this difference (3) over the same denominator (3): $\dfrac{3}{3}$	Any fraction whose numerator and denominator are the same number (other than 0) equals 1.

Thus: $\dfrac{4}{3} - \dfrac{1}{3} = \dfrac{3}{3} = 1$.

EXAMPLE

Two like fractions whose difference is greater than *one*:

$$\frac{33}{16} - \frac{15}{16} =$$

SOLUTION

Problem	Step 1	Step 2	Step 3
$\dfrac{33}{16} - \dfrac{15}{16} =$	$\dfrac{33-15}{16}$	$\dfrac{18}{16}$	$\dfrac{18}{16} = 1\dfrac{2}{16}$ $= 1\dfrac{1}{8}$
	Subtract the numerators: $33 - 15 = 18$	Place the difference (18) over the same denominator (16): $\dfrac{18}{16}$	Change the improper fraction to a mixed number: $\dfrac{18}{16} = 1\dfrac{2}{16}$ $= 1\dfrac{1}{8}$

Thus: $\dfrac{33}{16} - \dfrac{15}{16} = \dfrac{18}{16} = 1\dfrac{2}{16} = 1\dfrac{1}{8}$.

PRACTICE EXERCISE 87

Subtract each of the following fractions. Write the difference in its simplest form.

1. $\dfrac{6}{5} - \dfrac{1}{5} =$

2. $\dfrac{5}{3} - \dfrac{1}{3} =$

3. $\dfrac{9}{4} - \dfrac{3}{4} =$

4. $\dfrac{22}{10} - \dfrac{11}{10} =$

5. $\dfrac{17}{8} - \dfrac{1}{8} =$

6. $\dfrac{13}{7} - \dfrac{4}{7} =$

7. $\dfrac{15}{12} - \dfrac{1}{12} =$

8. $\dfrac{25}{16} - \dfrac{5}{16} =$

9. $\dfrac{43}{32} - \dfrac{11}{32} =$

8.4 Mixed Numbers Can Also Be Subtracted

Subtracting **mixed** numbers is similar to adding mixed numbers. Remember, you added the fractions, **then added** the whole numbers, and combined the two sums into a *mixed* number. The **same method** is used for subtracting mixed numbers except that the operation is subtraction. **Look at** this example:

EXAMPLE

Mr. Jacono, **a s**hopkeeper, had bought $5\frac{3}{4}$ dozen shirts and sold $3\frac{1}{4}$ dozen. How many dozen still remain?

SOLUTION

To answer **the** question you must subtract what he sold from what he bought.

$$5\frac{3}{4} - 3\frac{1}{4}$$

Following the procedure outlined, the whole numbers and the fractions are treated individually **and the result** is combined. The solution follows:

Problem	Step 1	Step 2	Step 3
$\begin{aligned} 5\frac{3}{4} \\ -3\frac{1}{4} \\ \hline \end{aligned}$	$\begin{aligned} 5\frac{3}{4} \\ -3\frac{1}{4} \\ \hline \frac{2}{4} \end{aligned}$	$\begin{aligned} 5\frac{3}{4} \\ -3\frac{1}{4} \\ \hline 2\frac{2}{4} \end{aligned}$	$2\frac{2}{4} = 2\frac{1}{2}$
	Subtract the fractions: $\dfrac{3-1}{4} = \dfrac{2}{4}$	Subtract the whole numbers: $5 - 3 = 2$	Simplify the fraction $\dfrac{2}{4}$: $\dfrac{2}{4} = \dfrac{1}{2}$

Thus: $5\frac{3}{4} - 3\frac{1}{4} = 2\frac{2}{4} = 2\frac{1}{2}$.

PRACTICE EXERCISE 88

Subtract. Write each difference in its simplest form.

1. $6\dfrac{3}{7}$
   $-2\dfrac{1}{7}$

2. $8\dfrac{3}{4}$
   $-3\dfrac{1}{4}$

3. $9\dfrac{5}{9}$
   $-\dfrac{1}{9}$

4. $14\dfrac{1}{2}$
   $-\ 9$

5. $4\dfrac{5}{6}$
   $-1\dfrac{1}{6}$

6. $5\dfrac{11}{16}$
   $-\dfrac{7}{16}$

7. $10\dfrac{2}{3}$
   $-\ 3\dfrac{1}{3}$

8. $18\dfrac{11}{12}$
   $-\ 7\dfrac{5}{12}$

9. $16\dfrac{5}{6}$
   $-\ 4$

10. $7\dfrac{11}{16}$
    $-3\dfrac{3}{16}$

8.5 Exchanging in Subtracting Fractions

EXAMPLE

You use $\dfrac{3}{8}$ of a yard of felt to decorate your jeans. If you bought 1 yd. of felt, how much do you have left?

SOLUTION

You must subtract the fraction $\frac{3}{8}$ from the whole number 1. The subtraction method is illustrated below.

Problem	Step 1	Step 2
$\begin{array}{r} 1 \\ -\frac{3}{8} \\ \hline \end{array}$	$\begin{array}{r} 1 = \frac{8}{8} \\ -\frac{3}{8} = \frac{3}{8} \\ \hline \end{array}$	$\begin{array}{r} 1 = \frac{8}{8} \\ -\frac{3}{8} = \frac{3}{8} \\ \hline \frac{5}{8} \end{array}$
	Since $\frac{3}{8}$ cannot be subtracted from 0, change the whole number 1 to 8ths. $1 = \frac{8}{8}$ This is called exchanging.	Subtract the fractions: $\frac{8}{8} - \frac{3}{8} = \frac{5}{8}$

Thus: $1 \text{ yd.} - \frac{3}{8} \text{ yd.} = \frac{5}{8} \text{ yd.}$

EXAMPLE

A bag of flour contains 4 cups of flour. Marilyn uses $1\frac{1}{4}$ cups of flour making a cake. How many cups of flour still remain?

SOLUTION

$$4 - 1\frac{1}{4} = ?$$

Problem	Step 1	Step 2	Step 3
4 $-1\frac{1}{4}$ ___	$4 = 3 + \frac{4}{3}$ $-1\frac{1}{4}$ ___	$3\frac{4}{4}$ $-1\frac{1}{4}$ ___ $\frac{3}{4}$	$3\frac{4}{4}$ $-1\frac{1}{4}$ ___ $2\frac{3}{4}$
	Write the whole number 4 as 3 + 1, exchanging: $1 = \frac{4}{4}$ Thus: $4 = 3 + \frac{4}{4}$	Subtract the fractions: $\frac{4-1}{4} = \frac{3}{4}$	Subtract the whole numbers: $3 - 1 = 2$

Thus: $4 - 1\frac{1}{4} = 2\frac{3}{4}$.

PRACTICE EXERCISE 89

Subtract.

1. $\begin{array}{r} 1 \\ -\frac{7}{8} \\ \hline \end{array}$

2. $\begin{array}{r} 1 \\ -\frac{2}{5} \\ \hline \end{array}$

3. $\begin{array}{r} 3 \\ -\frac{3}{4} \\ \hline \end{array}$

4. $\begin{array}{r} 5 \\ -1\frac{1}{2} \\ \hline \end{array}$

5. $\begin{array}{r} 6 \\ -\frac{7}{12} \\ \hline \end{array}$

6. $\begin{array}{r} 1 \\ -\frac{3}{16} \\ \hline \end{array}$

7. $\begin{array}{r} 2 \\ -\frac{5}{8} \\ \hline \end{array}$

8. $\begin{array}{r} 13 \\ -5\frac{1}{6} \\ \hline \end{array}$

9. $\begin{array}{r} 37 \\ -22\frac{3}{7} \\ \hline \end{array}$

10. $\begin{array}{r} 44 \\ -29\frac{2}{3} \\ \hline \end{array}$

11. $\begin{array}{r} 1 \\ -\frac{5}{32} \\ \hline \end{array}$

12. $\begin{array}{r} 37 \\ -9\frac{7}{12} \\ \hline \end{array}$

8.6 Exchanging in Subtracting Mixed Numbers (Like Denominators)

EXAMPLE

A truck radiator holds $6\frac{2}{5}$ gallons of solution. To $4\frac{4}{5}$ gallons of antifreeze how many gallons of water must be added to fill the radiator?

SOLUTION

$6\frac{2}{5} - 4\frac{4}{5} = ?$

Problem	Step 1	Step 2
$6\frac{2}{5}$ $-4\frac{4}{5}$	$6\frac{2}{5} = 5 + 1 + \frac{2}{5} =$ $5 + \frac{5}{5} + \frac{2}{5} = 5\frac{7}{5}$ $-4\frac{4}{5} \qquad = -4\frac{4}{5}$	$5\frac{7}{5}$ $-4\frac{4}{5}$ $1\frac{3}{5}$
	Since it is impossible to subtract $\frac{4}{5}$ from $\frac{2}{5}$, rewrite the top mixed number $6\frac{2}{5}$ as $5 + 1 + \frac{2}{5}$. *Exchange:* $1 = \frac{5}{5}$ Thus: $5 + 1 + \frac{2}{5} =$ $5 + \frac{5}{5} + \frac{2}{5}$ Add the fractions: $5\frac{7}{5}$	Subtract the fractions: $\frac{7}{5} - \frac{4}{5} = \frac{3}{5}$ Subtract the whole numbers: $5 - 4 = 1$

Thus: $6\frac{2}{5}$ gal. $- 4\frac{4}{5}$ gal. $= 1\frac{3}{5}$ gal.

EXAMPLE

$$15\frac{5}{12} - 7\frac{11}{12} = ?$$

SOLUTION

Problem	Step 1	Step 2
$15\frac{5}{12}$ $-\ 7\frac{11}{12}$	$15\frac{5}{12} = 14+1+\frac{5}{12} =$ $14+\frac{12}{12}+\frac{5}{12} = 14\frac{17}{12}$ $-7\frac{11}{12}$ $\qquad -7\frac{11}{12}$	$14\frac{17}{12}$ $-\ 7\frac{11}{12}$ $7\frac{6}{12} = 7\frac{1}{2}$
	Since $\frac{11}{12}$ cannot be subtracted from $\frac{5}{12}$, rewrite the top mixed number $15\frac{5}{12}$ as $14+1+\frac{5}{12}$. Exchange: $1=\frac{12}{12}$ Thus: $14+1+\frac{5}{12} =$ $14+\frac{12}{12}+\frac{5}{12}$ Add the fractions: $14\frac{17}{12}$	Subtract the fractions: $\frac{17}{12} - \frac{11}{12} = \frac{6}{12}$ Subtract the whole numbers: $14 - 7 = 7$ Combine the result and simplify: $7\frac{6}{12} = 7\frac{1}{2}$

Thus: $15\frac{5}{12} - 7\frac{11}{12} = 7\frac{1}{2}$.

PRACTICE EXERCISE 90

Subtract.

1. $\begin{array}{r} 4\frac{1}{3} \\ -3\frac{2}{3} \\ \hline \end{array}$

2. $\begin{array}{r} 6\frac{1}{5} \\ -\frac{3}{5} \\ \hline \end{array}$

3. $\begin{array}{r} 8\frac{3}{8} \\ -4\frac{7}{8} \\ \hline \end{array}$

4. $\begin{array}{r} 12\frac{1}{4} \\ -3\frac{3}{4} \\ \hline \end{array}$

5. $\begin{array}{r} 13\frac{2}{7} \\ -7\frac{5}{7} \\ \hline \end{array}$

6. $\begin{array}{r} 33\frac{1}{10} \\ -9\frac{9}{10} \\ \hline \end{array}$

7. $\begin{array}{r} 21\frac{3}{16} \\ -13\frac{13}{16} \\ \hline \end{array}$

8. $\begin{array}{r} 73\frac{11}{15} \\ -68\frac{14}{15} \\ \hline \end{array}$

9. $\begin{array}{r} 104\frac{17}{32} \\ -59\frac{31}{32} \\ \hline \end{array}$

8.7 Word Problems

Before you begin, go back to the General Instructions in Chapter 2, section 2.7, to review the methods used to solve any word problems. The specific instructions for solving problems in subtraction (Chapter 3, section 3.5) are the same for solving problems in subtracting fractions.

PRACTICE EXERCISE 91

Solve these word problems, reducing fractions where possible.

1. Juan invests of his savings in Treasury bonds. What fraction of his savings did he not invest?

2. If a trip to Washington, D.C., takes 4 h. by bus and only $1\frac{1}{2}$ h. by plane, how many hours are saved in traveling by plane?

3. Lucinda weighs $114\frac{3}{4}$ lb. now, but last year she weighed $106\frac{1}{4}$ lb. How many pounds has she gained?

4. If rainfall for the month measured $2\frac{3}{8}$ in. this year and only $1\frac{7}{8}$ in. for the same month last year, then what was the increase?

5. If one machine can complete a job in $3\frac{5}{6}$ h. while a newer machine takes only 2 h., how many hours are save by using the newer machine?

6. The trip to work is $12\frac{3}{10}$ mi. Mary Lou drives $7\frac{1}{10}$ mi. before running out of gas. How many miles further must she travel to work?

7. You drank $\frac{2}{8}$ of a can of soda. Mary drank $\frac{1}{8}$ of the can. What part of the soda is left?

8. You intend to study 4 h. for a test. After studying $1\frac{5}{6}$ h., you take a break. How many hours do you have left to study?

9. What is the length of the third side of a triangle if the sum of the other two sides is $39\frac{1}{2}$ in. and its entire perimeter is 58 in.? (Remember, the perimeter is the sum of all three sides of the triangle.)

10.   From a 16-ft. piece of lumber a carpenter cuts three pieces measuring $4\frac{11}{16}$ ft., $5\frac{3}{16}$ ft. and $3\frac{5}{16}$ ft. How many feet still remain? (Waste from cutting should not be considered.)

8.8 Subtracting Fractions with Unlike Denominators

All of the fractions and mixed numbers you have subtracted so far have had *like denominators*. When they have the same denominator, you keep the denominator and subtract only the numerators. See the following example:

$$\frac{7}{8} - \frac{2}{8} = \frac{7-2}{8} = \frac{5}{8}$$

How do you subtract two fractions which have unlike denominators like these?

$$\frac{2}{3} - \frac{1}{6} = ?$$

You know how to subtract fractions which have like denominators, so let's see how you can relate the problem of unlike denominators to the one you already know.

First, you find the *lowest common denominator* (L.C.D.) for the two fractions. (Review section 6.6 in Chapter 6 if it is necessary.) Second, you proceed as you did in adding fractions. Look at the method used to solve the following example.

EXAMPLE

From a candle $\frac{2}{3}$ its original size, Maria burned another $\frac{1}{6}$. What part of the candle still remains?

SOLUTION

To find the part that still remains, subtract $\frac{1}{6}$ from $\frac{2}{3}$ or

$$\frac{2}{3} - \frac{1}{6} = ?$$

Problem	Step 1	Step 2	Step 3
$\frac{2}{3}$ $-\frac{1}{6}$	L.C.D. = 6	$\frac{2\times2}{3\times2}=\frac{4}{6}$ $-\frac{1}{6}=\frac{1}{6}$	$\frac{4}{6}$ $-\frac{1}{6}$ $\frac{3}{6}=\frac{1}{2}$
	The L.C.D. for the fractions is 6, since 6 is divisible by both 3 and 6.	Change each fraction into an equivalent fraction whose denominator is 6: $\frac{2}{3}=\frac{4}{6}$ $\frac{1}{6}=\frac{1}{6}$	Subtract the fractions: $\frac{4}{6}-\frac{1}{6}=$ $\frac{4-1}{6}=\frac{3}{6}$ Simplify the fraction: $\frac{3}{6}=\frac{1}{2}$

Thus: $\frac{2}{3}-\frac{1}{6}=\frac{1}{2}$.

EXAMPLE

Find the difference when $\frac{3}{5}$ is subtracted from $\frac{2}{3}$.

SOLUTION

$$\frac{2}{3} - \frac{3}{5} = ?$$

Problem	Step 1	Step 2	Step 3
$\dfrac{2}{3}$ $-\dfrac{3}{5}$	L.C.D. = 15	$\dfrac{2\times 5}{3\times 5} = \dfrac{10}{15}$ $\dfrac{3\times 3}{5\times 3} = \dfrac{9}{15}$	$\dfrac{2}{3} = \dfrac{10}{15}$ $-\dfrac{3}{5} = \dfrac{9}{15}$ $\dfrac{1}{15}$
	The L.C.D. for the two fractions is 15. Both 3 and 5 divide evenly into 15.	Change each fraction into an equivalent fraction whose denominator is 15: $\dfrac{2}{3} = \dfrac{10}{15}$ $\dfrac{3}{5} = \dfrac{9}{15}$	Subtract the fractions: $\dfrac{10}{15} - \dfrac{9}{15} =$ $\dfrac{10-9}{15} = \dfrac{1}{15}$

Thus: $\dfrac{2}{3} - \dfrac{3}{5} = \dfrac{1}{15}$.

PRACTICE EXERCISE 92

Subtract.

1. $\dfrac{1}{2} - \dfrac{2}{5} =$

2. $\dfrac{1}{2} - \dfrac{1}{4} =$

3. $\dfrac{3}{4} - \dfrac{1}{6} =$

4. $\dfrac{3}{4}$
 $-\dfrac{2}{3}$

5. $\dfrac{5}{8}$
 $-\dfrac{1}{2}$

6. $\dfrac{3}{4}$
 $-\dfrac{3}{5}$

7. $\dfrac{5}{6} - \dfrac{2}{3} =$ 8. From $\dfrac{3}{5}$ subtract $\dfrac{3}{10}$. 9. $\dfrac{13}{16} - \dfrac{1}{2} =$

10. Find the difference when $\dfrac{3}{16}$ is subtracted from $\dfrac{7}{8}$.

8.9 Subtracting Mixed Numbers

The procedures for subtracting fractions with unlike denominators can be used in subtracting mixed numbers when the fractional parts have different denominators. After subtracting the fractional part of the mixed numbers, you then subtract the whole numbers and combine the two results. The following example will illustrate the method for you.

EXAMPLE

From a piece of lumber $52\dfrac{1}{2}$ in. long, Mr. Gonzalez cut $24\dfrac{1}{8}$ in. How many inches still remain? (Waste from cutting is not to be considered.)

SOLUTION

Subtracting $24\dfrac{1}{8}$ from $52\dfrac{1}{2}$ will result in the desired solution. Note this illustration:

Problem	Step 1	Step 2	Step 3
$52\dfrac{1}{2}$ $-24\dfrac{1}{8}$	L.C.D. = 8	$52\dfrac{1}{2} = 52\dfrac{4}{8}$ $-24\dfrac{1}{8} = 24\dfrac{1}{8}$	
	The L.C.D. for 2 and 8 is 8.	Change $\dfrac{1}{2}$ to 8ths: $\dfrac{1}{2} = \dfrac{4}{8}$	Subtract the fractions: $\dfrac{4}{8} - \dfrac{1}{8} =$ $\dfrac{4-1}{8} = \dfrac{3}{8}$ Subtract the whole numbers: $52 - 24 = 28$

Thus: $52\dfrac{1}{2}$ in. $- 24\dfrac{1}{8}$ in. $= 28\dfrac{3}{8}$ in.

PRACTICE EXERCISE 93

Subtract and simplify.

1. $5\dfrac{5}{8}$
    $-3\dfrac{3}{16}$

2. $8\dfrac{7}{8}$
    $-3\dfrac{5}{12}$

3. $10\dfrac{7}{10}$
    $-\ 4\dfrac{1}{4}$

4. $7\dfrac{2}{3}$
    $-3\dfrac{1}{2}$

5. $6\dfrac{1}{2}$
    $-4\dfrac{3}{8}$

6. $5\dfrac{2}{5}$
    $-2\dfrac{1}{8}$

7. $4\dfrac{7}{12}$
    $-\ 3\dfrac{1}{3}$

8. $61\dfrac{5}{8}$
    $-47\dfrac{9}{16}$

9. $156\dfrac{3}{10}$
    $-148\dfrac{1}{4}$

8.10 Exchanging in Subtracting Mixed Numbers (Unlike Denominators)

EXAMPLE

Jack found that he now weighs $156\dfrac{1}{4}$ lb. while last year he weighed $148\dfrac{1}{2}$ lb. How many pounds has he gained?

SOLUTION

You must find the difference between $156\dfrac{1}{4}$ and $148\dfrac{1}{2}$ to discover the number of pounds Jack has gained:

$$156\dfrac{1}{4} - 148\dfrac{1}{2} = ?$$

Problem	Step 1	Step 2	Step 3
$156\dfrac{1}{4}$ $-148\dfrac{1}{2}$	$156\dfrac{1}{4}=156\dfrac{1}{4}$ $-148\dfrac{1}{2}=148\dfrac{2}{4}$	$156\dfrac{1}{4}=155+\dfrac{4}{4}+\dfrac{1}{4}$ $=155\dfrac{5}{4}$ $-148\dfrac{2}{4}=148\dfrac{2}{4}$	$155\dfrac{5}{4}$ $-148\dfrac{2}{4}$ $7\dfrac{3}{4}$
	Change each fraction to the L.C.D. (4): $\dfrac{1}{2}=\dfrac{2}{4}$	Since you can't subtract $\dfrac{1}{4}-\dfrac{2}{4}=?$, write the top mixed number $156\dfrac{1}{4}$ as $155+1+\dfrac{1}{4}$. Exchange: $1=\dfrac{4}{4}$ Thus: $155+1+\dfrac{1}{4}=$ $155+\dfrac{4}{4}+\dfrac{1}{4}$ Add the fractions: $155\dfrac{5}{4}$	Subtract the fractions: $\dfrac{5}{4}-\dfrac{2}{4}=\dfrac{3}{4}$ Subtract the whole numbers: $155-148=7$

Thus: $156\dfrac{1}{4}$ lb. $-148\dfrac{1}{2}$ lb. $=7\dfrac{3}{4}$ lb.

EXAMPLE

Traveling from New York to Washington, D.C. takes $5\dfrac{1}{2}$ h. by car and only $3\dfrac{5}{6}$ h. by train. How many fewer hours does it take traveling by train?

SOLUTION

You must find the difference between $5\dfrac{1}{2}$ and $3\dfrac{5}{6}$ to find the answer.

Problem	Step 1	Step 2	Step 3
$5\dfrac{1}{2}$ $-3\dfrac{5}{6}$	$5\dfrac{1}{2}=5\dfrac{3}{6}$ $-3\dfrac{5}{6}=3\dfrac{5}{6}$	$5\dfrac{3}{6}=4+\dfrac{6}{6}+\dfrac{3}{6}$ $=4\dfrac{9}{6}$ $-3\dfrac{5}{6}=3\dfrac{5}{6}$	$4\dfrac{9}{6}$ $-3\dfrac{5}{6}$ $1\dfrac{4}{6}=1\dfrac{2}{3}$
	Change $\dfrac{1}{2}$ to 6ths: the L.C.D. for both fractions: $\dfrac{1}{2}=\dfrac{3}{6}$	Since you can't subtract $\dfrac{3}{6}-\dfrac{5}{6}=?$, write the top mixed number $5\dfrac{3}{6}$ as $4+1+\dfrac{3}{6}$. Exchange: $1=\dfrac{6}{6}$ Thus: $4+1+\dfrac{3}{6}=$ $4+\dfrac{6}{6}+\dfrac{3}{6}$ Add the fractions: $4\dfrac{9}{6}$	Subtract the fractions: $\dfrac{9}{6}-\dfrac{5}{6}=\dfrac{4}{6}$ Subtract the whole numbers: $4-3=1$ Simplify the fraction: $1\dfrac{4}{6}=1\dfrac{2}{3}$

Thus: $5\dfrac{1}{2}$ h. $-3\dfrac{5}{6}$ h. $=1\dfrac{2}{3}$ h.

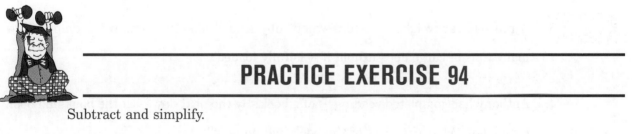

PRACTICE EXERCISE 94

Subtract and simplify.

1. $3\dfrac{1}{2}$
    $-\dfrac{3}{4}$

2. $9\dfrac{7}{12}$
    $-4\dfrac{3}{4}$

3. $8\dfrac{1}{5}$
    $-2\dfrac{1}{2}$

4. $3\dfrac{1}{6}$

   $-2\dfrac{2}{5}$
   ___

5. $6\dfrac{2}{7}$

   $-5\dfrac{2}{3}$
   ___

6. $9\dfrac{1}{16}$

   $-5\dfrac{5}{8}$
   ___

7. $4\dfrac{2}{5} - 2\dfrac{1}{2} =$

8. $5\dfrac{1}{3} - \dfrac{1}{2} =$

9. $5\dfrac{3}{8} - 4\dfrac{2}{3} =$

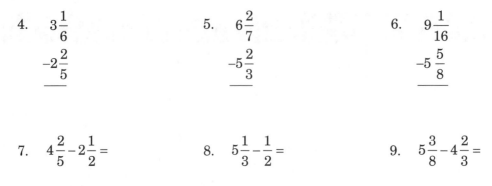

8.11 More Difficult Word Problems

Look over the problems in Practice Exercise 91 before doing this exercise. Be sure you understand the previous ones thoroughly, so that you will have little difficulty with these more advanced word problems. To understand a problem better, draw a diagram.

PRACTICE EXERCISE 95

1. Mrs. Washington found that she had $\dfrac{7}{8}$ lb. of butter in the refrigerator. After she baked a cake which required $\dfrac{1}{2}$ lb. of butter, how much butter still remained?

2. From a roll of upholstery fabric $24\dfrac{1}{8}$ yd. long, a piece measuring $12\dfrac{3}{4}$ yd. is cut. How many yards of fabric remain?

3. A railroad car weighs $47\dfrac{7}{8}$ tons when fully loaded and $12\dfrac{3}{4}$ tons when empty. How many tons of cargo does it hold if it is fully loaded?

4. Anthony had completed reading $\dfrac{3}{5}$ of a book. By the next day $\dfrac{5}{8}$ of the book had been completed. What new part of the book had he read?

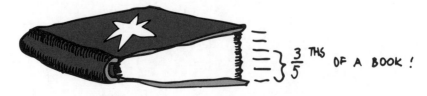

5. Beulah walked $\frac{9}{16}$ mi. while Bertha walked $\frac{2}{3}$ mi. What is the difference in the distances walked?

6. To make a single strand of jewelry chain $1\frac{5}{6}$ ft. is needed. A double strand requires $3\frac{1}{3}$ ft. of chain. How many more feet of chain are needed to make a double strand rather than a single strand?

7. From a board 16 ft. long, a carpenter cut lengths of 6 ft., $5\frac{5}{8}$ ft., $2\frac{1}{2}$ ft., and $1\frac{11}{16}$ ft. How much of the original board remains?

8. If $58\frac{1}{8}$ ft. of fencing are required to enclose a triangular plot of ground, what is the length of the third side of the triangle if the other two sides measure $19\frac{3}{16}$ ft. and $20\frac{11}{32}$ ft.?

9. Mr. Dostanza finds the dimensions of a rectangular-shaped room are $13\frac{1}{2}$ ft. by $14\frac{7}{8}$ ft. If he buys 64 ft. of molding to place around the entire room, how many feet, if any, are left over? (Hint: Draw a diagram)

10. A mixture of five blends of coffee must weigh 231 lb. How many pounds of Blend E are needed if the four other blends weigh $45\frac{1}{8}$ lb., $49\frac{1}{2}$ lb., $32\frac{7}{8}$ lb., and $75\frac{3}{4}$ lb.?

TERM YOU SHOULD REMEMBER

Exchanging   Replacing one thing by another.

REVIEW OF IMPORTANT IDEAS

Some of the most important ideas in Chapter 8 were:

 To subtract *like fractions* you subtract the numerators and place the difference over the same denominator.

 To subtract *unlike fractions* you change each fraction to a fraction with the lowest common denominator and then subtract the numerators. Place the difference over the lowest common denominator.

 To subtract *mixed numbers* you subtract the fractions and you subtract the whole numbers separately. Combine each difference into a mixed number.

CHECK WHAT YOU HAVE LEARNED

This chapter completed the second operation dealing with fractions, namely, the subtraction of fractions. The chapter test allows you to check on your understanding of the material in the chapter. Try it to see how well you have mastered the material. Good luck!

A Score of	Means That You
21–23	Did very well. You can move to Chapter 9.
18–20	Know this material except for a few points. Reread the sections about the ones you missed.
15–17	Need to check carefully on the sections you missed.
0–14	Need to review the chapter again to refresh your memory and improve your skills.

Questions	Are Covered in Section
1	8.1
2	8.2
3, 4	8.3
5–7, 11	8.4
8, 12	8.5
9, 10	8.6
13, 14	8.7
15, 16	8.8
17, 21	8.9
18, 19	8.10
20, 22, 23	8.11

Chapter Test 8

Subtract. Write **each** answer in its simplest form below each question.

1. $\dfrac{3}{5}$
    $-\dfrac{1}{5}$
    ___

2. $\dfrac{7}{10}$
    $-\dfrac{1}{10}$
    ___

3. $\dfrac{4}{3}$
    $-\dfrac{1}{3}$
    ___

4. $\dfrac{13}{8}$
    $-\dfrac{3}{8}$
    ___

5. $5\dfrac{2}{3}$
    $-3\dfrac{1}{3}$
    ___

6. $7\dfrac{3}{4}$
    $-6\dfrac{1}{4}$
    ___

7. $9\dfrac{5}{16}$
    $-\dfrac{4}{16}$
    ___

8. 1
    $-\dfrac{3}{8}$
    ___

9. $6\dfrac{1}{3}$
    $-\dfrac{2}{3}$
    ___

10. $7\dfrac{1}{8}$
    $-3\dfrac{5}{8}$
    ___

11. $14\dfrac{4}{7}$
    $-\ 3$
    ___

12. 8
    $-5\dfrac{3}{5}$
    ___

13. Margaret bought $2\dfrac{1}{4}$ lb. of potatoes. She used $1\dfrac{3}{4}$ lb. in a stew. How many pounds of potatoes still remain?

14. A carpenter has a board that is $\dfrac{7}{8}$ in. thick. If she planes off $\dfrac{1}{8}$ in., how thick will the board be?

15. $\dfrac{11}{12}$
    $-\dfrac{3}{4}$
    ___

16. $\dfrac{7}{9}$
    $-\dfrac{1}{2}$
    ___

17. $8\dfrac{5}{8}$
    $-7\dfrac{2}{5}$
    ___

18. $7\dfrac{1}{2}$
$-5\dfrac{3}{4}$
———

19. $19\dfrac{1}{4}$
$-\ 6\dfrac{3}{7}$
———

20.   In a recent dry spell, the water level of Lost Pond dropped from $36\dfrac{1}{4}$ ft. to $19\dfrac{2}{3}$ ft. How much did the water level drop?

21.   From the sum of $4\dfrac{3}{8}$ and $1\dfrac{1}{3}$, subtract $3\dfrac{1}{16}$.

22.   Magdalena had two packages of chopped meat, one weighing $2\dfrac{1}{2}$ lb. and another weighing $6\dfrac{3}{4}$ lb. If she used $3\dfrac{5}{8}$ lb. for a dish of lasagna, how many pounds still remain?

23.   From a roll of wallpaper $24\dfrac{7}{8}$ yd. long, a piece measuring $12\dfrac{3}{4}$ yd. is cut. How many yards of wallpaper remain?

HOLD IT!

These problems are a general review of all that you have learned up to this point. If you get all the answers correct, you are really making progress and on your way to success in mathematics.

1.   386
× 47
———

2. $28\overline{)483}$

3. $\dfrac{1}{2}$
$\dfrac{3}{8}$
$+1\dfrac{1}{4}$
———

4. $4\dfrac{2}{5}$

   $-1\dfrac{1}{3}$

5. $98\dfrac{7}{8}$

   $-\dfrac{15}{16}$

6. $38\dfrac{1}{7}$

   $42\dfrac{20}{21}$

   $+\ \dfrac{1}{3}$

7. Agnes made a two-piece suit that required $1\dfrac{1}{2}$ yd. for the top and $2\dfrac{3}{4}$ yd. for the skirt. How many yards of material did she use?

8. $\quad4397$
   $\quad2143$
   $+8067$

9. John has agreed to work 12 h. weekly. After working $8\dfrac{1}{8}$ h., how many additional hours must he work?

10. When Sam joined a weight control program, he weighed $206\dfrac{1}{4}$ lb. During July he lost $6\dfrac{7}{8}$ lb. and in August he lost $5\dfrac{3}{4}$ lb. What was Sam's weight at the beginning of September?

MATHEMATICAL MAZE

This maze contains 13 mathematical words. Some are written horizontally ↔, some vertically ↕, some diagonally ↗ ↘; some go in two or three directions. When you locate a word in the maze, draw a ring around it. The word TIMES has already been done for you. The solution is below.

```
A   M   B   C   D   E   S   F   G   H   F
I   J   E   K   S   M   R   E   T   L   R
T   A   M   D   N   O   P   Q   M   R   A
N   E   S   T   I   U   V   T   O   I   C
E   R   C   W   X   A   C   Y   D   Z   T
I   A   A   U   B   U   N   A   E   M   I
T   C   D   O   D   E   F   G   H   I   O
O   I   R   O   K   E   L   M   N   X   N
U   P   R   P   Q   P   R   O   P   E   R
Q   P   S   R   O   S   I   V   I   D   T
```

WORDS OF:

4 letters	*5 letters*	*6 letters*	*7 letters*	*8 letters*
MODE	MIXED	PROPER	DIVISOR	FRACTION
MEAN	TERMS	REDUCE	PRODUCT	QUOTIENT
AREA	TIMES	MEDIAN		

ANSWERS

1. 18,142

2. 17 R 7 or $17\frac{7}{28} = 17\frac{1}{4}$

3. $2\frac{1}{8}$

4. $3\frac{1}{15}$

5. $97\frac{15}{16}$

6. $81\frac{3}{7}$

7. $4\frac{1}{4}$ yd.

8. 14,607

9. $3\frac{7}{8}$ h.

10. $193\frac{5}{8}$ lb.

Once again, remember that you should not allow your mistakes to make you angry or frustrated. Go back over the material you haven't mastered and see where your difficulty is. Keep trying and eventually you must succeed.

Multiplying and Dividing Fractions

Multiplying Fractions

Multiplying fractions is an operation in mathematics we ordinarily use each day in our life. We see signs advertising sales of $\frac{1}{3}$ to $\frac{1}{2}$ off, $\frac{1}{2}$ price, or $\frac{3}{4}$- or $\frac{1}{2}$-acre lots for sale. Each of these statements suggests an example in multiplying fractions which will be explained in this chapter. Check your skill in multiplying fractions in the pretest which follows.

Pretest 9A—See What You Know and Remember

Do these examples as carefully as possible. Some are more difficult than others so try your best. Write the answers below each question, reducing fractions if possible.

1. $\frac{1}{5} \times \frac{2}{3} =$

2. $\frac{1}{2} \times \frac{4}{5} =$

3. $\frac{4}{7} \times \frac{5}{6} =$

4. $\frac{3}{4} \times \frac{8}{9} =$

5. $\frac{2}{3} \times 6 =$

6. $2 \times \frac{3}{4} =$

7. $10 \times \frac{9}{10} =$

8. Find the product of $\frac{3}{5}$ and $\frac{1}{6}$.

9. Margarita found a recipe which called for $\frac{1}{2}$ cup of sugar. If she wanted to cut the recipe in half, how much sugar would she require?

10. $3\frac{1}{2} \times 7 =$ 11. $6 \times 7\frac{1}{2} =$ 12. $\frac{7}{8} \times 1\frac{1}{4} =$

13. $\frac{3}{4}$ of $2\frac{1}{8} =$ 14. $\frac{2}{9}$ of $1\frac{1}{2} =$ 15. $1\frac{1}{4} \times 1\frac{3}{5} =$

16. $2\frac{1}{8} \times 1\frac{1}{5} =$ 17. $4\frac{1}{3} \times 2\frac{1}{2} =$ 18. $\frac{3}{4} \times \frac{5}{6} \times \frac{1}{10} =$

19. $\frac{2}{3} \times 6 \times \frac{1}{8} =$

20. Carmen rode a bicycle $3\frac{1}{2}$ mi. each day for 4 days. How many miles did she go?

21. Mary uses $2\frac{1}{2}$ cups of sugar for one cake. How much sugar will be required to bake three cakes?

22. A pattern to make a sleeveless shirt requires $1\frac{1}{2}$ yd. of fabric and a pattern to make a skirt requires $2\frac{1}{3}$ yd. of fabric. How much fabric is required if you want to make 3 shirts and 3 skirts?

Now turn to the back of the book to check your answers. Add up all that you had correct.

A Score of	Means That You
20–22	Did very well. You can move to the second half of this chapter, "Dividing Fractions."
18–19	Know this material except for a few points. Read the sections about the ones you missed.
14–17	Need to check carefully on the sections you missed.
0–13	Need to work with this part of the chapter to refresh your memory and improve your skills.

Questions	Are Covered in Section
1	9.1
2–4, 8	9.2
5–7	9.3
18, 19	9.4
10–17	9.5
9, 20–22	9.6

9.1 The Product of Two Fractions

Cookbooks have recipes which serve 4, 6, 8, or more people. A particular recipe requires $\frac{1}{3}$ teaspoon of salt for a dish serving 4 people. Suppose you wish to make this for only 2 people. Since you are cooking for half as many people, you will then cut each quantity in half. You will have to know one-half of one-third, or $\frac{1}{2}$ of $\frac{1}{3}$. This is pictured below. The rectangle is divided into three equal parts and one of these parts is $\frac{1}{3}$ of the whole.

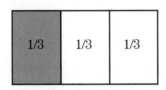

Another line is drawn horizontally, dividing the rectangle into two parts, and is shaded differently. One of the parts is $\frac{1}{2}$ of the whole.

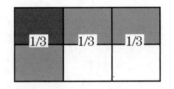

The rectangle is divided into 6 equal parts, one of which is shaded darker than the others. This darker section represents $\frac{1}{6}$ of the rectangle. You can now conclude that

$$\frac{1}{2} \text{ of } \frac{1}{3} = \frac{1}{6}$$

This problem can be done more quickly and efficiently by using mathematical methods:

Problem	Step 1	Step 2
$\frac{1}{2} \times \frac{1}{3} =$	$\frac{1}{2} \times \frac{1}{3} = \frac{1}{}$	$\frac{1}{2} \times \frac{1}{3} = \frac{1}{6}$
	Multiply the numerators: $1 \times 1 = 1$	Multiply the denominators: $2 \times 3 = 6$

Thus: $\frac{1}{2} \times \frac{1}{3} = \frac{1}{6}$.

Since the answer to $\frac{1}{2}$ of $\frac{1}{3}$ and $\frac{1}{2} \times \frac{1}{3}$ is the same, $\frac{1}{6}$, you can conclude that the word *of* indicates the operation of multiplication.

PRACTICE EXERCISE 96

Multiply. Be sure all answers are written in their simplest form. Answers to this test and subsequent practice exercises begin on page 455.

1. $\frac{1}{3} \times \frac{1}{8} =$

2. $\frac{3}{5} \times \frac{6}{7} =$

3. $\frac{1}{2} \times \frac{1}{2} =$

4. $\frac{2}{3} \times \frac{1}{4} =$

5. $\frac{2}{5} \times \frac{5}{6} =$

6. $\frac{3}{4} \times \frac{6}{7} =$

7. $\frac{1}{2}$ of $\frac{4}{5} =$
    (Write it as a multiplication example first.)

8. $\frac{1}{6} \times \frac{2}{3} =$

9. $\frac{1}{2}$ of $\frac{2}{3} =$

10. $\frac{3}{4}$ of $\frac{1}{8} =$

11. $\frac{7}{8} \times \frac{2}{3} =$

12. $\frac{2}{3} \times \frac{3}{4} =$

9.2 Cancellation

In the problems just completed you multiplied numerators together, multiplied denominators together, and wrote the products as a fraction. The fraction in the answer was then simplified, if possible. For example:

$$\frac{1}{3} \times \frac{3}{8} = \frac{3}{24} \text{ and } \frac{3}{24} = \frac{1}{8}$$

Another method of doing this same example would be to *cancel* (or "cross-out") any numerator or any denominator which can be divided by the same factor.

Problem	Step 1	Step 2
$\dfrac{1}{3} \times \dfrac{3}{8} =$	$\dfrac{1}{\overset{1}{\cancel{3}}} \times \dfrac{\overset{1}{\cancel{3}}}{8} =$	$\dfrac{1}{1} \times \dfrac{1}{8} = \dfrac{1 \times 8}{1 \times 8} = \dfrac{1}{8}$
	Divide any numerator and any denominator by one number which will divide evenly into both. This is called *cancellation*. 3 divides into both 3s: $3 \div 3 = 1$ $3 \div 3 = 1$	Multiply the numerators: $1 \times 1 = 1$ Multiply the denominators: $1 \times 8 = 8$

Thus: $\dfrac{1}{3} \times \dfrac{3}{8} = \dfrac{1}{8}$.

EXAMPLE

The J & R Realty Corp. was selling individual lots of $\dfrac{4}{9}$ acre. The Barkan family decided to purchase a lot and build a house covering $\dfrac{3}{4}$ of the lot. How much acreage will the house cover?

SOLUTION

The words $\dfrac{3}{4}$ of the lot indicate that the problem is a multiplication example.

Since the lot is $\dfrac{4}{9}$ of an acre, the answer to the problem will be found by the calculation,

$$\frac{3}{4} \text{ of } \frac{4}{9} \text{ or } \frac{3}{4} \times \frac{4}{9} = ?$$

This is illustrated below.

Problem	Step 1	Step 2	Step 3
$\dfrac{3}{4} \times \dfrac{4}{9} =$	$\dfrac{3}{\cancel{4}_1} \times \dfrac{\cancel{4}^1}{9} =$	$\dfrac{\cancel{3}^1}{1} \times \dfrac{1}{\cancel{9}_3} =$	$\dfrac{1}{1} \times \dfrac{1}{3} = \dfrac{1}{3}$
	Divide each 4 in the numerator and the denominator (*cancellation*) by 4: $4 \div 4 = 1$ $4 \div 4 = 1$	Divide the 3 in the numerator and the 9 in the denominator by 3 (*cancellation*): $3 \div 3 = 1$ $9 \div 3 = 3$	Multiply the numerators: $1 \times 1 = 1$ Multiply the denominators: $1 \times 3 = 3$

Thus: $\dfrac{3}{4} \times \dfrac{4}{9} = \dfrac{1}{3}$.

PRACTICE EXERCISE 97

Multiply. Use the method of cancellation when you can.

1. $\dfrac{2}{3} \times \dfrac{3}{5} =$

2. $\dfrac{3}{4} \times \dfrac{2}{3} =$

3. $\dfrac{4}{5}$ of $\dfrac{1}{2} =$

4. $\dfrac{3}{8} \times \dfrac{10}{21} =$

5. $\dfrac{5}{6}$ of $\dfrac{12}{25} =$

6. $\dfrac{3}{5} \times \dfrac{7}{9} =$

7. $\dfrac{6}{7} \times \dfrac{3}{4} =$

8. $\dfrac{2}{5} \times \dfrac{5}{6} =$

9. $\dfrac{2}{3}$ of $\dfrac{1}{6} =$

9.3 Finding Parts of Whole Numbers

EXAMPLE

Jim went shopping for a CD and saw the following ad in the store window. He found a CD usually selling for $16. What did it cost during the sale?

**CDs
1/2 PRICE**

SOLUTION

Jim must find $\frac{1}{2}$ of $16. Recall that $\frac{1}{2}$ of $16 is written as $\frac{1}{2} \times \$16$. This is pictured as:

The cost of the CD during the sale is $\frac{1}{2}$ of $16 = $8. This problem is illustrated as:

Problem	Step 1	Step 2	Step 3
$\frac{1}{2} \times \$16 =$	$\frac{1}{2} \times \frac{16}{1} =$	$\frac{1}{\cancel{2}} \times \frac{\overset{8}{\cancel{16}}}{1} =$	$\frac{1}{1} \times \frac{8}{1} = \frac{8}{1} = 8$
	Write the whole number (16) as a fraction $\left(\frac{16}{1} \right)$ *Any whole number can be written as a fraction with 1 as the denominator*	Use cancellation: $2 \div 2 = 1$ $16 \div 2 = 8$	Multiply the numerators: $1 \times 8 = 8$ Multiply the denominators: $1 \times 1 = 1$

Thus: $\frac{1}{2} \times \$16 = \frac{8}{1} = \$8$.

It does not matter whether you multiply a fraction by a whole number or a whole number by a fraction as shown in the following example:

EXAMPLE

Find the product of 10 and $\frac{2}{3}$.

SOLUTION

Problem	Step 1	Step 2
$10 \times \dfrac{2}{3} =$	$\dfrac{10}{1} \times \dfrac{2}{3} =$	$\dfrac{10}{1} \times \dfrac{2}{3} = \dfrac{20}{3} = 6\dfrac{2}{3}$
	Write the whole number (10) as a fraction $\left(\dfrac{10}{1}\right)$	Multiply the numerators: $10 \times 2 = 20$ Multiply the denominators: $1 \times 3 = 3$ Change $\dfrac{20}{3}$ into a mixed number.

Thus: $10 \times \dfrac{2}{3} = \dfrac{20}{3} = 6\dfrac{2}{3}$.

PRACTICE EXERCISE 98

Multiply. Write each answer in its simplest form.

1. $6 \times \dfrac{5}{6} =$

2. Find $\dfrac{3}{8}$ of 50.

3. Find $\dfrac{2}{3}$ of 60.

4. $20 \times \dfrac{2}{3} =$

5. $\dfrac{1}{8} \times 2 =$

6. $\dfrac{4}{5} \times 12 =$

7. $\dfrac{3}{5} \times 15 =$

8. $\dfrac{2}{3} \times 8 =$

9. $28 \times \dfrac{2}{7} =$

9.4 Multiplying Three or More Fractions

This problem isn't any more difficult than those illustrated already. Recall that to use *cancellation* you look for a number that divides evenly into any numerator and any denominator. Look at the following example to see how this method applies to multiplying three fractions.

EXAMPLE

Multiply: $\dfrac{5}{6} \times \dfrac{3}{7} \times \dfrac{2}{5} = ?$

SOLUTION

Problem	Step 1	Step 2	Step 3	Step 4
$\dfrac{5}{6} \times \dfrac{3}{7} \times \dfrac{2}{5} =$	$\dfrac{\overset{1}{\cancel{5}}}{6} \times \dfrac{3}{7} \times \dfrac{2}{\underset{1}{\cancel{5}}} =$	$\dfrac{1}{\underset{3}{\cancel{6}}} \times \dfrac{3}{7} \times \dfrac{\overset{1}{\cancel{2}}}{1} =$	$\dfrac{1}{\underset{1}{\cancel{3}}} \times \dfrac{\overset{1}{\cancel{3}}}{7} \times \dfrac{1}{1} =$	$\dfrac{1}{3} \times \dfrac{1}{7} \times \dfrac{1}{1} = \dfrac{1}{7}$
	Cancel: $5 \div 5 = 1$ $5 \div 5 = 1$	Cancel: $6 \div 2 = 3$ $2 \div 2 = 1$	Cancel: $3 \div 3 = 1$ $3 \div 3 = 1$	Multiply the numerators: $1 \times 1 \times 1 = 1$ Multiply the denominators: $1 \times 7 \times 1 = 7$

Thus: $\dfrac{5}{6} \times \dfrac{3}{7} \times \dfrac{2}{5} = \dfrac{1}{7}$.

PRACTICE EXERCISE 99

1. $\dfrac{1}{2} \times \dfrac{2}{3} \times \dfrac{3}{5} =$

2. $\dfrac{3}{4} \times \dfrac{1}{2} \times \dfrac{8}{9} =$

3. $\dfrac{3}{4} \times 2 \times \dfrac{5}{6} =$
   (*Hint:* To solve, write the whole number 2 as a fraction $\dfrac{2}{1}$).

4. $\dfrac{7}{8} \times \dfrac{3}{5} \times 4 =$

9.5 Multiplying Mixed Numbers

None of the examples presented up to this point deals with mixed numbers. Multiplying mixed numbers should not cause you any difficulty if you have mastered the early portion of this chapter.

To multiply mixed numbers you change each mixed number into an improper fraction and proceed in the same way as you would to multiply any other fractions.

EXAMPLE

A page measures $8\frac{7}{8}$ in. in length. How many inches are there in $\frac{1}{2}$ the page?

SOLUTION

You must find $\frac{1}{2}$ of $8\frac{7}{8}$ in., or $\frac{1}{2} \times 8\frac{7}{8}$ in., since *of* implies multiplication. Look at the problem below.

Problem	Step 1	Step 2
$\frac{1}{2} \times 8\frac{7}{8} =$	$\frac{1}{2} \times \frac{71}{8} =$	$\frac{1}{2} \times \frac{71}{8} = \frac{71}{16} = 4\frac{7}{16}$
	Change the mixed number $8\frac{7}{8}$ into an improper fraction: $8\frac{7}{8} = \frac{71}{8}$	Multiply the numerators: $1 \times 71 = 71$ Multiply the denominators: $2 \times 8 = 16$ Change the improper fraction $\frac{71}{16}$ to a mixed number: $\frac{71}{16} = 4\frac{7}{16}$

Thus: $\frac{1}{2} \times 8\frac{7}{8}$ in. $= \frac{71}{16} = 4\frac{7}{16}$ in.

EXAMPLE

Multiply $2\frac{1}{8}$ by $1\frac{1}{3}$. Write the product in simplest form.

SOLUTION

To solve this problem you must find the product:

$$2\frac{1}{8} \times 1\frac{1}{3} = ?$$

Problem	Step 1	Step 2	Step 3
$2\frac{1}{8} \times 1\frac{1}{3} =$	$\frac{17}{8} \times \frac{4}{3} =$	$\frac{17}{\underset{2}{\cancel{8}}} \times \frac{\overset{1}{\cancel{4}}}{3} =$	$\frac{17}{2} \times \frac{1}{3} = \frac{17}{6} = 2\frac{5}{6}$
	Change each mixed number into an improper fraction: $2\frac{1}{8} = \frac{17}{8}$ $1\frac{1}{3} = \frac{4}{3}$	Cancel: $8 \div 4 = 2$ $4 \div 4 = 1$	Multiply the numerators: $17 \times 1 = 17$ Multiply the denominators: $2 \times 3 = 6$ Change the improper fraction to a mixed number: $\frac{17}{6} = 2\frac{5}{6}$

Thus: $2\frac{1}{8} \times 1\frac{1}{3} = \frac{17}{6} = 2\frac{5}{6}$.

PRACTICE EXERCISE 100

Multiply. Write each answer in its simplest form.

1. $\dfrac{4}{5} \times 8\dfrac{3}{4} =$

2. $2\dfrac{3}{5} \times 5 =$

3. $5\dfrac{1}{3} \times 3\dfrac{1}{4} =$

4. $2\dfrac{1}{2} \times 4\dfrac{1}{4} =$

5. $2\dfrac{2}{5} \times \dfrac{2}{3} =$

6. $\dfrac{1}{4} \times 4\dfrac{1}{5} =$

7. $2\dfrac{2}{3} \times 12 =$

8. $3\dfrac{1}{2} \times 4\dfrac{2}{3} =$

9. $7\dfrac{1}{5} \times 6\dfrac{3}{4} =$

10. $1\dfrac{2}{3} \times 4\dfrac{1}{5} =$

11. $3\dfrac{1}{2} \times 5\dfrac{1}{2} =$

12. $3\dfrac{1}{4} \times 4\dfrac{1}{5} =$

9.6 Word Problems

Before we begin word problems, let's renew your skills in mathematics. As you work on problems you have probably noted that you need a system or way of "setting up" the problem. Here is one system that has been devised; see if it helps you.

STEPS TO SOLVING WORD PROBLEMS

1. Read the problem through to get the main idea.
2. Reread the problem for details:
   a. What terms or symbols are used?
   b. What facts are given?
   c. What is to be found?
3. Decide the process or method that will solve the problem.
4. Work the problem.
5. Check that your answer is logical and correct.

Use your knowledge and the above suggestions to solve this problem:

EXAMPLE

A large rectangular tract of land measures $8\frac{1}{4}$ miles by $7\frac{1}{3}$ miles. Swamp covers $\frac{3}{4}$ of the land. What is the area of the swamp?

SOLUTION
1. The main idea of the problem is *area* (review Chapter 4, section 4.8).
2. a. The key symbols and terms for a wall are rectangle, length, width, and the formula for area.
   b. The facts are that the length is $8\frac{1}{4}$ mi., the width is $7\frac{1}{3}$ mi., and the swamp covers $\frac{3}{4}$ of the land.
   c. The unknown facts are the number of square miles (area) the tract contains and the size of the swamp.
3. In this problem, you require the formula for the area of the rectangle.
   Area = Length × Width
   Area of the tract = $8\frac{1}{4}$ mi. × $7\frac{1}{3}$ mi.
4. When you work the problem, you must know how to multiply fractions or mixed numbers.

I. Area of the tract = $8\dfrac{1}{4} \times 7\dfrac{1}{3}$

Area of the tract = $\dfrac{33}{4} \times \dfrac{22}{3}$

Area = $\dfrac{\overset{11}{\cancel{33}}}{\underset{2}{\cancel{4}}} \times \dfrac{\overset{11}{\cancel{22}}}{\underset{1}{\cancel{3}}}$

Area = $\dfrac{121}{2}$

Area of the tract = $60\dfrac{1}{2}$ mi.²

II. Area of the swamp = area of the tract $\times \dfrac{3}{4}$

Area of the swamp = $60\dfrac{1}{2} \times \dfrac{3}{4}$

Area of the swamp = $\dfrac{121}{2} \times \dfrac{3}{4}$

Area of the swamp = $\dfrac{363}{8}$

Area of the swamp = $45\dfrac{3}{8}$ mi.²

5. The answer is logical because the total area of the tract is approximately 60 mi.²
and the swamp is $\dfrac{3}{4}$ of that area. Thus, our result should be less than 60 mi.²,
which it is.

This illustrates one method you can use to solve a problem. Another approach may be to set up a diagram that fits the example and use the diagram to help you solve the problem. There are many different methods to employ when you are faced with a problem and the need for a solution. Try your skills on the practice exercise which follows.

PRACTICE EXERCISE 101

1. A case of soda contains 24 bottles. If Mrs. Smith ordered $\dfrac{2}{3}$ of the case, how many bottles will she receive?

2. The trip to work each day and back is $8\dfrac{7}{8}$ mi. If Mr. Goris does this 5 days a week, how many miles does he travel?

3. Mr. Clark had to cut 24 pieces of wood, each $7\frac{1}{16}$ in. long, from a 16 ft. board. How many inches remained after the pieces were cut? (*Hint:* You must convert 16 feet to inches.)

4. Mr. and Mrs. Young decide to save a portion of their salaries for next year's vacation trip. They find that they can save only $\frac{1}{10}$ of their combined weekly salaries of $925.
    a. How much do they save each week?
    b. How much will they save at the end of a year (52 weeks = 1 year)?

5. If $3\frac{7}{8}$ yd. of material are needed to make a dress, how many yards will be needed to complete an order for 300 dresses?

6. Find the area of a rectangle whose dimensions are $6\frac{3}{5}$ ft. by $8\frac{4}{11}$ ft.

7. Sam can do a certain job in $3\frac{3}{4}$ h. while John can do the same job in $\frac{1}{3}$ of that time. How many hours does it take John to do the job?

8. Find the area of a triangle whose base is $4\frac{5}{8}$ in. and whose height is $9\frac{1}{2}$ in. The formula for the area of a triangle is $A = \frac{1}{2}b \times h$ where A represents area, b represents the base, and h the height.

9. Mrs. Jackson made 36 cupcakes for a church dinner. Some of the cupcakes were chocolate and some vanilla. If $\frac{2}{3}$ of them were vanilla, how many chocolate cupcakes were there?

10. If $2\frac{7}{8}$ yd. of material is needed for a dress and $2\frac{1}{16}$ yd. for a blouse, then how much material is required if Jane makes 2 dresses and 4 blouses?

TERMS YOU SHOULD REMEMBER

Cancellation   The removal of a common factor from the numerator and denominator of a fraction.
Area   The measure of a flat, open surface or space.
Product   The result obtained in a multiplication example.

REVIEW OF IMPORTANT IDEAS

Some of the most important ideas thus far in Chapter 9 have been:

To multiply fractions, you multiply the numerators, multiply the denominators, and write the result as a fraction.

The word *of* indicates the operation of multiplication. For example: two-thirds *of* 6 is $\frac{2}{3} \times 6$.

Cancellation is used when multiplying fractions if a numerator and a denominator contain a common factor.

Finding the area of a geometric figure requires you to learn the area formula associated with the figure. For example: $A = L \times W$ is the area formula for a rectangle.

CHECK WHAT YOU HAVE LEARNED

This test gives you a chance to see if you have been successful in multiplying fractions. Try hard for a satisfactory grade.

A Score of	Means That You
20–22	Did very well. You can move to the next part of this chapter, "Dividing Fractions."
18–19	Know this material except for a few points. Reread the sections about the ones you missed.
14–17	Need to check carefully on the sections you missed.
0–13	Need to review this part of the chapter again to refresh your memory and improve your skills.

Chapter Test 9A

Write your answers below each question.

1. $\dfrac{2}{5} \times \dfrac{1}{3} =$

2. $\dfrac{1}{3} \times \dfrac{6}{7} =$

3. $\dfrac{5}{8} \times \dfrac{6}{7} =$

4. $\dfrac{3}{5} \times \dfrac{10}{21} =$

5. $\dfrac{3}{4} \times 8 =$

6. $3 \times \dfrac{5}{6} =$

7. $12 \times \dfrac{7}{12} =$

8. Find the product of $\dfrac{4}{5}$ and $\dfrac{1}{12}$.

9. Solange's favorite recipe calls for $\dfrac{1}{3}$ cup of sugar. If she wanted to cut the recipe in thirds, how much sugar would she require?

10. $4\dfrac{1}{2} \times 9 =$

11. $10 \times 6\dfrac{1}{2} =$

12. $\dfrac{5}{6} \times 1\dfrac{1}{6} =$

13. $\dfrac{2}{3}$ of $2\dfrac{1}{8} =$

14. $\dfrac{2}{21} \times 3\dfrac{1}{2} =$

15. $2\dfrac{1}{4} \times 1\dfrac{7}{9} =$

16. $3\dfrac{1}{4} \times 1\dfrac{1}{5} =$

17. $3\dfrac{1}{3} \times 2\dfrac{1}{3} =$

18. $\dfrac{2}{3} \times \dfrac{5}{6} \times \dfrac{1}{15} =$

19. $\dfrac{2}{5} \times 10 \times \dfrac{1}{8} =$

20. Daisy walks $1\dfrac{1}{2}$ mi. 6 days a week to the train station. How many miles a week does she walk to the station?

21. Mike wishes to bake 3 apple pies. If $3\frac{1}{2}$ lb. of apples are used for each pie, how many pounds in all are needed?

22. Nadia needs $1\frac{1}{2}$ ft. of steel tubing to make a toy car and $2\frac{1}{4}$ ft. to make a toy train. How many feet of steel tubing are needed if Nadia makes 4 toy cars and 5 toy trains?

Dividing Fractions

Division is our last operation dealing with fractions. It requires the same intensive study you have been doing for the other operations. Division of fractions will help you to estimate the number of floor tiles, each measuring $\frac{9}{16}$ ft.^{2}, needed to cover a floor with dimension of 9 ft. × 12 ft. or any other size. It will allow you to discover how many trips you need to make to remove $7\frac{1}{2}$ tons of gravel if your truck holds only $2\frac{1}{2}$ tons. It is an important operation with many useful applications.

   Do you still have your skill in dividing fractions? Here's your chance to find out. Take the pretest which follows.

Pretest 9B—See What You Know and Remember

Work these examples as carefully as possible. Some are more difficult than others. Write the answers, in simplified form, below each question.

1. $9 \div \frac{1}{3} =$

2. $20 \div \frac{4}{5} =$

3. $5 \div \frac{3}{8} =$

4. $18 \div 2\frac{1}{4} =$

5. $4 \div 2\frac{5}{8} =$

6. $6 \div 6\frac{3}{4} =$

7. $15 \div 1\frac{9}{16} =$

8. $\frac{3}{4} \div \frac{2}{3} =$

9. $\frac{5}{8} \div \frac{15}{16} =$

10. $\frac{7}{8} \div \frac{7}{32} =$

11. $\frac{2}{3} \div \frac{5}{12} =$

12. $\frac{4}{5} \div 7 =$

13. $\dfrac{9}{10} \div 12 =$

14. How many packages of candy, each containing $\dfrac{3}{4}$ lb., can be filled from 45 lb. of candy?

15. Mrs. Chung bought $8\dfrac{1}{3}$ yd. of cloth for $50. How much did she pay for each yard?

16. $2\dfrac{3}{4} \div 3 =$

17. $3\dfrac{1}{5} \div 4 =$

18. $17\dfrac{1}{2} \div 10 =$

19. $1\dfrac{7}{8} \div \dfrac{15}{32} =$

20. $1\dfrac{2}{3} \div \dfrac{3}{4} =$

21. $3\dfrac{5}{6} \div \dfrac{2}{3} =$

22. $\dfrac{5}{16} \div 2\dfrac{2}{3} =$

23. $\dfrac{7}{10} \div 5\dfrac{3}{5} =$

24. $6\dfrac{3}{4} \div 1\dfrac{1}{8} =$

25. $4\dfrac{1}{6} \div 4\dfrac{3}{8} =$

26. $5\dfrac{4}{5} \div 2\dfrac{1}{2} =$

27. $2\dfrac{11}{32} \div 1\dfrac{1}{4} =$

28. Ten girls are to share $3\dfrac{3}{4}$ lb. of cake equally. How much cake does each girl get?

29. How many $4\dfrac{1}{2}$ in. lengths of ribbon can be cut from a piece $13\dfrac{1}{2}$ in. long?

30. An athlete participating in the long jump makes three leaps of $22\dfrac{1}{2}$ ft., 24 ft., and $23\dfrac{5}{8}$ ft. What is the average length of his three jumps?

Now turn to the back of the book to check your answers. Add up all that you had correct.

A Score of	Means That You
27–30	Did very well. You can move to Chapter 10.
24–26	Know the material except for a few points. Read the sections about the ones you missed.
20–23	Need to check carefully on the sections you missed.
0–19	Need to work with this part of the chapter to refresh your memory and improve your skills.

Questions	Are Covered in Section
8–11	9.7
1–7, 12, 13	
16–27	9.8
14, 15, 28–30	9.9

9.7 The Quotient of Two Fractions

EXAMPLE

Diana buys $\frac{1}{2}$ lb. of cashew nuts and asks the salesclerk to put the nuts into $\frac{1}{4}$ lb. bags. How many bags does Diana receive from the salesclerk?

SOLUTION

The way to solve this example is to divide $\frac{1}{2}$ by $\frac{1}{4}$, since you are trying to determine the number of quarters there are in one half.

$$\frac{1}{2} \div \frac{1}{4} = ?$$

can be illustrated by the following diagram. Start with $\frac{1}{2}$ by dividing the whole into two equal parts with a solid line. Shade in $\frac{1}{2}$ of the whole.

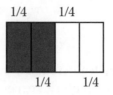

Divide the whole into 4 equal parts. You see that there are two quarters in the shaded portion equal to $\frac{1}{2}$. Thus,

$$\frac{1}{2} \div \frac{1}{4} = 2$$

While this is a suitable method for finding the answer to simple problems, what will you do when you are faced with more difficult ones? Obviously, an easier method must be found for dealing with all types of division examples involving fractions.

From the previous section on multiplying fractions you know that $\frac{1}{2} \times \frac{4}{1} = 2$. This problem has two elements which are the same as $\frac{1}{2} \div \frac{1}{4} = 2$, namely, the dividend $\frac{1}{2}$ and the quotient 2. The divisor $\frac{1}{4}$ is the *reciprocal* of $\frac{4}{1}$. The *reciprocal* of a fraction is the fraction obtained by interchanging its numerator and denominator.

The reciprocal of $\frac{2}{3}$ is $\frac{3}{2}$.

The reciprocal of $\frac{5}{12}$ is $\frac{12}{5}$.

The reciprocal of 6 is $\frac{1}{6}$. (Remember: $6 = \frac{6}{1}$).

You notice that the division example $\frac{1}{2} \div \frac{1}{4} = 2$ and the multiplication example $\frac{1}{2} \times \frac{4}{1} = 2$ result in the same solution. This gives you a hint on how to divide fractions. All you have to do is to write the reciprocal of the divisor, change the operation to multiplication, and then proceed, using the method for multiplying fractions. Look at the following example:

Problem	Step 1	Step 2	Step 3
$\dfrac{1}{2} \div \dfrac{1}{4} =$	$\dfrac{1}{2} \times \dfrac{4}{1} =$	$\dfrac{1}{\cancel{2}_{1}} \times \dfrac{\cancel{4}^{2}}{1} =$	$\dfrac{1}{1} \times \dfrac{2}{1} = \dfrac{2}{1} = 2$
	Write the reciprocal of the divisor $\left(\dfrac{1}{4}\right) \rightarrow \dfrac{4}{1}$ Change the division operation to multiplication.	Use cancellation: $4 \div 2 = 2$ $2 \div 2 = 1$	Multiply the numerators: $1 \times 2 = 2$ Multiply the denominators: $1 \times 1 = 1$

Thus: $\dfrac{1}{2} \div \dfrac{1}{4} = 2$.

"Write the reciprocal of the divisor" is sometimes called "invert the divisor." That expression may be familiar to you.

EXAMPLE

$$\frac{5}{6} \div \frac{1}{3} = ?$$

SOLUTION

Problem	Step 1	Step 2	Step 3
$\dfrac{5}{6} \div \dfrac{1}{3} =$	$\dfrac{5}{6} \times \dfrac{3}{1} =$	$\dfrac{5}{\cancel{6}_{2}} \times \dfrac{\cancel{3}^{1}}{1} =$	$\dfrac{5}{2} \times \dfrac{1}{1} = \dfrac{5}{2} = 2\dfrac{1}{2}$
	Invert the divisor $\left(\dfrac{1}{3} \text{ becomes } \dfrac{3}{1}\right)$ and multiply.	Use cancellation: $6 \div 3 = 2$ $3 \div 3 = 1$	Multiply the numerators: $5 \times 1 = 5$ Multiply the denominators: $2 \times 1 = 2$ Simplify the fraction: $\dfrac{5}{2} = 2\dfrac{1}{2}$

Thus: $\dfrac{5}{6} \div \dfrac{1}{3} = 2\dfrac{1}{2}$.

PRACTICE EXERCISE 102

Divide. Use cancellation when you can and write each answer in its simplest form.

1. $\dfrac{3}{4} \div \dfrac{1}{8} =$

2. $\dfrac{2}{3} \div \dfrac{1}{4} =$

3. $\dfrac{1}{2} \div \dfrac{2}{3} =$

4. $\dfrac{1}{2} \div \dfrac{1}{4} =$

5. $\dfrac{3}{4} \div \dfrac{1}{2} =$

6. $\dfrac{2}{3} \div \dfrac{4}{9} =$

7. $\dfrac{5}{8} \div \dfrac{1}{2} =$

8. $\dfrac{1}{3} \div \dfrac{1}{4} =$

9. $\dfrac{7}{8} \div \dfrac{7}{12} =$

10. $\dfrac{1}{3} \div \dfrac{1}{6} =$

11. $\dfrac{1}{8} \div \dfrac{1}{2} =$

12. $\dfrac{2}{3} \div \dfrac{1}{2} =$

9.8 Dividing Mixed Numbers

The division of mixed numbers is handled in the same way as the multiplication of mixed numbers. You change the mixed number or whole number to a fraction and then proceed as you learned in the previous section.

Remember: *Change the mixed number to a fraction and then invert the divisor and multiply.* Use the example that follows as a model.

EXAMPLE

A recipe calls for $\dfrac{1}{4}$ lb. of meat per person. Mrs. Jackson buys 3 lb. of meat. How many people will this feed?

SOLUTION

The solution to this example can be found by dividing 3 by $\dfrac{1}{4}$, or

$$3 \div \dfrac{1}{4} = ?$$

Problem	Step 1	Step 2	Step 3
$3 \div \frac{1}{4} =$	$\frac{3}{1} \div \frac{1}{4} =$	$\frac{3}{1} \times \frac{4}{1} =$	$\frac{3}{1} \times \frac{4}{1} = \frac{12}{1} = 12$
	Write the whole number as a fraction: $3 = \frac{3}{1}$	Invert the divisor $\left(\frac{1}{4} \text{ becomes } \frac{4}{1} \right)$ and multiply.	Multiply the numerators: $3 \times 4 = 12$ Multiply the denominators: $1 \times 1 = 1$

Thus: $3 \div \frac{1}{4} = 12$.

EXAMPLE

How many lengths of $1\frac{7}{8}$ in. can be cut from an aluminum rod measuring $7\frac{1}{2}$ in.? (Waste from cutting is not considered.)

SOLUTION

The solution to this problem is to divide $7\frac{1}{2}$ by $1\frac{7}{8}$ or

$$7\frac{1}{2} \div 1\frac{7}{8} = ?$$

Picturing this example on a ruler, like that drawn below, you see that there are *four* lengths of $1\frac{7}{8}$ in. in a line $7\frac{1}{2}$ in. in length.

Thus: $7\frac{1}{2} \div 1\frac{7}{8} = 4$.

The shorter arithmetic method requires you to change each of the mixed numbers to improper fractions first and then to continue as you did before.

Problem	Step 1	Step 2	Step 3	Step 4
$7\dfrac{1}{2} \div 1\dfrac{7}{8} =$	$\dfrac{15}{2} \div \dfrac{15}{8} =$	$\dfrac{15}{2} \times \dfrac{8}{15} =$	$\dfrac{\overset{1}{\cancel{15}}}{\underset{1}{\cancel{2}}} \times \dfrac{\overset{4}{\cancel{8}}}{\underset{1}{\cancel{15}}} =$	$\dfrac{1}{1} \times \dfrac{4}{1} = \dfrac{4}{1} = 4$
	Change each mixed number to an improper fraction: $7\dfrac{1}{2} = \dfrac{15}{2}$ $1\dfrac{7}{8} = \dfrac{15}{8}$	Invert the divisor $\dfrac{15}{8} \to \dfrac{8}{15}$ and multiply.	Use cancellation: $15 \div 15 = 1$ $15 \div 15 = 1$ and $8 \div 2 = 4$ $2 \div 2 = 1$	Multiply the numerators: $1 \times 4 = 4$ Multiply the denominators: $1 \times 1 = 1$

Thus: $7\dfrac{1}{2}$ in. $\div 1\dfrac{7}{8}$ in. $= 4$.

EXAMPLE

$$4 \div 6\dfrac{2}{3} = ?$$

SOLUTION

Problem	Step 1	Step 2	Step 3	Step 4
$4 \div 6\dfrac{2}{3} =$	$\dfrac{4}{1} \div \dfrac{20}{3} =$	$\dfrac{4}{1} \times \dfrac{3}{20} =$	$\dfrac{\overset{1}{\cancel{4}}}{1} \times \dfrac{3}{\underset{5}{\cancel{20}}} =$	$\dfrac{1}{1} \times \dfrac{3}{5} = \dfrac{3}{5}$
	Write each as an improper fraction: $4 = \dfrac{4}{1}$ $6\dfrac{2}{3} = \dfrac{20}{3}$	Invert the divisor $\dfrac{20}{3} \to \dfrac{3}{20}$ and multiply.	Use cancellation: $4 \div 4 = 1$ $20 \div 4 = 5$	Multiply the $1 \times 3 = 3$ and $1 \times 5 = 5$

Thus: $4 \div 6\dfrac{2}{3} = \dfrac{3}{5}$.

PRACTICE EXERCISE 103

Divide.

1. $6 \div \dfrac{2}{3} =$

2. $10 \div \dfrac{5}{6} =$

3. $4 \div \dfrac{1}{2} =$

4. $3 \div 1\dfrac{1}{2} =$

5. $1 \div 2\dfrac{1}{8} =$

6. $7 \div 2\dfrac{1}{3} =$

7. $4\dfrac{2}{5} \div 5\dfrac{1}{2} =$

8. $1\dfrac{2}{3} \div 2\dfrac{3}{5} =$

9. $10\dfrac{2}{3} \div 5\dfrac{1}{3} =$

10. $4\dfrac{5}{6} \div 3 =$

11. $6\dfrac{2}{3} \div 4 =$

12. $3\dfrac{7}{12} \div \dfrac{2}{3} =$

9.9 Word Problems

Reread section 9.6 of this chapter to be sure you are familiar with the *five* steps necessary to solve a word problem. When you feel sure you know these steps thoroughly, you will be ready to solve these practice exercises.

PRACTICE EXERCISE 104

1. How many pieces can be cut from a 27 in. aluminum pipe if each piece measures $3\dfrac{3}{8}$ in.?

2. How many bags of seed weighing $2\dfrac{1}{2}$ lb. each can be made from a sack weighing 60 lb.?

3. If a board $8\dfrac{3}{4}$ ft. long is cut into 7 equal parts, how long is each part?

4. Mr. Oak purchased $8\frac{1}{2}$ acres of land and plans to build 17 houses, each on an equal amount of land. How many acres will each plot of land contain?

5. Mr. Small has 5 crates of oranges which total $182\frac{1}{2}$ lb. Each crate contains an equal weight and no crate can exceed 35 lb. when shipped.

   a. Are the crates within the limit of 35 lb. each?

   b. How many pounds are they above or below the weight of 35 lb. each?

6. Find the average of the following *three* heights jumped by a pole vaulter in a recent track meet: $12\frac{1}{2}$ ft., $14\frac{7}{8}$ ft., and $14\frac{3}{4}$ ft.

7. If a dress requires $2\frac{9}{16}$ yd. fabric, how many dresses could be made if you had 82 yd. of fabric?

8. A steel bar measures $12\frac{1}{4}$ ft. in length. How many rods $1\frac{3}{4}$ ft. long can be cut from it?

9. A vegetable garden contains 45 ft.^{2} If the length of the rectangular garden is $5\frac{5}{8}$ ft., what is its width?

10. A recipe calls for $1\frac{1}{3}$ tablespoons of butter and serves 6 people. If you wish to cook this recipe for yourself, how much butter should you use?

TERMS YOU SHOULD REMEMBER

Quotient   The quantity resulting from dividing one quantity by another.

Invert   To interchange the numerator and denominator of a fraction. (Flip the fraction.)

Average   Most often referred to as the *arithmetic mean*.

REVIEW OF IMPORTANT IDEAS

Some of the most important ideas in this part of Chapter 9 were:

To divide fractions, invert the divisor, change the division sign to a multiplication sign, and proceed as you did when multiplying fractions.

When dividing mixed numbers, change each mixed number to a fraction, invert the divisor, and proceed as you did when multiplying fractions.

To find either a length or width of a rectangle, given one of these dimensions and the area, you divide the area by one of the given dimensions. In symbols:

$$L = \frac{A}{W} \text{ or } W = \frac{A}{L}$$

CHECK WHAT YOU HAVE LEARNED

This test will allow you to see how your skill in computing is progressing. Try to achieve a satisfactory grade.

A Score of	Means That You
27–30	Did very well. You can move to Chapter 10.
24–26	Know this material except for a few points. Reread the sections about the ones you missed.
20–23	Need to check carefully on the sections you missed.
0–19	Need to review this part of the chapter again to refresh your memory and improve your skills.

Questions	Are Covered in Section
8–11	9.7
1–7, 12, 13, 16–27	9.8
14, 15, 28–30	9.9

Chapter Test 9B

Write your answers below each question.

1. $8 \div \dfrac{1}{2} =$

2. $24 \div \dfrac{3}{4} =$

3. $7 \div \dfrac{4}{5} =$

4. $12 \div 1\dfrac{1}{5} =$

5. $3 \div 2\dfrac{5}{6} =$

6. $6 \div 3\dfrac{3}{4} =$

7. $18 \div 1\dfrac{5}{16} =$

8. $\dfrac{2}{3} \div \dfrac{3}{4} =$

9. $\dfrac{3}{4} \div \dfrac{15}{16} =$

10. $\dfrac{5}{9} \div \dfrac{5}{27} =$

11. $\dfrac{5}{6} \div \dfrac{7}{18} =$

12. $\dfrac{2}{5} \div 9 =$

13. $\dfrac{9}{10} \div 15 =$

14. How many cans of nuts, each containing $\dfrac{5}{6}$ lb., can be filled from 40 lb. of nuts?

15. Jean bought $7\dfrac{1}{2}$ yd. of fabric for \$60. How much did she pay for one yard?

16. $3\dfrac{2}{5} \div 4 =$

17. $2\dfrac{1}{4} \div 3 =$

18. $12\dfrac{1}{2} \div 20 =$

19. $1\dfrac{5}{8} \div \dfrac{13}{16} =$

20. $1\dfrac{3}{4} \div \dfrac{4}{5} =$

21. $3\dfrac{5}{8} \div \dfrac{3}{4} =$

22. $\dfrac{7}{20} \div 3\dfrac{1}{3} =$

23. $\dfrac{3}{10} \div 5\dfrac{2}{5} =$

24. $6\dfrac{2}{3} \div 1\dfrac{1}{9} =$

25. $3\dfrac{3}{4} \div 4\dfrac{1}{6} =$

26. $4\dfrac{1}{5} \div 1\dfrac{2}{3} =$

27. $2\dfrac{17}{32} \div 2\dfrac{1}{4} =$

28. Nine girls want to share $5\dfrac{1}{4}$ pounds of candy equally. How much candy should each girl get?

29. How many $3\dfrac{1}{2}$ in. pieces of ribbon can be cut from a piece $17\dfrac{1}{2}$ in. long?

30. Three track stars run a race in times of $39\frac{1}{2}$ s., $38\frac{3}{4}$ s., and $39\frac{1}{8}$ s. What is the average time for this event?

HOLD IT!

Here is your chance to see how your skills in computing are returning to you. This exercise reviews some of the work covered so far. Try hard for a perfect score.

1. $3{,}432 + 732 + 43{,}970 + 9 =$

2. Find the difference of 5,000 and 2,739.

3. Add: $3\frac{7}{8} + 4\frac{1}{6}$.

4. $37\overline{)3774}$

5. $\begin{array}{r} 673 \\ \times\ 45 \\ \hline \end{array}$

6. $\begin{array}{r} 193\frac{1}{2} \\ -146\frac{3}{8} \\ \hline \end{array}$

7. $6 \times 1\frac{1}{4} =$

8. $3\frac{5}{8} \div 1\frac{1}{16} =$

9. If $3\frac{5}{8}$ yd. of fabric are needed to make a dress, how many yards are needed to make 96 dresses?

10. If $13\frac{9}{16}$ ft. are cut from a 16 ft. board, how many feet still remain?

ANSWERS

1. 48,143

2. 2,261

3. $8\frac{1}{24}$

4. 102

5. 30,285

6. $47\frac{1}{8}$

7. $7\frac{1}{2}$

8. $3\frac{7}{17}$

9. 348 yd.

10. $2\frac{7}{16}$ ft.

Decimals–
What's the Point?

Decimals play a major role in our daily lives. The most use of decimals by the average person is in relation to money. Machine shop operators, parts inspectors, and other skilled persons concerned with more precise measurements than one can read from a ruler need to know decimals very well.

You have certainly seen many instances where decimals have an important role. Baseball players have batting averages of .283; pitchers have earned run averages of 3.54; track races are timed in 9.1 s.; and distances between cities are written as 38.7 mi.; all are examples of decimals.

You probably know other examples of the same type. Decimals are no more difficult to use than the fractions you have completed. As a matter of fact, decimals may be easier for you to learn now that you understand fractions.

These next four chapters deal with this type of number, its meaning, its operations, and its relationship to fractions.

Pretest 10–See What You Know and Remember

The pretest will determine whether you know the meaning of decimals—how to read them, how to write them, and how to round them off. Try hard to do correctly as many of these problems as you can. Some are more difficult than others so do not be discouraged. Write your answers in the space provided.

1. The rectangle below has been divided into 10 equal parts. Represent as a decimal the part of the rectangle that has been shaded.

2. Express as a decimal the equivalents for each of the following fractions.

   a. $\dfrac{6}{10}$ b. $\dfrac{48}{100}$ c. $\dfrac{162}{1,000}$ d. $\dfrac{5}{100}$ e. $\dfrac{22}{1,000}$

3. Change these decimals to proper fractions written in their simplest form.

   a. .39   b. .8   c. .625   d. .04   e. .002

4. Place a decimal point in the number 1125 so that the following sentence makes sense: "Susan, who is in my adult education class, weighs *1125* lb."

5. Mr. Brown earns $100 a day. He spends $86 and saves the rest. Express, as a decimal, the part of his salary he saves.

6. Eight-thousandths is written as

   a. 8.000   b. .008   c. .0008   d. .8000

7. a. Write, in words, the meaning of the number: .2

   b. Write, in words, the meaning of the number: 3.07

8. Write in figures, the number represented by:

   a. forty-four hundredths

   b. three and two-tenths

   c. three thousand four and sixty-three thousandths

9. Place one of the symbols >, <, or = within the parentheses to make each of the following statements true:

   a. .8 (   ) .4   b. .50 (   ) .5   c. .070 (   ) .70

   d. .1 (   ) .0925   e. 2.0 (   ) 2.009

10. Arrange each decimal in order of size with the smallest number first.

   a. .6, .06, .65

   b. 7.37, 7.037, 7.307

   c. 2.55, 2.505, 2.5

11. Round off each of the following numbers to the indicated place.

   a. .87 to the nearest tenth

   b. 11.434 to the nearest hundredth

   c. 3.1416 to the nearest thousandth

   Now turn to the back of the book to check your answers. Add up all that you had correct. Count by the number of separate answers, not by the number of questions. In this pretest there were 11 questions, but 30 separate answers.

A Score of	Means That You
27–30	Did very well. You can move to Chapter 11.
24–26	Know this material except for a few points. Read the sections about the ones you missed.
20–23	Need to check carefully on the sections you missed.
0–19	Need to work with this chapter to refresh your memory and improve your skills.

Questions	Are Covered in Section
1–3, 5	10.1
4, 6	10.2
7, 8	10.3
9, 10	10.4
11	10.5

10.1 Introducing Decimals

You have previously learned that each digit in a numeral has two values—*face* and *place* values. In the numeral 234, the digits 2, 3, and 4 have *face* values equivalent to 2, 3, and 4 units respectively. But because of their position in the number, they have a second value. Look at the illustration which follows.

In the numeral 234,

2 represents 2 hundreds
3 represents 3 tens
4 represents 4 ones

Each digit has a place value which is always 10 times the place value of the digit immediately to its right and one-tenth as much as the place value of the digit immediately to its left. In tabular form it looks like this:

	TEN THOUSANDS	THOUSANDS	HUNDREDS	TENS	ONES
etc. ← ←					
	$10 \times 10 \times 10 \times 10 \times 1 =$ 10,000	$10 \times 10 \times 10 \times 1 =$ 1,000	$10 \times 10 \times 1 = 100$	$10 \times 1 = 10$	1
→			2	3	4

Clearly, 234 belongs in the three columns to the right, as shown below the table, marked with an ⟶. The 2 is in the hundreds column, the 3 is in the tens column, and the 4 is in the ones column.

You know that there are other place values farther to the left than are indicated on the chart. Hundred-thousands, millions, etc., are all columns to the left of those listed. Each of these columns has a place value that is 10 times the column to its right. Thousands are 10 times the hundreds column, hundreds are 10 times the tens column, and so on.

Let us consider the relationship of any column to the one on its right. The column to the right is one-tenth or $\frac{1}{10}$ of the column on its left. This is illustrated in the following table using the number 234.875:

10,000	1,000	100	10	1	$\frac{1}{10}$	$\frac{1}{100}$	$\frac{1}{1,000}$	$\frac{1}{10,000}$
		2	3	4 .	8	7	5	
...TEN THOUSANDS	THOUSANDS	HUNDREDS	TENS	ONES .	TENTHS	HUNDREDTHS	THOUSANDTHS	TEN-THOUSANDTHS...

Each place has a value one-tenth of that to the left. Each place has a value ten times the place to the right.

The place value of the column to the right of the ones column is one-tenth of its place value and is called the *tenths* column. This column represents fractional values and each column to its right also represents fractional values.

Looking at it again, you see that the place value of the column to the right of the ones column must be multiplied by 10 and the product must be the place value of the ones column, or 1. Clearly, the problem becomes

$$10 \times ? = 1$$

The answer is $\frac{1}{10}$, since $10 \times \frac{1}{10} = 1$.

Thus, in the numeral 234.875, at the top of the table, eight-tenths = .8 = $\frac{8}{10}$. This is a *one-place* decimal. It is written as .8 and read as 8 tenths.

PRACTICE EXERCISE 105

Write each of the following in decimal form.

1. $\dfrac{6}{10}$ 2. $\dfrac{3}{10}$ 3. $\dfrac{1}{10}$ 4. $\dfrac{9}{10}$ 5. $\dfrac{7}{10}$

Write each of the following in fractional form.

6. .5   7. .2   8. .8   9. .1   10. .4

Moving to the next column to the right, you have

$$10 \times ? = \frac{1}{10}$$

The solution to this example is $\dfrac{1}{100}$, and the column to the right of the tenths column is called the *hundredths* column. Other columns are found the same way and are called thousandths, ten-thousandths, hundred-thousandths, etc. You may continue indefinitely. You should also notice that as you move to the right values are shrinking. One-thousandth is much less than one-tenth:

$$\frac{1}{1,000} < \frac{1}{10}$$

The *decimal point* (.) separates the whole-number portion from the fractional portion. Thus, in the decimal numeral 234.875, illustrated in the preceding table, 234 is the whole-number part of the numeral and 875 is the fractional part.

$$.8 = \frac{8}{10}$$

$$.87 = \frac{87}{100}$$

$$.875 = \frac{875}{1,000}$$

EXAMPLE

During the first two weeks in September $\frac{32}{100}$ in. of rain fell. Write this as a decimal.

SOLUTION

$\frac{32}{100} = .32$

This is a two-place decimal. It is written as .32 and read as 32 hundredths.

PRACTICE EXERCISE 106

Write each of the following in decimal form.

1. $\frac{37}{100}$
2. $\frac{463}{1,000}$
3. $\frac{99}{100}$
4. $\frac{4}{100}$

5. $\frac{36}{1,000}$
6. $\frac{7,834}{10,000}$
7. $\frac{834}{10,000}$
8. $\frac{1}{1,000}$

Write each of the following in fraction form.

9. .53
10. .25
11. .08
12. .032

13. .205
14. .1034
15. .0078
16. .005

10.2 Reading and Writing Decimals

In the table showing our decimal system (page 283), you notice that numbers to the right of the decimal point are common fractions whose denominators are powers of 10, such as 10, 100, 1,000, etc.

A *one-place* decimal is a common fraction whose denominator is 10, like $\frac{6}{10}$, which is written as .6 and read as 6 tenths.

A *two-place* decimal is a common fraction whose denominator is 100, like $\frac{6}{100}$. It is written as .06 and read as 6 hundredths.

When a decimal has *three places* to the right of the decimal point, it is read as thousandths. Therefore, .006 is read as 6 thousandths.

Notice that the zeros in the number are quite important since they help place the digit or digits in their proper place. Without them, each number in the previous illustrations would look like .6 and each would be read as 6 tenths.

EXAMPLE

How is the decimal .25 read and written?

SOLUTION
1. The numeral .25 is a *two-place* decimal.
2. Read it as: 25 hundredths.
3. Write it as: twenty-five hundredths.

EXAMPLE

In the men's 60-yd. dash, the time of the winner was 5.8 s. Read and write the decimal.

SOLUTION
1. 5.8 contains a whole number and a one-place decimal.
2. Read it as: 5 *and* 8 tenths (*and* represents the decimal point).
3. Write it as: five and eight-tenths.

EXAMPLE

Some instruments measure as accurately as one-thousandth of an inch. Write this as a decimal.

SOLUTION
Since thousandths is the third place in the decimal, you place the 1 in that place and fill the missing spaces with zeros. Thus: one-thousandth = .001.

PRACTICE EXERCISE 107

Write in words the meaning of each decimal.

1. .9

2. .27

3. .05

4. 6.1

5. 3.141

6. .0051

7. $.05\frac{1}{4}$

8. 7.02

Write each of the following as a decimal.

9. 6 and 9 hundredths

10. 3 thousandths

11. 3 and 5 tenths

12. 312 and 4 hundredths

13. $4\frac{3}{4}$ hundredths

14. 93 thousandths

15. 7 and 5 ten-thousandths

16. 40 and 43 hundredths

Write each of the following as a decimal.

17. nine-tenths

18. twenty-five hundredths

19. two hundred seven thousandths

20. three and fourteen thousandths

21. thirty-seven and three hundredths

22. nine hundred and nine thousandths

10.3 Which Fraction Equals Which Decimal?

Decimals which terminate or end can be written as a fraction. To write a decimal as a fraction is as easy as reading a decimal.

EXAMPLE

Write the decimal .8 as a fraction in its simplest form.

SOLUTION

Problem	Step 1	Step 2
.8 =	8 tenths =	$\frac{8}{10} = \frac{4}{5}$
	Read the decimal.	Write the decimal as a fraction. Simplify the fraction.

Thus: $.8 = \frac{4}{5}$

EXAMPLE

Mr. Washington reads a dimension on a blueprint as .875 in. What measurement on a standard ruler is this equivalent to?

SOLUTION

Problem	Step 1	Step 2
.875 =	875 thousandths =	$\dfrac{875}{1,000} = \dfrac{7}{8}$
	Read the decimal.	Write the decimal as a fraction. Simplify the fraction.

Thus: .875 in. $= \dfrac{7}{8}$ in.

PRACTICE EXERCISE 108

Express each of these examples as a fraction in its simplest form (lowest terms).

1. .5

2. .25

3. .625

4. .75

5. .34

6. .125

7. .9

8. .12

9. .08

Express each of these examples as a mixed number in its simplest form.

10. 3.7

11. 5.75

12. 12.1

13. 6.375

14. 1.875

15. 123.5

10.4 Comparing Decimals

EXAMPLE

When purchasing a sheet of aluminum, Mr. Green had to choose between two different thicknesses. One sheet measured .09 in. while the other was .11 in. Mr. Green desired the thicker of the two sheets of aluminum. Which did he choose?

SOLUTION

In this problem you must compare the sizes of the two decimals .09 and .11. If you change each decimal to a fraction you will see that it is easy to answer the question. Look below:

$$.09 = \frac{9}{100}$$

$$.11 = \frac{11}{100}$$

Since $\frac{11}{100} > \frac{9}{100}$, it follows that .11 > .09.

   Comparing decimals can be done in a simpler way as shown in the following problem.

EXAMPLE

Compare the decimals .5 and .505 to determine the smaller of the two.

SOLUTION

In the previous problem, the two decimals, .09 and .11, were changed to fractions and then compared. This was easily done since both decimals had denominators of 100. In this example, .5 has a denominator of 10 and .505 has a denominator of 1,000. This can be solved in this way:

Problem	Step 1	Step 2
Compare .5 and .505.	$.5 = .500 = \dfrac{500}{1,000}$ $.505 = \dfrac{505}{1,000}$	$\dfrac{500}{1,000} < \dfrac{505}{1,000}$ or $.500 < .505$ or $.5 < .505$
	.505 is a three-place decimal. Change .5 to a three-place decimal by adding two zeros to the right of the decimal: $.5 = .500$ Change both decimals to fractions.	Compare the fractions.

Thus: $.5 < .505$

EXAMPLE

Arrange these numbers in order of size with the smallest first: .2, .22, .022

SOLUTION
1. .2 is a *one*-place decimal
2. .22 is a *two*-place decimal
3. .022 is a *three*-place decimal
4. The number .022 has the largest number of decimal places; therefore each of the other decimals will have to become three-place decimals by adding zeros.

Problem	Step 1	Step 2
Compare .2, .22, .022.	$.2 = .200 = \dfrac{200}{1,000}$ $.22 = .220 = \dfrac{220}{1,000}$ $.022 = .022 = \dfrac{22}{1,000}$	$\dfrac{22}{1,000} < \dfrac{200}{1,000} < \dfrac{220}{1,000}$ or $.022 < .2 < .22$
	Since .022 is a *three*-place decimal, change both .2 and .22 to three-place decimals also. To .2 add two zeros: .2 = .200. To .22 add one zero: .22 = .220. Convert each decimal to an equivalent common fraction; all will have the same denominator.	Compare the fractions. Write the equivalent decimals.

Thus: $.022 < .2 < .22$

PRACTICE EXERCISE 109

Change each of these decimals to *equivalent* ones expressed as thousandths.

1. .43   2. .04   3. .1

Choose the decimals which have the same value.

4. .56, .506, .560   5. .05, .055, .05000, .050

Select the larger of the two numbers.

6. .82 and .8   7. .95 and .92   8. .71 and .69

9. .004 and .04   10. .1 and .01   11. 4.01 and 3.9

Arrange the numbers in order of size, with the smallest one first.

12. .077, .07, .7   13. 3.01, 3.001, 3.1

14. 1.3, 1.03, 1.31   15. .001, .0001, .1

10.5 Rounding Off Decimals

EXAMPLE

Ms. Black works as a payroll clerk and must figure the amount of money to be withheld for Social Security taxes for each employee. While computing the tax for an employee, she figured out that she had to deduct $74.156 from his salary. How much money was deducted?

SOLUTION

$74.156 is a sum of money greater than $74.15 and less than $74.16 or

$$\$74.15 < \$74.156 < \$74.16$$

Ms. Black realized that the first place past the decimal point represents tenths and, in this case, tenths of a dollar, or dimes. The second place represents hundredths of a dollar, or pennies. The third place represents thousandths of a dollar and is called mills, but there is no coin which is equivalent to a mill. Ms. Black had a decision to make—deduct $74.15 or $74.16. The process involved is called *rounding off a decimal*.

Since the digit in the thousandths place is 5 or greater than 5, the digit in the hundredths place is raised to 6. Thus, $74.16 was deducted from the employee's salary.

This illustrates *rounding off a decimal* to the *nearest hundredth*. All of the digits to the left of that place remain the same, the 7, the 4, and the 1. The digit in the hundredths place is changed upward to 6 because the digit in the thousandths place is 5 or greater than 5. Look at this illustration:

Problem	Step 1	Step 2
Round off $74.156 to the nearest cent. (Hundredths).	$74.1<u>5</u>6	$74.16
	Underline the digit in the hundredths place (5).	All the digits to the left of 5 remain the same: $74.1__. Since the digit in the thousandths place (6) is 5 or greater than 5, the digit in the hundredths place is increased by one.

Thus: $74.156 ≈ $74.16

NOTE

The symbol ≈ means approximately equal to. It is not correct to write $74.156 = $74.16, but it is correct to write $74.156 ≈ $74.16.

EXAMPLE

Suppose you had the decimal $7.152 and you wished to round it off to the nearest cent.

SOLUTION

Problem	Step 1	Step 2
Round off $7.152 to the nearest cent.	$7.1<u>5</u>2	$7.15
	Underline the digit in the hundredths place (5).	All the digits to the left of 5 remain the same: $ 7.1_ Since the digit in the thousandths place (2) is less than 5, the digit in the hundredths place remains the same.

Thus: $7.152 ≈ $7.15

PRACTICE EXERCISE 110

Round off each of the following decimals to the nearest tenth.

1. .34   2. 2.381   3. 3.07   4. .67

Round off each of the following decimals to the nearest hundredth.

5. 4.345   6. .768   7. 2.796   8. .871

Round off each of the following decimals to the nearest thousandth.

9. 3.1416   10. .6784   11. 5.5555

TERMS YOU SHOULD REMEMBER

Decimal   A number that uses a decimal point to represent a fraction, every decimal place indicating a power of 10.

Decimal point   A period placed to the left of a decimal.

Round off   Approximate the value according to designated rules.

REVIEW OF IMPORTANT IDEAS

Some of the most important ideas in Chapter 10 were:

To read a decimal—
1. Read the whole number portion.
2. Read the decimal point as *and*.
3. Read the number to the right of the decimal point and then name the position of the last digit on the right (.04 is read as 4 hundredths since the 4 is in the hundredths place).

To write a decimal—
1. Write the whole number portion first.
2. Place the decimal point.
3. Place the last digit in the number in the place corresponding to the position stated (to write 3 thousandths, place the 3 in the third place in the decimal and fill in with zeros: .003).

To compare decimals—
1. Determine which of the numbers you are comparing has the greater number of decimal places.
2. Change each of the decimals to an equivalent one by adding zeros to the right of the number so that each decimal has the same number of places.
3. Change each decimal to an equivalent common fraction.
4. Compare the resulting numbers.

To round off decimals—
1. Underline the digit in the decimal place to which you wish to round off the number.
2. Write down all the digits to the left of the underlined digit as part of the answer.
3. Check the digit to the right of the underlined digit:
   a. If this digit is less than 5, the underlined digit remains the same.
   b. If this digit is 5 or greater than 5, the underlined digit is increased by one.

CHECK WHAT YOU HAVE LEARNED

Good luck on the chapter test. See if you can achieve a satisfactory grade.

In counting up your answers, remember that there were 30 separate answers in this test.

A Score of	Means That You
27–30	Did very well. You can move to Chapter 11.
24–26	Know this material except for a few points. Reread the sections about the ones you missed.
20–23	Need to check carefully one the sections you missed.
0–19	Need to review the chapter again to refresh your memory and improve your skills.

Questions	Are Covered in Section
1–3, 5	10.1
4, 6	10.2
7, 8	10.3
9, 10	10.4
11	10.5

Chapter Test 10

Solve these problems, reducing fractions to their lowest terms. Write your answers in the space provided.

1. This rectangle has been divided into 10 equal parts. Represent as a decimal the part of the rectangle that has been shaded.

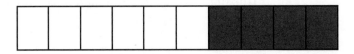

2. Express as a decimal the equivalents for each of the following fractions:

    a. $\dfrac{5}{10}$ b. $\dfrac{34}{100}$ c. $\dfrac{612}{1,000}$ d. $\dfrac{4}{100}$ e. $\dfrac{87}{1,000}$

3. Change these decimals to proper fractions written in their simplest form.

    a. .93   b. .6   c. .875   d. .06   e. .004

4. Place a decimal point in the number 2146 so that the following statement makes sense. "John, who is in my adult education class, weighs *2146* lb."

5. Mr. Greene earns $10,000 a month. He spends $9,600 and saves the rest. Express, as a decimal, the part of his salary he saves.

6. Six-thousandths is written as

    a. 6.000   b. .006   c. .06   d. .6000

7. a. Write, in words, the meaning of the decimal: .4

    b. Write, in words, the meaning of the decimal numeral: 7.03

8. Write, in figures, the number represented by:

    a. thirty-six hundredths

    b. two and three-tenths

    c. four thousand three and thirty-six thousandths

9. Place one of the symbols >, <, or = within the parentheses to make each of the following statements true.

    a. .7 (   ) .5   b. .60 (   ) .6   c. .090 (   ) .90

    d. .4 (   ) .3925   e. 4.0 (   ) 4.001

10. Arrange the decimals in order of size with the smallest number first.

    a. .5, .05, .56

    b. 8.34, 8.034, 8.304

    c. 5.22, 5.202, 5.2

11. Round off each of these numbers to the indicated place.

    a. .78 to the nearest tenth

    b. 14.212 to the nearest hundredth

    c. 4.1518 to the nearest thousandth

    You have come a long way since you began to study and relearn the mathematics contained in this book. It has been fairly difficult for you to get to this point. You should be very proud of yourself. You still have a way to go before you can say you are proficient in mathematics. You have started on that climb; keep going!

Adding and Subtracting Decimals

11

Adding Decimals

You have completed the introduction to decimals and have seen the relationship between the decimal form and the fraction form of a number. You have been successful in this area and you are ready to start the first fundamental operation associated with decimals, adding decimals.

Before you begin, though, take the pretest which follows, to see how much of the material covered in this topic is familiar to you.

Pretest 11A–See What You Know and Remember

Do as many problems as you can. Try as hard as you can. Some may be harder than others for you. Write each answer in the space provided.

1.   .2
  +.7

2.   .15
  +.04

3.   .76
  +.23

4.   .07
  +.02

5.   .8
  +.6

6.   .3
    .5
  +.7

7.   3.6
  +2.1

8.   2.134
  + .35

9.   13.018
  + 7.703

10. 1.3 + 9.2 + 13 =   11.   4.5   12.   .013
                                   .07   +.006
                                +20.457   _____

13. $.95 + $1.48 =

14. Bob went to the supermarket on Thursday to buy some bread and butter. The butter cost $1.98 and the bread cost $1.19. What was the total cost of his purchase?

15. How many miles did Sonia drive her car if she drove 14 mi. on Monday, 23.8 mi. on Tuesday, 5.3 mi. on Wednesday, and 3.25 mi. on Thursday?

16. The Capital Men's Shop had a sale on men's clothing. A suit sold for $374.95, a hat for $20.00, shoes for $66.99, and shirts for $26.00 each. How much money will it cost if you bought a suit, a pair of shoes, and one shirt?

Now turn to the back of the book to check your answers. Add up all that you had correct.

A Score of	Means That You
15–16	Did very well. You can proceed to the second half of this chapter, "Subtracting Decimals."
13–14	Know this material except for a few points. Read the sections about the ones you missed.
11–12	Need to check carefully on the sections you missed.
0–10	Need to work with this part of the chapter to refresh your memory and improve your skills.

Questions	Are Covered in Section
1–4, 12, 13	11.1
5–11	11.2
14–16	11.3

11.1 Adding Decimals

When adding decimals, you must make sure you are adding tenths to tenths, hundredths to hundredths, and so on. To do this, make sure to line up the decimal points in a vertical column. Then add the numbers as you would add whole numbers.

EXAMPLE

What is the overall length of the steel template drawn below?

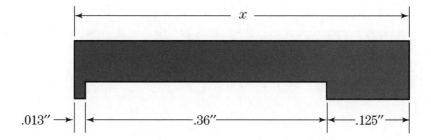

SOLUTION

Problem	Step 1	Step 2
.013 + .360 + .125 =	.013 .360 + .125	.013 .360 + .125 .498 in.
	Place the decimals in vertical columns, lining up the decimal points directly below those of the addends.	Add the numbers.

Thus: .013 in. + .360 in. + .125 in. = .498 in.

PRACTICE EXERCISE 111

Add these examples.

1. .8
    +.1

2. .17
    +.21

3. .86
    +.04

4. .06
    +.45

5. .325
    .242
    +.051

6. .08
    .14
    .23
    +.46

7. .504
    +.308

8. .46
    +.30

What is the significance of zeros in decimals?
Zeros to the right of a decimal point *do not* affect the value of the number. For instance:

$.5 or $50 are both 50 cents.

Zeros to the right of the decimal point and to the left of the other digits in the number *do* affect the value of the number. For instance:

$.01 and $.1 represent two different quantities.

$.01 = 1 penny while $.1 = 1 dime.

The zero is used as a place holder and cannot be omitted.

EXAMPLE

The thickness of a piece of paper is .025 in. and the thickness of a strand of hair is .019 in. If one is placed on top of the other, how much will be their combined thickness?

SOLUTION

Problem	Step 1	Step 2
.025 + .019 =	.025 + .019 —— .	.025 + .019 —— .044
	Line up the numbers by the decimal points. Place the decimal point in the answer below that of the addends.	Add the numbers. Be sure to include the zero. Remember: .044 ≠ (is not equal to) .44.

Thus: .025 in. + .019 in. = .044 in.

PRACTICE EXERCISE 112

Add. Be sure to include the zero if necessary.

1. .05
    +.03
    ——

2. .037
    +.045
    ——

3. .006
    +.045
    ——

4. .0013
    .0121
    + .0504
    ——

5. .009
    .027
    .031
    +.016
    ——

6. .04
    .02
    +.05
    ——

Sometimes, decimals to be added contain different numbers of digits. The numbers will then be uneven on the right-hand side of the number when they are lined up in vertical columns. This is illustrated in the following example:

EXAMPLE

Find the sum of .3205, .27, and .121.

SOLUTION

Problem	Step 1	Step 2	Step 3
.3205 + .27 + .121 =	.3205 .27 + .121 ───── .	.3205 .2700 + .1210 ───── .7115	.3205 .27 + .121 ───── .7115
	Line up the decimals and place the decimal point in the answer.	Add up the numbers. You can annex zeros to the right of each Number without changing its value. Thus: .27 = .2700 .121 = .1210	OR Since the zeros are understood to be there, just add the numbers without annexing the zeros.

Thus: .3205 + .27 + .121 = .7115

PRACTICE EXERCISE 113

Add.

1. .1034
    +.246
    ─────

2. .037
    .03
    .41
    +.002
    ─────

3. .274
    .29
    +.16
    ─────

4. .345
    .4
    .05
    +.19
    ─────

5. .48
    .1
    +.332
    ─────

6. .1
    .793
    +.07
    ─────

11.2 Adding Mixed Decimals

EXAMPLE

The Smith family plans a trip across country this summer and wishes to travel no more than 300 mi. each day. Checking the mileage between cities, they see that the mileage from Jackson to Lakeland is 151.2 miles, from Lakeland to Netcong is 87.3 miles, and from Netcong to Erie is 43.6 miles. Can the Smiths travel from Jackson to Erie and still be within their 300-mile limit?

SOLUTION

Adding mixed decimals is the same as adding decimals in section 11.1. Place the addends in vertical columns so that the decimal points are lined up one below the other. Look at this example:

Problem	Step 1	Step 2
151.2 + 87.3 + 43.6 = ?	151.2 87.3 + 43.6	151.2 87.3 + 46.6 —— 282.1
	Line up the addends and place the decimal point in the answer.	Add the numbers.

Thus: 151.2 mi. + 87.3 mi. + 43.6 mi. = 282.1 mi.

This sum can be verified by changing each of the decimals to fractions and then adding.

Jackson to Lakeland $\qquad 151.2 = 151\dfrac{2}{10}$

Lakeland to Netcong $\qquad 87.3 = 87\dfrac{3}{10}$

Netcong to Erie $\qquad + \ 43.6 = 43\dfrac{6}{10}$

$$282.1 \quad 281\frac{11}{10} = 281 + 1\frac{1}{10} =$$

$$282\frac{1}{10} = 282.1 = 282.1 \text{ mi.}$$

EXAMPLE

Find the sum of .73, .67, .2, and .259.

SOLUTION

Problem	Step 1	Step 2
.73 + .67 + .2 + .259 =	.73 .67 .2 + .259 .	.73 .67 .2 + .259 1.859
	Line up the addends and place the decimal point in the answer.	Add the numbers.

Thus: .73 + .67 + .2 + .259 = 1.859

EXAMPLE

Find the sum of 403.2, 45.96, 16, and .829.

SOLUTION

Problem	Step 1	Step 2
403.2 + 45.96 + 16 + .829 =	403.2 45.96 16. + .829 .	403.2 45.96 16. + .829 465.989
	Line up the addends and place the decimal point in the answer. Remember that the number 16 is a whole number and the decimal point is at its far right.	Add the numbers.

Thus: 403.2 + 45.96 + 16 + .829 = 465.989

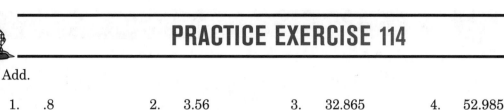

PRACTICE EXERCISE 114

Add.

1.   .8
    +.4

2.   3.56
    +4.73

3.   32.865
    + 2.97

4.   52.985
    + .06

Arrange these problems in column form before adding. Be sure to place the decimal points one under the other.

5.   .345 + .4 + .05 + 1.9 =

6.   5.8 + 10 + .432 =

7. $72.87 + $3.95 + $1.99 =

8.   103.456 + 7.932 + 19.57 + .12 =

11.3 Word Problems Requiring the Addition of Decimals

Adding decimals occurs frequently in various industries. One such problem is illustrated below.

EXAMPLE

Mr. Castro works in a metal shop. One day he received a blueprint drawing of a metal rod, pictured below, which he had to make.

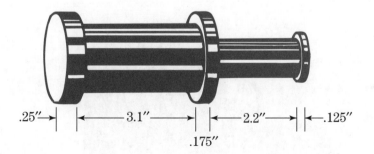

He had to find the overall length of the metal rod. What is the length?

SOLUTION
To find the length of the metal rod he had to find the sum of its parts. Thus:

Problem	Step 1	Step 2
.25 + 3.1 + .175 + 2.2 + .125 +	.25 3.1 .175 2.2 + .125 .	.25 3.1 .175 2.2 + .125 5.850 or 5.85
	Line up the addends and place the decimal point in the answer.	Add the numbers.

Thus: .25 in. + 3.1 in. + .175 in. + 2.2 in. + .125 in. = 5.85 in.

PRACTICE EXERCISE 115

1. Mr. O'Leary is a traveling salesman and in one week covered these distances. On Monday he traveled 73.4 mi.; Tuesday, 55.8 mi.; Wednesday, 113.9 mi.; Thursday, 93.7 mi.; and Friday, 27.5 mi. Find the total number of miles he traveled during the week.

2. A driver was permitted to carry a maximum truck load of 1,500 lbs. He loaded 7 crates weighing 138.4 lb., 258 lb., 212.9 lb., 371.6 lb., 157.9 lb., 197.2 lb., and 164 lb.

   a. Had he exceeded the maximum truck load of 1,500 lb.?
   b. If not, what was the weight of his load?

3. Ms. Chung is a social worker and receives a mileage allowance for her car. In one month she traveled the following weekly distances: 383.7 mi., 412.4 mi., 139.9 mi., and 273.8 mi. How many miles did she travel during the month?

4. Jenny shopped in a supermarket and bought the following priced items: $1.78, $2.58, $1.18, $1.38, $.58, and $8.86. How much money does Jenny need to pay for her purchases?

5. A rectangular plot of land measures 43.87 ft. along its length and 31.9 ft. along its width. What is the perimeter of this plot of land?

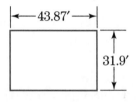

6. What is the length of this rod?

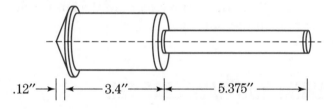

7. Find the perimeter of a triangle drawn below.

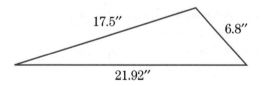

8. A sales slip was made for the following items. What was the total charge?

1 tennis shirt	$44.98
1 pair of tennis shorts	47.70
1 can of balls	5.99
1 tennis racket	189.99

9. What is the total width of four books whose individual widths are 2.73 in., 0.67 in., 1.2 in. and 3.259 in.?

10. The world's indoor record for the two-mile run is 8 min. 4.69 s. Mike ran the same distance in 18.9 s. more. What was his total time?

REVIEW OF IMPORTANT IDEAS

Some of the most important ideas in this part of Chapter 11 were:

Adding decimals is similar to adding whole numbers.

The decimal point separates the whole number portion of the number from the fractional part.

To add decimal numbers—
1. Place the decimals in vertical columns with the decimal points directly underneath each other.
2. Add the numbers together as you would add whole numbers.
3. Place the decimal point in the answer directly below the decimal point in the addends.

CHECK WHAT YOU HAVE LEARNED

The following test lets you see how well you learned the ideas thus far presented in Chapter 11.

A Score of   Means That You

15–16	Did very well. You can move on to the second part of the chapter, "Subtracting Decimals."
13–14	Know the material except for a few points. Reread the sections about the ones you missed.
11–12	Need to check carefully the sections you missed.
0–10	Need to review this part of the chapter again to refresh your memory and improve your skills.

Questions   Are Covered in Section

1–4, 12, 13	11.1
5–11	11.2
14–16	11.3

Chapter Test 11A

Write each answer in the space provided. Use a calculator to check your answers after you have completed the entire chapter test.

1.   .6
    +.3
    ———

2.   .12
    +.06
    ———

3.   .67
    +.32
    ———

4.   .03
    +.06
    ———

5.   .9
    +.8
    ———

6.   .7
    .6
    +.4
    ———

7.   6.3
    +1.2
    ———

8.   2.431
    + .35
    ———

9.   14.017
    + 7.704
    ———

10. 1.4 + .86 + 14 =

11.   8.37
     .03
    4.9
   + .002
    ———

12.   .0301
    +.0006
    ———

13. $2.59 + $.48 =

14. Martin went to the delicatessen on Tuesday to buy cole slaw and a carton of milk. The cole slaw cost $2.59 and the container of milk cost $1.87. What was the total cost of his purchase?

15. How many miles did Michael ride his bicycle if he rode 8.3 miles on Monday, 4.1 miles on Tuesday, 12.7 miles on Wednesday, and 2.46 miles on Thursday?

16. The Cape Cod Hardware Store had a second anniversary sale.

	Regular Price	Sale Price
Garden cart	$295.00	$249.50
Garden pruner	48.95	42.95
Garden shovel	44.95	39.99
Garden edge border	39.98	35.50
Garden fence kit	49.94	47.90
Plastic sprinkler can	12.95	9.95

Mrs. Maxson purchased a garden cart, a garden shovel, a plastic sprinkler can, and a garden fence kit. What was the total amount of her purchase?

Subtracting Decimals

To answer the question "How much remains after part is used?" you need to subtract. Change from a $10 bill after a purchase is made depends on subtraction. Finding the difference between any two quantities is another example of the operation. It is extremely important to know how to subtract decimals.

The pretest that follows will allow you an opportunity to see how good your skills are in subtracting decimals.

Pretest 11B—See What You Know and Remember

Some of these problems are more difficult than others. Do as many as you can. Write each answer in the space provided.

1. .8
    $-$.2
    ⎯⎯

2. .69
    $-$.41
    ⎯⎯

3. $7.6 - 3 =$

4. .39
    $-$.32
    ⎯⎯

5. 9.78
    $-$7.65
    ⎯⎯

6. 4.09
    $-$1.03
    ⎯⎯

7. 8.5
    $-$2.6
    ⎯⎯

8. $58.9 - 4.6 =$

9. $5 - .7 =$

10. 6.84
    $-$ .58
    ⎯⎯

11. 91.3
    $-$73.9
    ⎯⎯

12. .503
    $-$.39
    ⎯⎯

13. 6.51
    $-$2.47
    ⎯⎯

14. .084
    $-$.053
    ⎯⎯

15. 1.255
    $-$.165
    ⎯⎯

16. $1.03 - .75 =$

17. 39.29
    $-$ 4.7
    ⎯⎯

18. $5 - 3.2 =$

19. $19.8 - .32 =$

20. $.73 - .648 =$

21. Ramona had a piece of wood 8.75 ft. long. If she cut off a piece 3.25 ft. long to use for a shelf, how long was the piece which remained?

22. In the 1896 Olympics, Thomas E. Burke won the 100-meter dash in 12.0 s. In the 1968 Olympics, James Hines won the same race in 9.9 s. How much faster than Burke was Hines?

23. Edna weighed 117.65 lb. when summer began. When summer ended in September, she weighed 121.5 lb. How many pounds did Edna gain?

24. The gas tank in Mrs. Watts' car holds 20 gal. It took 16.3 gal. to fill the tank. How much gas was in the tank before it was filled?

25. Carol bought a book for $19.95 and a game for $14.99. She gave the clerk a $50 bill. How much change did she get?

Now turn to the back of the book to check your answers. Add up all that you had correct.

A Score of	Means That You
23–25	Did very well. You can proceed to Chapter 12.
20–22	Know the material except for a few points. Read the sections about the ones you missed.
17–19	Need to check carefully on the sections you missed.
0–16	Need to work with the chapter to refresh your memory and improve your skills.

Questions	Are Covered in Section
1, 2, 4, 12, 14, 20	11.4
3, 5–11, 13, 15–19	11.5
21–25	11.6

11.4 Subtracting Decimals

EXAMPLE

The thickness of a piece of paper is approximately .025 in. How much larger is this than the thickness of a strand of hair measuring .019 in.?

SOLUTION

You subtract .019 in. from .025 in. to get the solution to this example. To subtract (.025 − .019), you place the decimals in vertical columns with one decimal point directly underneath the other. Place the decimal point in the answer, or *difference*, directly below the decimal points in the numbers you are subtracting. Subtract the numbers as you would subtract whole numbers. See the following illustration:

Problem	Step 1	Step 2
.025 − .019 =	.025 −.019 —— .	11 .025̸ −.019 —— .006
	Place the decimals in vertical columns, lining up the decimal points. Place the decimal point in the answer directly below the other decimal points.	Subtract the numbers. Be sure to include the zeros. (Remember: .006 ≠ [is not equal to] .6.)

Thus: .025 in. − .019 in. = .006 in.

PRACTICE EXERCISE 116

Subtract. Be sure to indicate zeros when necessary.

1.   .9
    −.3
    ‾‾‾

2.   .48
    −.27
    ‾‾‾‾

3.   .311
    −.287
    ‾‾‾‾‾

4.   .431
    −.278
    ‾‾‾‾‾

5.   .1134
    −.0879
    ‾‾‾‾‾‾

6.   .373
    −.298
    ‾‾‾‾‾

7.   .302
    −.288
    ‾‾‾‾‾

8.   .151
    −.146
    ‾‾‾‾‾

9.   .995
    −.495
    ‾‾‾‾‾

EXAMPLE

Caridad won the race in .53 min. Her closest competitor ran the same race in .549 min. How much faster was Caridad?

SOLUTION

Problem	Step 1	Step 2	Step 3
.549 − .53 =	.549 −.53 ‾‾‾‾ .	.549 −.530 ‾‾‾‾‾ .	.549 −.530 ‾‾‾‾‾ .019
	Line up the decimal points so that one is directly below the other. Place the decimal point in the answer below the others.	.53 = .530 (since zeros to the right of the decimal number do not change its value)	Subtract the numbers. Be sure to include the zero.

Thus: .549 min. − .53 min. = .019 min.

EXAMPLE

Find the difference between .54 and .259.

SOLUTION

Problem	Step 1	Step 2	Step 3
.54 − 259 =	.54 −.259 ——— .	.540 −.259 ——— .	.540 −.259 ——— .281
	Line up the decimal points so that one is directly below the other. Place the decimal point in the answer below the others.	.54 = .540 (since zeros to the right of the decimal number do not change its value)	Subtract the numbers.

Thus: .54 − .259 = .281

PRACTICE EXERCISE 117

Subtract.

1.   .547
    −.29
    ————

2.   .43
    −.278
    ————

3.   .647
    −.19
    ————

4.   .6
    −.273
    ————

5.   .187
    −.099
    ————

6.   .07
    −.0691
    ————

7.   .482
    −.27
    ————

8.   .5644
    −.38
    ————

9.   .302
    −.28
    ————

11.5 Subtracting Mixed Decimals

Mixed decimals are subtracted in the same way as decimals in section 11.4.

EXAMPLE

Mr. Gonzalez had a credit of $8.95 in a men's clothing store. He bought a T-shirt for $15.98 and socks for $6.98. How much additional money must he pay?

SOLUTION

Mr. Gonzalez received a sales slip with the following computations:

A&R MEN'S STORE	
1 T-shirt	$15.98
2 pair of socks	6.98
	$22.96
Credit	8.95
To Be Paid	$14.01

From this information on the sales slip, Mr. Gonzalez determined that he had to pay an additional amount of $14.01 to cover the cost of his purchases. See the following subtraction example:

Problem	Step 1	Step 2
$22.96 – $8.95 =	$22.96 – 8.95	$22.96 – 8.95 ——— $14.01
	Line up the decimal points and place the decimal point in the answer.	Subtract the numbers. Be sure to include the zero in the answer.

Thus: $22.96 – $8.95 = $14.01.

Subtraction examples are checked by adding the *difference* to the number directly above the line. This sum must equal the top number. In this example you have:

$$\begin{array}{r} \$22.95 \\ -\ 8.95 \\ \hline \$14.01 \text{ (difference)} \end{array} \qquad \text{and} \qquad \begin{array}{r} \$8.95 \\ +\ 14.01 \text{ (difference)} \\ \hline \$22.96 \end{array}$$

Thus, you see that the $14.01 answer checks.

PRACTICE EXERCISE 118

Subtract.

1.	$5.43 – 2.59	2.	3.11 –2.87	3.	3.73 –2.98

4.	56.44 – .38	5.	38.16 – 9.47	6.	3.002 –2.97

7.	2.83 – .5737	8.	$113.43 – 87.95	9.	1.87 – .99

EXAMPLE

Beulah left work with $5 and spent $3.79 on her way home. How much money did she have when she arrived home?

SOLUTION

Problem	Step 1	Step 2	Step 3
$5 – $3.79 =	$5. – 3.79 —— .	$5.00 – 3.79 —— .	$5.00 – 3.79 —— $1.21
	Line up the decimals and place the decimal point in the answer. The decimal point in 5 is on its right: 5.	Annex 2 zeros to the right of the decimal point: 5. = 5.00	Subtract the numbers.

Thus: $5 – $3.79 = $1.21

EXAMPLE

Find the remainder when .6 is subtracted from 4.

SOLUTION

Problem	Step 1	Step 2	Step 3
4 − .6 =	4. − .6 ────── .	4.0 − .6 ────── .	4.0 − .6 ────── 3.4
	The decimal point in a whole number is on its right: 4.	Annexing zeros to the right of the decimal point doesn't change the value of the number: 4. = 4.0	Subtract the numbers.

Thus: 4 − .6 = 3.4

PRACTICE EXERCISE 119

Subtract. Place the decimal point in the numbers where it is needed. Remember to include zeros when necessary.

1. $4 − $1.23 =

2. 6 − .8 =

3. 4.9 − 2 =

4. 5.87 − .9 =

5. 113.4 − 87 =

6. 37 − 4.21 =

7. $6 − $1.95 =

8. Subtract .273 from 6.

9. From $10 subtract $6.91.

11.6 Word Problems Requiring the Subtraction of Decimals

The subtraction of decimals is necessary at many different times in our everyday life.

EXAMPLE

The radio announcer on Station WGLX reported that the barometer had fallen from 30.2 in. to 28.8 in. since 4:00 p.m. How many inches had it dropped?

SOLUTION

This example requires you to find the difference in the two barometer readings. The operation is subtraction, so subtract 28.8 in. from 30.2 in. See the following example.

Problem	Step 1	Step 2
30.2 − 28.8 =	30.2 −28.8	29 1 3̶0̶.2 −28.8 1.4
	Line up the decimals and place the decimal point in the answer.	Subtract the decimals.

Thus: 30.2 in. − 28.8 in. = 1.4 in.

PRACTICE EXERCISE 120

1. Two sheets of aluminum have thicknesses of .028 in. and .032 in. By how much do they differ in size?

2. Before the Reeds left New York for Cape Cod, the automobile odometer read, 4,543.7 mi. Upon their arrival, the odometer read 4,830.4 mi. How many miles is it from New York to Cape Cod?

3. Ms. Ramirez purchased a jacket for $59.95, a cap for $12.95, and a pair of shoes for $32.95. How much money did she receive in change if she gave the cashier $110?

4. The winner in a horse race ran the mile in 1.46 min. The second-place horse completed the mile in 1.51 min. By how many minutes faster was the winner?

5. Mr. Garcia measured a metal rod and found its length to be 5.85 in. If metal shrinks .005 in. during the winter months, what will its new length be?

6. A woman's club in Baltimore collected $564.75 from a garage sale. What was the profit if their expenses were $35.95?

7. If the perimeter of a triangular plot of ground is 34.671 ft. and two of its sides measure 12.7 ft. and 15.386 ft., find the length of this third side.

8. Farmer Appleby owns two different-shaped fields. One field is triangular and its dimensions are 138.2 yd., 170.6 yd., and 92.4 yd. The other is rectangular in shape and its dimensions are 89.6 yd. by 47.8 yd.

   a. What is the perimeter of the triangular field?

   b. What is the perimeter of the rectangular field?

   c. Which field has the longer perimeter?

   d. How many yards larger is it?

9. The gasoline tank in Mr. Alfaro's automobile holds 24 gal. If Mr. Alfaro fills the tank with 13.4 gal., how many gallons were in the tank before he filled it?

10. In the diagram below, find the missing dimension x.

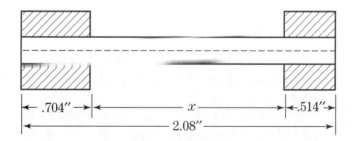

REVIEW OF IMPORTANT IDEAS

Some of the most important ideas in this part of Chapter 11 were:

 Subtracting decimals is almost identical to subtracting whole numbers.

 The decimal point is always placed to the right of a whole number.

 Annexing zeros to the right of the decimal number will not change its value.

 To subtract decimals—
1.   Place the number in a vertical column with one decimal point directly beneath the other.
2.   Place the decimal point in the answer directly below the decimal points in the numbers above.
3.   Subtract the numbers as you would whole numbers.

CHECK WHAT YOU HAVE LEARNED

Good luck on this exercise. These examples will test your understanding of subtracting decimals.

A Score of	Means That You
23–25	Did very well. You can move on to Chapter 12.
20–22	Know this material except for a few points. Reread the sections about the ones you missed.
17–19	Need to check carefully the sections you missed.
0–16	Need to review this part of the chapter again to refresh your memory and improve your skills.

Questions	Are Covered in Section
1, 2, 4, 12, 14, 20	11.4
3, 5–11, 13, 15–19	11.5
21–25	11.6

Chapter Test 11B

Write each answer in the space provided. Use a calculator to check your answers after you have completed the entire chapter test.

1. .9
    −.3
    ‾‾‾

2. .78
    −.36
    ‾‾‾‾

3. $8.4 - 5 =$

4. .59
    −.55
    ‾‾‾‾

5. 8.79
    −5.67
    ‾‾‾‾‾

6. 9.04
    −1.03
    ‾‾‾‾‾

7. 5.6
    −2.8
    ‾‾‾‾

8. $85.9 - 4.6 =$

9. $7 - .5 =$

10. 6.88
    − .59
    ‾‾‾‾‾

11. 31.1
    −13.9
    ‾‾‾‾‾

12. .407
    −.29
    ‾‾‾‾

13. 9.53
    −6.49
    ‾‾‾‾‾

14. .048
    −.037
    ‾‾‾‾‾

15. 1.345
    − .265
    ‾‾‾‾‾

16. $3.04 - .79 =$

17. 28.37
    − 5.6
    ‾‾‾‾‾

18. $7 - 3.8 =$

19. $27.5 - .38 =$

20. $.83 - .752 =$

21. Bob had an oil truck which held 350 gal. of oil. It took 273.2 gal. to fill the tank. How much oil was in the tank before it was filled?

22. Rafael cut a piece of wood 5.25 ft. long from a longer piece measuring 15.75 ft. How long a piece of wood remained?

23. Maria bought a CD for $16.99 and a book for $8.75. She gave the clerk a $50 bill. How much change did she get?

24. Last year's winner ran the women's 60-yard dash in 7.0 s. If the Madison Square Garden record is 6.6 s., how many seconds slower was last year's winner?

25. At the close of the year Daisy weighed 98.5 lb. and in February she weighed 103.2 lb. How many pounds did she gain during January?

Multiplying Decimals

Multiplying decimals is another basic operation you must know if you are to be proficient with mathematics. You use multiplication to find the total cost when you purchase 2 quarts of milk at $1.61 a quart or 3 blueberry muffins at $0.93 each. Cashiers are constantly required to use this skill.

To see how well you know this material, take the pretest which follows.

Pretest 12–See What You Know and Remember

Be careful doing these problems. Try to get as many correct as you can. Write each answer in the space provided.

1. .2
    × 4

2. 2.3
    × 3

3. 2.63
    × 2

4. .8
    ×.5

5. 4.03
    × 2

6. .197
    × 2

7. .23
    × .4

8. 27
    ×.08

9. 4.6
    ×1.8

10 .67
    ×8.4

11. .78
     ×.96

12. 9.3
     ×.09

13. .704
    $\times$ .06
    ——

14. .29
    $\times$.07
    ——

15. a. $.4 \times 100 =$
    b. $1.2 \times 10 =$
    c. $3.84 \times 1,000 =$

16. 32.6
    $\times$ 1.3
    ——

17. 48.7
    $\times$ .68
    ——

18. 8.02
    $\times$ .49
    ——

19. $59.13
    $\times$ 78
    ——

20. Mr. Michaels bought a chicken which weighed 3.75 lb. and cost $2.69 a pound. How much did he pay for the chicken?

21. Eduardo decides to bake a chocolate layer cake. The recipe calls for .5 lb. of chocolate. If he triples the recipe, how many pounds of chocolate will he need?

22. A piece of paper is .008 in. thick. If a package contains 250 of these sheets, how thick is the package?

23. An automobile can travel 24 mi. on a gallon of gasoline. How far can it travel on .8 of a gallon?

   Now turn to the back of the book to check your answers. Add up all that you had correct. Count by the number of separate answers, not by the number of questions. In this pretest there were 23 questions, but 25 separate answers.

A Score of	Means That You
23–25	Did very well. You can proceed to Chapter 13.
20–22	Know this material except for a few points. Read the sections about the ones you missed.
16–19	Need to check carefully on the sections you missed.
0–15	Need to work with this chapter to refresh your memory and improve your skills.

Questions	Are Covered in Section
1–3, 5, 6, 8, 19	12.1
4, 7, 9–14, 16–18	12.2
15	12.3
20–23	12.4

12.1 Multiplying a Decimal by a Whole Number

EXAMPLE

How much money does it cost to purchase 2 lb. of apples at $1.29 a pound?

SOLUTION

To answer that question requires the operation of either addition or multiplication. You could add the cost of each pound of apples or multiply the price of one pound by 2. These methods are shown below:

$$\begin{array}{cc}
\textit{Addition} & \textit{Multiplication} \\
\end{array}$$

$$\begin{array}{c}
\$1.29 \\
+\ 1.29 \\
\hline
\$2.58
\end{array}
\qquad
\begin{array}{c}
\$1.29 \\
\times\quad 2 \\
\hline
\$2.58
\end{array}
\qquad \text{or} \quad \frac{129}{100} \times 2 = \frac{258}{100} = 2.58$$

EXAMPLE

Find the product of .176 and 3.

SOLUTION

$$\begin{array}{cc}
\textit{Addition} & \textit{Multiplication} \\
\end{array}$$

$$\begin{array}{c}
.176 \\
.176 \\
+.176 \\
\hline
.528
\end{array}
\qquad
\begin{array}{c}
.176 \\
\times\quad 3 \\
\hline
.528
\end{array}
\qquad \text{or} \quad \frac{176}{1,000} \times 3 = \frac{528}{1,000} = .528$$

To multiply a decimal by a whole number you multiply the numbers as you would multiply whole numbers. The decimal in the *product* contains *the same number of places as the number of decimal places in the multiplicand*. See the following example:

EXAMPLE

How long must a metal rod measure if you wish to cut 6 rods of 5.85 in. each?

SOLUTION

Problem	Step 1	Step 2
5.85 × 6	5.85 × 6 ——— 3510	5.85 (multiplicand) × 6 (multiplier) ——— 35.10 (product)
	Multiply the numbers: 585 × 6 = 3510	5.85 has 2 decimal places. Place the decimal point in 3510 counting 2 places from right to left. Thus: 35.10 = 35.1

Thus: 5.85 in. × 6 = 35.1 in.

PRACTICE EXERCISE 121

The products in problems 1–10 have been completed but the decimal point has been left out of the answer. Place the decimal point where it belongs.

1. 200 × .05 = 1000

2. 40 × .30 = 1200

3. 4 × .5 = 20

4. 4.62 × 4 = 1848

5. .525 × 6 = 3150

6. 2 × .2 = 4

7. .50 × 5 = 250

8. 1.9 × 5 = 95

9. .2415 × 3 = 7245

10. .06 × 6 = 36

Find the product of each of the following examples. Place the decimal point in the answer.

11.   .125 × 500 =   12.   5.25 × 4 =

13.   14.5 × 12 =   14.   .4 × 5 =

15.   48.32 × 6 =   16.   5 × .96 =

17.   3 × .04 =   18. $14.38 × 7 =

19. $146.30 × 22 =   20. $62.50 × 8 =

12.2 Multiplying Decimal Numbers

EXAMPLE

If a metal rod costs $.43 an inch, how many dollars would a rod 35.1 in. long cost?

SOLUTION

To answer this question you multiply 35.1 in. by $.43, or

$$\begin{array}{r} 35.1 \\ \times\$.43 \\ \hline \end{array}$$

Follow the same procedure as illustrated in the previous section. You multiply as you would with whole numbers. You must then place the decimal point in the product. To decide where to place the decimal point, let's do this same example by writing each decimal in fraction form and then multiplying. Thus:

$$35.1 \times \$.43 =$$

$$35\frac{1}{10} \times \frac{43}{100} =$$

$$\frac{351}{10} \times \frac{43}{100} = \frac{15093}{1,000} = 15\frac{93}{1,000}$$

$15.093 or $15.09 rounded off to the nearest cent

Since the product of $35.1 \times .43$ using fractions is 15.093, then this same product should be obtained if you use decimals. Thus:

$$
\begin{array}{ll}
35.1 & \text{(multiplicand)} \\
\underline{\times\ .43} & \text{(multiplier)} \\
15.093 & \text{(product)}
\end{array}
$$

The multiplicand is a *one*-place decimal, the multiplier is a *two*-place decimal, and the product is a *three*-place decimal. This leads to the conclusion that you place the decimal point in the product, counting from right to left, the same number of places as the total decimal places in the multiplicand and multiplier. See this illustration:

Problem	Step 1	Step 2
8.72 ×0.48	8.72 × 0.48 ———— 6976 3488 ———— 41856	8.72 (two-place decimal) × 0.48 (two-place decimal) ———— 6976 3488 ———— 4.1856 (four-place decimal)
	Multiply the numbers: $0.48 \times 8.72 = 41856$	Place the decimal point in the product, counting 4 places from the right. Place it between the 4 and 1.

Thus: $0.48 \times 8.72 = 4.1856$

Notice that the decimal points do not need to be underneath one another before you can multiply. This is different from adding or subtracting. Check this method out on these two examples.

EXAMPLE

Find the product of $4.56 \times .12$.

SOLUTION (using fractions)

$$4.56 \times .12 =$$

$$4\frac{56}{100} \times \frac{12}{100} =$$

$$\frac{456}{100} \times \frac{12}{100} = \frac{5472}{10,000} = .5472$$

or solution using decimals

Problem	Step 1	Step 2	
4.56 × .12	4.56 × .12 ――― 912 456 ――― 5472	4.56 × .12 ――― 912 456 ――― .5472	(two-place decimal) (two-place decimal) (four-place decimal)
	Multiply the numbers: 456 × 12 = 5472	The sum of the decimal places in the multiplicand and multiplier is 4. Place the decimal point, starting from the right, 4 places to the left.	

Thus: $4.56 \times .12 = .5472$

EXAMPLE

What is the product of .003 and .01?

SOLUTION

Problem	Step 1	Step 2	
.003 × .01	.003 × .01 ――― 3	.003 × .01 ――― .00003	(three-place decimal) (two-place decimal) (five-place decimal)
	Multiply the numbers: 3 × 1 = 3	The sum of the decimal places in the multiplicand and multiplier is 5. Place 4 zeros to the left of 3 as placeholders. The result is .00003.	

Thus: $.003 \times .01 = .00003$

PRACTICE EXERCISE 122

The products in problems 1–10 have been completed but the decimal points have been left out of the answer. Place the decimal point where it belongs. Include zeros as place holders where necessary.

1. $4.62 \times .4 = 1848$ 2. $5.25 \times .6 = 3150$

3. $.2415 \times .3 = 7245$ 4. $.06 \times .6 = 36$

5. $48.32 \times .06 = 28992$ 6. $.5 \times .96 = 480$

7. $.03 \times .04 = 12$

In examples 8–10 round off the answer to the nearest cent after placing the decimal point.

8. $\$14.38 \times .07 = \$10066$ 9. $\$1,463 \times .222 = \$324786$

10. $\$62.50 \times 1.1 = \$68750$

Find the products in each of the following examples. Place the decimal point in the answer and add zeros when necessary.

11. $1.73 \times .8 = 1384$ 12. $37.42 \times 3.6 = 134712$

13. $13.2 \times .43 = 5676$ 14. $3.14 \times .16 = 5024$

15. $.037 \times .01 = 37$

In examples 16–20 round off the answer to the nearest cent after placing the decimal point.

16. $\$36.42 \times .08 = \$29136$ 17. $\$149.78 \times .15 = \$224670$

18. $\$3,200 \times .055 = \$176000$ 19. $\$225.72 \times 3.6 = \$812592$

20. $\$83.75 \times .042 = \$351750$

12.3 Special Multipliers of 10, 100, 1,000, and So On

It is often necessary to multiply by the special numbers 10, 100, 1,000, etc. Shortcuts save considerable amounts of time so look at these special multipliers in the following examples:

$$
\begin{array}{r}
1.234 \\
\times\quad 10 \\
\hline
12.34\cancel{0}
\end{array}
\qquad\qquad
\begin{array}{r}
1.234 \\
\times\quad 100 \\
\hline
123.40\cancel{0}
\end{array}
\qquad\qquad
\begin{array}{r}
1.234 \\
\times\quad 1,000 \\
\hline
1,234.0\cancel{0}\cancel{0}
\end{array}
$$

Notice that the decimal point in the multiplicand moved *one* place to the right when multiplying by *10*, *two* places to the right when multiplying by *100*, and *three* places to the right when multiplying by *1,000*. Therefore, when multiplying by the special numbers 10, 100, 1,000, etc., the decimal point in the multiplicand moves a number of places to the right equal to the number of zeros there are in the multiplier.

Thus, multiplying by 10, which contains one zero, moves the decimal point in the multiplicand one place to the right. Multiplying by 100, which contains two zeros, moves the decimal point in the multiplicand two places to the right. Look at the following examples which illustrate the short cut for the special multipliers of 10, 100, 1,000, etc.

Problem	Step 1
14.32 × 10 =	14.32 × 10 = 14.3.2 = 143.2
	Move the decimal point in the multiplicand 14.32 one place to the right since the multiplier 10 contains one zero.

Problem	Step 1
.138 × 10,000 =	.138 × 10,000 = .1380. = 1,380
	Move the decimal in the multiplicand .138 four places to the right since the multiplier 10,000 contains four zeros. Annex a zero as a place holder.

Thus: 14.32 × 10 = 143.2

Thus: .138 × 10,000 = 1,380

PRACTICE EXERCISE 123

Multiply. The decimal point has been left out of the answer. Place the decimal point where it belongs. Put in zeros when necessary.

1. $10 \times 4.2 = 42$ 2. $13.625 \times 100 = 13625$

3. $1.4 \times 10 = 14$ 4. $1,000 \times 2.4 = 24$

5. $.247 \times 10 = 247$ 6. $3.1416 \times 100 = 31416$

7. $.62 \times 10,000 = 62$ 8. $10 \times .4 = 4$

9. $100 \times \$125 = \$125$ 10. $\$3.47 \times 10 = \$347$

12.4 Word Problems Requiring the Multiplication of Decimals

EXAMPLE

Mr. Thomas fills his car with 14.3 gal. of gasoline costing $2.29 a gallon. How much does the gasoline cost him?

SOLUTION
Each gallon of gasoline cost $2.29, so multiply to find the cost of 14.3 gal.

Problem	Step 1	Step 2
14.3 ×2.29	14.3 × 2.29 1287 286 286 32747	14.3 (one-place decimal) × 2.29 (two-place decimal) 1287 286 286 32.747 (three-place decimal)
	Multiply the numbers: $14.3 \times 2.29 = 32747$	Mark off the decimal point in the product, starting from the right, three places to the left.

Thus: $14.3 \times \$2.29 = \$32.747$ or $\$32.75$ rounded off to the nearest cent.

PRACTICE EXERCISE 124

1. Carl buys 3.5 yd. of material at the cost of $8.39 per yd. How much does the material cost?

2. John Lewis belongs to a pension plan in his union. Each week $57.69 is deducted from his paycheck to pay for this pension plan. How much does he contribute to the fund after a full year of work? (Remember: 52 weeks = 1 year.)

3. Rain fell in the city this year at an average of .641 in. per month. How many inches of rain fell during the year?

4. Ms. Pearl knows that she can drive 22.6 mi. for each gallon of gasoline. If the capacity of her gasoline tank is 11.8 gal., then how far can she travel without stopping to refuel?

5. If a school charges $1.75 for a student lunch, how much would it take in if 1,000 students buy lunch?

6. Find .07 of $48.45 rounded off to the nearest cent.

7. If a store owner sell 12 boxes of candy at 2 boxes for $1.50, how much money did she receive for her candy?

8. Find the area of a rectangular plot of ground whose dimensions are 26.8 ft. by 31.37 ft.

9. The length of a circle is called the *circumference*. It can be found by multiplying the diameter, indicated on the diagram as 14 in., by 3.14, which is the value called *pi* (π). Find the circumference of this circle.

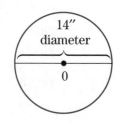

10. At $2.99 a pound, what is the cost of a chuck steak weighing 3.2 pounds?

REVIEW OF IMPORTANT IDEAS

Some of the most improtant ideas in Chapter 12 were:

Multiplying decimals is similar to multiplying whole numbers.

To multiply a decimal by either a whole number or a decimal—
1. Multiply the numbers in the same way as you would whole numbers.
2. Find the sum of the decimal places in the multiplier and the multiplicand.
3. Place the decimal point in the product, counting from right to left, the same number of places as the sum of the decimal places in the multiplier and the multiplicand.

To multiply by special numbers like 10, 100, 1,000, etc.—
1. Keep the same digits in the product as there are in the multiplicand.
2. Place the decimal point in the product as many places to the right of its position in the multiplicand as there are zeros in the multiplier. Add zeros as place holders when necessary.

CHECK WHAT YOU HAVE LEARNED

This test will reveal how well you have learned the material in this chapter. Remember, it's accuracy, not speed, which is most important.

In counting up your answers, remember that there were 25 separate answers on this test.

A Score of	Means That You
23–25	Did very well. You can move on to Chapter 13.
20–22	Know this material except for a few points. Read the sections about the ones you missed.
16–19	Need to check carefully the sections you missed.
0–15	Need to review the chapter again to refresh your memory and improve your skills.

Chapter Test 12

Try hard for a satisfactory grade. Use a calculator to check your answers after you have completed the entire chapter test.

1.   .3
  ×3

2.   3.2
  × 3

3.   3.53
  × 2

4.   .7
  ×.6

5.   3.04
  × 2

6.   .186
  × 3

7.   .32
  × .4

8.   34
  ×.07

9.   6.4
  ×1.8

10.   .76
  ×8.4

11.   .87
  ×.69

12.   9.5
  ×.07

13.   .407
  × .06

14.   .92
  ×.07

15   a.   .6 × 100 =
    b.   9.2 × 10 =
    c.   3.75 × 1,000 =

16.   44.8
  × 1.2

17.   84.7
  × .68

18.   9.02
  × .38

19. $39.43
  ×   87

20. Mr. Bates bought a steak weighing 2.7 lb. which cost $3.29 per lb. How much did he pay for the steak?

21. Cathy decides to cook a cheese fondue. The recipe calls for .5 cups of wine. If she triples the recipe, how much wine will she need?

22. A sheet of cardboard is .015 in. thick. If 150 sheets of cardboard are stacked one on top of the other, how thick is the stack?

23. An automobile can travel 18 mi. on a gallon of gasoline. How far can it travel on .6 of a gallon?

Dividing Decimals

The topic covered in this chapter has various applications in our lives. You buy gasoline for your car, travel a given distance, and want to know how many miles you have traveled for each gallon of gasoline purchased. This is a problem in dividing decimals. You are offered a job with a yearly salary and want to know how many dollars and cents you will earn monthly or weekly. This is another example of dividing decimals.

Your knowledge of this topic can be checked with the pretest which follows.

Pretest 13—See What You Know and Remember

Do as many of these problems as you can. Try hard for a high score. Write each answer in the space provided.

1. $4\overline{).84}$

2. $3\overline{)9.3}$

3. $2\overline{)1.2}$

4. $8\overline{)40.8}$

5. $4\overline{).16}$

6. $3\overline{)3.009}$

7. $4\overline{).084}$

8. $6\overline{).042}$

9. a. $8 \div 100 =$

    b. $1.32 \div 10 =$

    c. $54.2 \div 1,000 =$

10. Change $\dfrac{4}{5}$ to a decimal fraction.

11. Change $\dfrac{4}{9}$ to a two-place decimal.

12. Mrs. Ling used 12 gal. of gasoline to drive a distance of 146.4 mi. On the average, how many miles did she drive with 1 gal. of gasoline?

13. If 36 gumdrops weigh 11 oz., find the weight of a gumdrop to the nearest hundredth of an ounce.

14. $.3\overline{).9}$ 15. $.5\overline{).25}$ 16. $.7\overline{)4.9}$

17. $.8\overline{)64}$ 18. $.6\overline{).72}$ 19. $.03\overline{)15.6}$

20. $.4\overline{).002}$ 21. $.73\overline{)87.6}$ 22. $7.2\overline{)17.28}$

23. Divide 53,238 by .76.

24. Tomatoes sell for $1.59 per lb. How many pounds did Mrs. Romero buy if she paid $9.54 for her tomatoes?

25. Mr. Woo purchased $2\frac{2}{3}$ yd. of material for $18.59. What was the cost of 1 yd. of material?

26. If 2 lb. of potatoes cost $1.78, how many pounds can be bought for $10.68?

27. Jane worked a year earning $20,722. How much did she earn weekly? (52 weeks = 1 year)

28. Find the average of 56, 63, 72, 45, 79, and 62 to the nearest tenth.

29. Change $.06\frac{1}{4}$ to a decimal.

30. Change $.87\frac{1}{2}$ to a fraction in its simplest form.

Now turn to the back of the book to check your answers. Add up all that you had correct. Count by the number of separate answers, not by the number of questions. In this pretest there were 30 questions, but 32 separate answers.

A Score of	Means That You
29–32	Did very well. You can proceed to Chapter 14.
26–28	Know the material except for a few points. Read the sections about the ones you missed.
21–25	Need to check carefully on the sections you missed.
0–20	Need to work with this chapter to refresh your memory and improve your skills.

Questions	Are Covered in Section
1–4	13.1
5–8	13.2
10, 11	13.3
14–23	13.4
12, 13, 24, 26–28	13.5
9	13.6
30	13.7
29	13.8
25	13.9

13.1 Dividing a Decimal by a Whole Number

EXAMPLE

If .8 lb. of cashew nuts are distributed among 4 people, how many pounds does each one get?

SOLUTION

You must divide .8 by 4 or

$$4\overline{)\smash{.8}}$$

to find the solution to this problem. Let's use our knowledge of fractions to discover the method of dividing a decimal by a whole number.

$$.8 \div 4 =$$

$$\frac{8}{10} \div \frac{4}{1} =$$

$$\frac{\overset{2}{\cancel{8}}}{10} \times \frac{1}{\underset{1}{\cancel{4}}} = \frac{2 \times 1}{10 \times 1} = \frac{2}{10} = .2$$

$$\text{Thus: } 4\overline{)\,.8}^{\,.2}$$

This example illustrates that you divide as in whole numbers and then place the decimal point in the quotient directly above the decimal point in the dividend. See the following example:

EXAMPLE

Divide: $2.8 \div 4$.

SOLUTION

Problem	Step 1	Step 2
$4\overline{)2.8}$	$4\overline{)2.8}^{\,7}$	$4\overline{)2.8}^{\,.7}$
	Divide: $28 \div 4 = 7$	Place the decimal point in the quotient (7) above the decimal point in the dividend.

$$\text{Thus: } 4\overline{)2.8}^{\,.7}$$

PRACTICE EXERCISE 125

Divide.

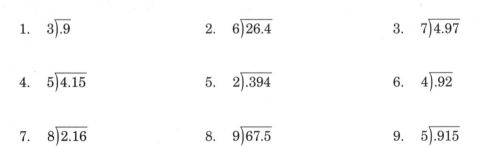

1. $3\overline{)\,.9}$

2. $6\overline{)26.4}$

3. $7\overline{)4.97}$

4. $5\overline{)4.15}$

5. $2\overline{)\,.394}$

6. $4\overline{)\,.92}$

7. $8\overline{)2.16}$

8. $9\overline{)67.5}$

9. $5\overline{)\,.915}$

13.2 Zero As a Place Holder in the Quotient

Zero is used to hold a place between the decimal point and the first digit in the quotient. Sometimes there may be more than one zero. See the two examples which follow.

EXAMPLE

Find the quotient of .16 divided by 16.

SOLUTION

Problem	Step 1	Step 2
$16\overline{)\,.16}$	$\overset{1}{16\overline{)\,.16}}$	$\overset{.01}{16\overline{)\,.16}}$
	Divide: $16 \div 16 = 1$ Place the 1 above the 6 in the dividend.	Place the decimal point in the quotient above the decimal point in the dividend. *Remember* to include the zero between the decimal point and the 1 as a place holder.

Thus: $\overset{.01}{16\overline{)\,.16}}$

EXAMPLE

$5\overline{)\,.025}$

SOLUTION

Problem	Step 1	Step 2
$5\overline{)\,.025}$	$\overset{5}{5\overline{)\,.025}}$	$\overset{.005}{5\overline{)\,.025}}$
	Divide: $25 \div 5 = 5$ Place the 5 in the quotient above the dividend.	Place the decimal point in the quotient above the decimal point in the dividend. Include two zeros between the decimal point and the 5 as place holders.

Thus:

PRACTICE EXERCISE 126

Divide. Remember to add zeros as place holders.

1. $3\overline{).09}$ 2. $5\overline{).45}$ 3. $9\overline{).081}$

4. $7\overline{).49}$ 5. $6\overline{).0042}$ 6. $4\overline{).052}$

7. $2\overline{).0034}$ 8. $7\overline{).518}$ 9. $6\overline{).036}$

10. $8\overline{).6464}$ 11. $5\overline{).2715}$ 12. $8\overline{).0736}$

13.3 Changing a Fraction into a Decimal

WITH NO REMAINDER

EXAMPLE

Juan decides to divide a gift of $15 among his 4 children. How much money does each one receive?

SOLUTION

Juan knows that he must divide 15 into 4 equal parts. This means that each child will get $\frac{15}{4}$ or $3\frac{3}{4}$ dollars. The decimal $.75 represents $\frac{3}{4}$ of a dollar; therefore, each child will receive $3.75.

One meaning of a fraction is that the line between the terms of the fraction means that the numerator is being divided by the denominator. In this example, $15 \div 4$ or $4\overline{)15}$. Look at the solution:

Problem	Step 1	Step 2	Step 3
$4\overline{)15}$	$4\overline{)15.}$	$4\overline{)15.00}$	$\begin{array}{r} 3.75 \\ 4\overline{)15.00} \end{array}$
	The decimal point is placed to the right of the whole number 15.	Since 4 cannot divide evenly into 15 add as many zeros to the right of the decimal point as you wish. Recall: $15 = 15. = 15.0 = 15.00$, etc.	Place the decimal point in the quotient. Divide: $1500 \div 4 = 75$

Thus: $\frac{15}{4} = 3.75$

PRACTICE EXERCISE 127

Find the decimal equivalent for each of these fractions by dividing the numerator by the denominator. Add zeros to the dividend until the problem works out evenly.

1. $\frac{2}{5}$ 2. $\frac{1}{4}$ 3. $\frac{1}{2}$

4. $\frac{1}{8}$ 5. $\frac{8}{10}$ 6. $\frac{5}{8}$

WITH THE REMAINDER ROUNDED OFF

Suppose you divide and divide and divide and the problem never works out evenly. What do you do? Instructions to these examples indicate how far you should divide. Look at this example:

EXAMPLE

Express the fraction $\frac{2}{3}$ as a decimal correct to the nearest hundredth.

SOLUTION

The instructions in the example tell you that you must round off the answer to the nearest hundredth. This means that you will have a two-place decimal as your answer. Since you must round off your decimal to the hundredth, you must know the value of the digit in the thousandths place so you can round off properly. Look how the following example is worked out:

Problem	Step 1	Step 2	Step 3
$3\overline{)2}$	$3\overline{)2.000}$	$.666$ $3\overline{)2.000}$	$.666 \approx .67$
	Place the decimal point to the right of 2 and add 3 zeros (thousandths place): 2 = 2.000	Divide. Find the quotient to one more place than the question calls for.	Round off to the nearest hundredth. Since the digit in the thousandths place (6) is more than 5, increase the digit in the hundredths place by 1: $.666 \approx .67$

Thus: $\frac{2}{3} \approx .67$ to the nearest hundredth.

EXAMPLE

Find the quotient of $1.10 divided by 3 correct to the nearest cent.

SOLUTION

Problem	Step 1	Step 2	Step 3
3)$1.10̄	3)$1.100̄	$ .366 3)$1.100 9 20 18 20 18	$.366 ≈ $.37
	Add 1 zero to the dividend: 1.10 = 1.100	Divide. Find the quotient to the thousandths place: .366	Round off to the nearest cent (hundredths). $.366 ≈ $.37

Thus: $1.10 ÷ 3 ≈ $.37 to the nearest cent.

PRACTICE EXERCISE 128

Find the decimal equivalent to the nearest tenth of:

1. $\frac{1}{3}$ 2. $\frac{5}{8}$ 3. $\frac{2}{7}$ 4. $\frac{1}{9}$

Find the decimal equivalent to the nearest hundredth of:

5. $\frac{4}{9}$ 6. $\frac{7}{8}$ 7. $\frac{2}{7}$ 8. $\frac{5}{6}$

9. Find correct to the nearest tenth: 4)3.5̄

10. If a 6-ft. board is cut into 9 equal parts, find the length of each part correct to the nearest hundredth of a foot.

13.4 Dividing a Decimal by a Decimal

Now that you have seen that dividing a decimal by a whole number was hardly more difficult than ordinary division, you are ready to tackle division of a decimal by a decimal. Look at this example.

EXAMPLE

A bag of candy weighs .6 lb. Into how many bags weighing .3 lb. each can it be divided?

SOLUTION

The problem becomes .6 ÷ .3 = ? Let's use fractions to find the solution before tackling the problem of dividing a decimal by a decimal.

$$
.6 \div .3 = \frac{.6}{.3}
$$

$$
= \frac{.6 \times 10}{.3 \times 10}
$$

$$
= \frac{6}{3} = 2
$$

As you see, the divisor (.3) was made into a whole number by multiplying by 10. You must multiply both terms of a fraction by the same number to maintain an equivalent fraction, so multiply the numerator (.6) by 10 also.

Since the divisor (.3) was multiplied by 10, the result is the whole number 3. You must multiply the dividend (.6) by 10 also to maintain an equivalent fraction. Since the problem is written in fractional form, it is easier to use the terms numerator and denominator.

$$
\text{Recall:} \quad \frac{\text{numerator}}{\text{denominator}} = \text{quotient}
$$

Thus you see that there is no special method for dividing a decimal by a decimal. All you have to do is change the divisor into a whole number and proceed as you did earlier in the chapter. Look at the following solution to this same example:

Problem	Step 1	Step 2
.3)‾.6	.3.)‾.6.	2. = 2 3)6.
	Multiply the divisor by 10, 100, 1,000, etc., to make it a whole number. In this case, multiply by 10: .3 × 10 = 3. Multiply the dividend also by 10: .6 × 10 = 6 Place the decimal point in quotient directly above its new position in the dividend.	Divide the whole number (3) into the dividend (6): 6 ÷ 3 = 2

Thus: .6 ÷ .3 = 2

EXAMPLE

.04)‾.008

SOLUTION

Be sure to change the divisor .04 into a whole number before dividing. Here is how:

Problem	Step 1	Step 2
.04)‾.008	.04.)‾.00.8	.2 4)‾.8
	Multiply the divisor (.04) by 100 to make it a whole number: .04 × 100 = .04. = 4 Multiply the dividend (.008) also by 100: .008 × 100 = .00.8 = .8 Place a decimal point in the quotient.	Divide: .8 ÷ 4 = .2

Thus: .008 ÷ .04 = .2

PRACTICE EXERCISE 129

Divide. The decimal point has been left out of the quotient. Place the decimal point where it belongs. Put in zeros when necessary.

1. $\overset{3}{.3\overline{).9}}$

2. $\overset{3}{.3\overline{).09}}$

3. $\overset{3}{.03\overline{).009}}$

4. $\overset{3}{.3\overline{).009}}$

5. $\overset{5}{.5\overline{).025}}$

6. $\overset{5}{.05\overline{).25}}$

7. $\overset{5}{.5\overline{)2.5}}$

8. $\overset{5}{.005\overline{).0025}}$

9. Divide: .7342 ÷ .002

10. Find the quotient of .034 ÷ .008.

13.5 Zero As a Place Holder in the Dividend

Before dividing, it might be necessary to use zero as a place holder in the dividend.

EXAMPLE

If a bus ride costs $1.25, how many rides can you take for $10?

SOLUTION

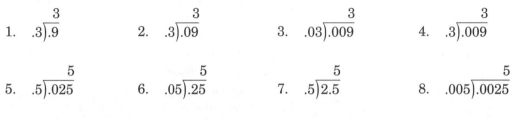

Problem	Step 1	Step 2
$1.25\overline{)10}$	$1.25.\overline{)10.00.}$	$\overset{8}{125\overline{)1000}}$
	Multiply the divisor (1.25) by 100 to make it a whole number: $1.25 \times 100 = 125$ Place the decimal point to the right of 10 in the dividend and multiply by 100: $10 \times 100 = 1000$	Divide: $1000 \div 125 = 8$

Thus: $10 ÷ $1.25 = 8 rides

EXAMPLE

$2.5 \div .005 =$

SOLUTION

Problem	Step 1	Step 2
$.005\overline{)2.5}$	$.005.\overline{)2.500.}$	$\begin{array}{r} 500 \\ 5\overline{)2,500} \end{array}$
	Multiply the divisor (.005) by 1,000: $.005 \times 1,000 = 5$ Multiply the dividend 2.5 also by 1,000: $2.5 \times 1,000 = 2.500.$ $= 2,500$	Divide: $2,500 \div 5 = 500$

Thus: $2.5 \div .005 = 500.$

PRACTICE EXERCISE 130

The answers to the following problems have been completed except for the placement of the decimal point in the quotient. Place the decimal point where it belongs. Annex zeros when necessary.

1. $25\overline{)62.5}$ with quotient 25

2. $.25\overline{)6.25}$ with quotient 25

3. $2.5\overline{)625}$ with quotient 25

4. $.025\overline{)6.25}$ with quotient 25

5. $.25\overline{)62.5}$ with quotient 25

6. $2.5\overline{).0625}$ with quotient 25

7. $.025\overline{)62.5}$ with quotient 25

8. $.25\overline{)625}$ with quotient 25

9. $.0025\overline{).625}$ with quotient 25

10. $.25\overline{).00625}$ with quotient 25

In this next group of problems, divide and place the decimal point in the quotient.

11. $5.6\overline{)4.592}$ 12. $.81\overline{)429.3}$ 13. $.98\overline{).8526}$ 14. $4.6\overline{)128.8}$

15. Find to the nearest tenth the quotient of $6.5 \div .261$.

13.6 Word Problems Requiring the Division of Decimals

EXAMPLE

Sonia, who works for the A.B.C. Dress Company as a bookkeeper, received the following bill in the mail.

M.J.R. FABRIC CORPORATION
Houston, Texas

To: A.B.C. Dress Company
    Los Angeles, California

Quantity	Description	Amount
42 yd.	No. 123	$ 377.58
26 yd.	No. 710	275.34
104 yd.	No. 62C	706.16
	Total	$1359.08

She has to check the price per yard of each item purchased and the total amount before she can pay the bill.

SOLUTION
You divide each amount by the quantity received to find the price per yard or

$$\frac{\text{amount}}{\text{quantity}} = \text{price per yard}.$$

This means that each of these separate division examples must be completed as shown below.

$$\begin{array}{r} \$\ \ 8.99 \\ 42\overline{)\$377.58} \\ \underline{336} \\ 41\,5 \\ \underline{37\,8} \\ 3\,78 \\ \underline{3\,78} \end{array} \qquad \begin{array}{r} \$\ 10.59 \\ 26\overline{)\$275.34} \\ \underline{26} \\ 15\,3 \\ \underline{13\,0} \\ 2\,34 \\ \underline{2\,34} \end{array} \qquad \begin{array}{r} \$\ \ 6.79 \\ 104\overline{)\$706.16} \\ \underline{624} \\ 82\,1 \\ \underline{72\,8} \\ 9\,36 \\ \underline{9\,36} \end{array}$$

Thus:

42 yd. at $8.99 per yd. = $377.58
26 yd. at $10.59 per yd. = $275.34
104 yd. at $6.79 per yd. = $706.16
$377.58 + $275.34 + $706.16 = $1359.08 Total

PRACTICE EXERCISE 131

1. Tomatoes sell for $1.69 a pound. How many pounds can be purchased for $8.45?

2. If $60 is to be divided equally among 80 people, what decimal fraction of a dollar does each person receive?

3. A board 12.8 ft. long is to be cut into 4 equal lengths. Discounting the loss of material for each cut, what will be the length of each piece?

4. A parents' association purchased a quantity of candy to sell at the annual school bazaar. If the total cost of the 450 boxes of candy was $751.50, what was the price for each box?

5. Find the average price for the following 5 items if their individual prices are $1.30, $2.56, $.50, $1.96, and $2.38.

6. If the area of a rectangular plot of ground is 263.22 in.^{2} and the length measures 21.4 in., find the measure of its width.

7. Find the number of miles a car travels on 1 gal. of gasoline if it requires 11.7 gal. to travel 289.2 mi. Round off the answer to the nearest tenth.

8.   Mr. Gonzalez drove 180.6 mi. in 3.8 h. Find his average rate of speed to the nearest tenth of a mile (speed = distance ÷ time).

9.   A strip of metal 16.7 in. long is to be cut into 5 equal parts. What is the length of each part? (Allow nothing for the cut of the shears).

10.   A machine performs an operation every .8 s. How many operations does it perform after working 28.8 s.?

11.   Mr. Kobetts purchased gasoline for his automobile for $16.60. If gasoline costs $2.29 per gallon, how many gallons did he purchase? The answer should be rounded off to the nearest tenth of a gallon.

12.   The St. Louis baseball team won 63 games and lost 72 games.
   a. Express the ratio of the number of games won divided by the total number of games played.
   b. Find the quotient to the nearest thousandth.

13.   Bananas sell for $.59 per pound. How many pounds did Mrs. Cappelletto buy if she paid $5.31 for her bananas?

13.7 Special Divisors of 10, 100, 1,000, and So On

EXAMPLE

A subscription to a sports magazine reads as follows:

52 issues for $35.36
104 issues for $64.48

Which of the two choices would you make if you wanted to spend the least amount of money for each copy?

SOLUTION
Divide the cost by the number of issues to find the price of one copy.
Therefore:

$$
\begin{array}{r}
\$\ \ .68 \\
52\overline{)\$35.36} \\
31\,2 \\
\hline
4\,16 \\
4\,16 \\
\hline
\end{array}
\qquad \text{and} \qquad
\begin{array}{r}
\$\ \ .62 \\
104\overline{)\$64.48} \\
62\,4 \\
\hline
2\,08 \\
2\,08 \\
\hline
\end{array}
$$

You were probably correct; it is cheaper to purchase 104 issues rather than 52 issues. For 52 issues you pay 68¢ per copy while for 104 issues you pay only 62¢ a copy, representing a savings of 6¢ per copy.

In Chapter 12 you learned to multiply by the special numbers of 10, 100, 1,000, etc. You see you must divide by these numbers too. Let's look for and learn a short method for dividing by them as you did in multiplication. To discover the short cut, look at the following examples:

1. $8 \div 10 =$

$$10\overline{)8.0}^{\ .8}$$

When we divide by 10, the decimal point in the dividend (8) moves *one place to the left* (.8.) so that its position in the quotient is now .8.

2. $8 \div 100 =$

$$100\overline{)8.00}^{\ .08}$$

When we divide by 100, the decimal point in the dividend (8) moves *two places to the left* (.08.) so that its position in the quotient is now .08.

3. $8 \div 1,000 =$

$$1,000\overline{)8.000}^{\ .008}$$

When we divide by 1,000, the decimal point in the dividend (8) moves *three places to the left* (.008.) so that its position in the quotient is now .008.

PRACTICE EXERCISE 132

The quotients in problems 1–12 have been completed but the decimal point has been left out of the answer. Place the decimal point where it belongs. Put in zeros when necessary.

1. $42 \div 10 = 42$

2. $1,362.5 \div 100 = 13625$

3. $14 \div 10 = 14$

4. $2,400 \div 1,000 = 24$

5. $2.47 \div 10 = 247$

6. $.456 \div 10 = 456$

7. $314.16 \div 100 = 31416$

8. $6,200 \div 10,000 = 62$

9. $4 \div 10 = 4$

10. $\$125.00 \div 100 = \$125$

11. $\$34.70 \div 10 = \$347$

12. $.456 \div 100 = 456$

13.8 Changing a Decimal into a Fraction

A few pages back you learned to change a fraction into a decimal. Now look at the opposite operation, changing a decimal into a fraction.

EXAMPLE

If two cans of corn cost $1.25, how many cans can you purchase for $5?

SOLUTION

If two cans of corn cost $1.25, then $.62$\frac{1}{2}$ is the cost of one can. To find the number of cans you can purchase, perform the following division example:

$$\$5 \div \$.62\frac{1}{2} = \qquad\qquad \text{or} \qquad\qquad .62.\frac{1}{2}\overline{)5.00.}$$

This problem is both complicated and difficult since you haven't attempted to solve any problem which contained $.62\frac{1}{2}$ as a divisor. What do you do?

The decimal $.62\frac{1}{2}$ is read as sixty two and one-half hundredths and is written

$$\frac{62\frac{1}{2}}{100}$$

as a fraction. Look at the following solution:

Problem	Step 1	Step 2	Step 3	Step 4
$\dfrac{62\frac{1}{2}}{100}$	$62\frac{1}{2} \div 100$	$\dfrac{125}{2} \div \dfrac{100}{1}$	$\dfrac{\overset{5}{\cancel{125}}}{2} \div \dfrac{1}{\underset{4}{\cancel{100}}} =$	$\dfrac{5 \times 1}{2 \times 4} = \dfrac{5}{8}$
	Rewrite in this form.	Change the mixed number $62\frac{1}{2}$ to $\dfrac{125}{2}$.	Invert the divisor $\dfrac{100}{1}$ and cancel.	Multiply.

Thus: $.62\frac{1}{2} = \frac{5}{8}$.

Therefore: $\$5 \div \$.62\frac{1}{2} =$

$$5 \div \frac{5}{8} =$$

$$\frac{5}{1} \times \frac{8}{5} = 8 \text{ cans}$$

The equality of $.62\frac{1}{2} = \frac{5}{8}$ resulted in a simple division example. Sometimes it is more convenient to use the fractional equivalent of a decimal when doing computations. One problem was just demonstrated. Look at the following example.

EXAMPLE

Change .025 to a fraction in its simplest form.

SOLUTION

Problem	Step 1	Step 2
$.025 =$	$\dfrac{25}{1,000}$	$\dfrac{25 \div 25}{1,000 \div 25} = \dfrac{1}{40}$
	Write the decimal as a fraction: $.025 = 25$ thousandths	Simplify the fraction: $\dfrac{25}{1,000} = \dfrac{1}{40}$

Thus: $.025 = \dfrac{1}{40}$

EXAMPLE

Change $.33\frac{1}{3}$ to a fraction in its simplest form.

SOLUTION

Problem	Step 1	Step 2	Step 3	Step 4
$.33\frac{1}{3} =$	$\dfrac{33\frac{1}{3}}{100}$	$33\frac{1}{3} \div 100 =$	$\dfrac{100}{3} \div \dfrac{100}{1} =$	$\dfrac{\overset{1}{\cancel{100}}}{3} \times \dfrac{1}{\underset{1}{\cancel{100}}} = \dfrac{1}{3}$
	Write as a fraction.	Rewrite in this form.	Change $33\frac{1}{3}$ to an improper fraction.	Invert the divisor and multiply.

Thus: $.33\frac{1}{3} = \dfrac{1}{3}$

You are now able to convert any decimal to a fraction or any fraction to a decimal. Choosing one rather than another usually depends upon the nature of the problem. There are some fractions and decimals which are used more frequently than others. These should be familiar to you and you should commit them to memory. Many of them are included in the next practice exercise.

PRACTICE EXERCISE 133

In the following table certain values are missing. Find the fraction or decimal equivalent and place it in the table.

	Fraction	Decimal Equivalent		Fraction	Decimal Equivalent
1.	$\frac{1}{2}$	.50	**11.**	$\frac{4}{5}$	
2.	$\frac{1}{3}$	$.33\frac{1}{3}$	**12.**		$.83\frac{1}{3}$
3.	$\frac{1}{4}$	.25	**13.**		$.62\frac{1}{2}$
4.		.20	**14.**		$.87\frac{1}{2}$
5.	$\frac{1}{6}$		**15.**	$\frac{7}{10}$	
6.		$.12\frac{1}{2}$	**16.**		.9
7.	$\frac{1}{10}$		**17.**		$.66\frac{2}{3}$
8.		.4	**18.**	$\frac{3}{4}$	
9.	$\frac{3}{5}$		**19.**	$\frac{3}{8}$	
10.		.3	**20.**		$.42\frac{6}{7}$

13.9 Rewriting Decimals with Fractions

An area of difficulty is the decimal and fraction combination like $.62\frac{1}{2}$. You saw that it can be changed into the fraction $\frac{5}{8}$. Can it be written as a decimal number without a fraction?

Look at the decimal number .625. What is its fractional value? It is $\frac{625}{1,000} = \frac{5}{8}$. Thus:

$$.62\frac{1}{2} = .625.$$

EXAMPLE

Is it true that $.06\frac{1}{4} = .0625$?

SOLUTION

$$.06\frac{1}{4} = \frac{6\frac{1}{4}}{100} = 6\frac{1}{4} \div 100$$

$$= \frac{25}{4} \div \frac{100}{1}$$

$$= \frac{\overset{1}{\cancel{25}}}{4} \times \frac{1}{\underset{4}{\cancel{100}}}$$

$$= \frac{1}{16}$$

$$.0625 = \frac{625}{10,000}$$

$$= \frac{625 \div 625}{10,000 \div 625}$$

$$= \frac{1}{16}$$

Thus: $.06\frac{1}{4} = .0625$

To rewrite a decimal containing a fraction like $\frac{1}{2}$, $\frac{1}{4}$, etc., replace the fraction with its decimal equivalent, $\frac{1}{2} = .5$, $\frac{1}{4} = .25$, etc. Write the decimal equivalent to the right of the original decimal but do not include another decimal point. Look at the following illustrations.

1. $.07\frac{1}{4} = .0725$

2. $.02\frac{1}{2} = .025$

3. $3.42\frac{1}{8} = 3.42125$

since $\frac{1}{4} = .25$

since $\frac{1}{2} = .5$

since $\frac{1}{8} = .125$

PRACTICE EXERCISE 134

Change each decimal to a common fraction or an improper fraction in its simplest form.

1. $.6 =$

2. $.04 =$

3. $.375 =$

4. $.07\frac{3}{4} =$

5. $.222 =$

6. $.15 =$

7. $1.2 =$

8. $.80 =$

9. $.62\frac{1}{2} =$

10. $.83\frac{1}{3} =$

11. If 3 cans of peas cost \$2, then how many cans of peas can be purchased for \$8?

12. If a towel requires $.66\frac{2}{3}$ yd. of material, then how much material would be needed to make 8 towels for the family?

13.10 Word Problems Combining Fractions and Decimals

EXAMPLE

The cost of a dozen cans of soda is \$7.20. What is the cost of $\frac{3}{4}$ of a dozen?

SOLUTION

To determine the cost of the purchase find $\frac{3}{4} \times \$7.20$ using:

Method 1	Method 2	Method 3
(all fractions)	(all decimals)	(fractions and decimals)

Method 1 (all fractions)

$$\$7.20 = 7\frac{20}{100} = 7\frac{1}{5}$$

$$\frac{3}{4} \times 7\frac{1}{5} =$$

$$\frac{3}{4} \times \frac{\overset{18}{\cancel{36}}}{5} = \frac{54}{10} = \$5.40$$
$$\qquad\; 2$$

Method 2 (all decimals)

$$\frac{3}{5} = .75$$

$$\begin{array}{r} \$7.20 \\ \times \quad .75 \\ \hline 3600 \\ 5040 \\ \hline \$5.4000 = \$5.40 \end{array}$$

Method 3 (fractions and decimals)

$$\frac{3}{4} \text{ of } \$7.20 =$$

$$\frac{3}{\underset{1}{\cancel{4}}} \times \frac{\overset{1.80}{\cancel{\$7.20}}}{1} = \$5.40$$

EXAMPLE

Yolanda bought $3\frac{1}{4}$ yd. of dress material on sale at \$3.96 per yd. How much did the material cost her?

SOLUTION

One yard of material costs \$3.96. To find the cost of $3\frac{1}{4}$ yards, multiply \$3.96 by $3\frac{1}{4}$ using:

Method 1	Method 2	Method 3
(all fractions)	(all decimals)	(fractions and decimals)

Method 1 (all fractions)

$$3\frac{1}{4} \times \$3.96 =$$

$$3\frac{1}{4} \times 3\frac{96}{100} =$$

$$\frac{13}{\cancel{4}} \times \frac{\cancel{396}}{100} = \frac{1287}{100}$$

$$= 12\frac{87}{100}$$

$$= \$12.87$$

Method 2 (all decimals)

$$3\frac{1}{4} = 3.25$$

$$\begin{array}{r} \$3.96 \\ \times \quad 3.25 \\ \hline 1980 \\ 792 \\ 1188 \\ \hline \$12.8700 = \$12.87 \end{array}$$

Method 3 (fractions and decimals)

$$3\frac{1}{4} \times \$3.96 =$$

$$\frac{13}{\cancel{4}} \times \frac{\cancel{\$3.96}^{.99}}{1} = \$12.87$$

The methods just illustrated can be used as you desire. It is up to you to choose that method which best fits the problem and best suits you. You should be aware of each method and be able to do any one of them at any time.

PRACTICE EXERCISE 135

Find the value of each of the following problems.

1. Find $\frac{1}{2}$ of $4.02.

2. $.04 \div \frac{4}{5} =$

3. $6.03 \times \frac{2}{3} =$

4. Find $\frac{3}{4}$ of .0056 in.

5. Find the quotient of $.62 \div \frac{5}{8}$ to the nearest hundredth.

6. If two pieces of chocolate cost $.59 then what is the cost of 17 pieces?

7. If a foot of wire costs $.12$\frac{1}{4}$, find the number of feet of wire purchased for $2.94.

8. If a baseball player gets a hit .275 times for each of the 80 times he was at bat, then how many hits did he get?

9. Baseball standings are represented by a three-place decimal. This decimal is found by forming a fraction of the number of games won divided by the total number of games played. In the Cape League standings which follow, find the 3-place decimal for each team. Orleans has been completed for you.

CAPE LEAGUE STANDINGS

	W	L
Orleans	19	11
Wareham	20	12
Cotuit	17	16
Harwich	18	17
Chatham	17	18
Falmouth	13	19
Yarmouth	11	22

League	W	L	Fraction	Decimal
Orleans	19	11	$\dfrac{19}{30}$	$\dfrac{19}{30} = 30\overline{)19.0000}\,^{.6333}$ $= .6333 \approx .633$
Wareham				
Cotuit				
Harwich				
Chatham				
Falmouth				
Yarmouth				

10. If a $\dfrac{3}{8}$ in. rivet weighs $.37\dfrac{1}{2}$ lb. how many rivets are there in a pile that weighs $.67\dfrac{1}{2}$ lb.?

REVIEW OF IMPORTANT IDEAS

Some of the most important ideas in Chapter 13 were:

Dividing decimals is almost identical to dividing whole numbers.

Division by a decimal is accomplished after the divisor is made into a whole number.

To divide a decimal by a whole number—
1. Divide as you would with whole numbers.
2. Place the decimal point in the quotient directly above the decimal point in the dividend.

To divide a decimal by a decimal—
1. Multiply the divisor by a multiple of 10 so that it becomes a whole number.
2. Multiply the dividend by the same number used to multiply the divisor.
3. Proceed as you did when dividing a decimal by a whole number.

To divide by the special numbers of 10, 100, 1,000, etc.—
1. Keep the same digits in the quotient as there are in the dividend.
2. Move the decimal point in the dividend as many places to the left as there are zeros in the divisor.

To change a fraction into a decimal—
1. Place the decimal point in the dividend (fraction's numerator) at the extreme right of the number.
2. Annex as many zeros to the right of the decimal point in the dividend as desired.
3. Divide until there is no remainder, or to one decimal place more than the question calls for. Round off to the desired number of decimal places.

To change a decimal to a fraction—
1. Write the numerator of the fraction having the same digits as the decimal without the decimal point.
2. Write the denominator as a multiple of 10, 100, 1,000, etc., according to the number of digits in the decimal.
3. Simplify the resulting fraction if possible.

When doing computations with both decimals and fractions—
1. Change all the numbers to fractions and then proceed.
2. Change all the numbers to decimals and then proceed.
3. Do the computations without changing to the same type of terms.

CHECK WHAT YOU HAVE LEARNED

The examples in this chapter test will test your progress in dividing decimals. Try to get a satisfactory grade. Use a calculator to check your answers after you have completed the entire chapter test.

In counting up your answers, remember that there were 32 separate answers in this test.

A Score of	*Means That You*
29–32	Did very well. You can move on to Chapter 14.
26–28	Know this material except for a few points. Reread the sections about the ones you missed.
21–25	Need to check carefully the sections you missed.
0–20	Need to review the chapter again to refresh your memory and improve your skills.

Questions	*Are Covered in Section*
1–4	13.1
5–8	13.2
10, 11	13.3
14–23	13.4
12–13, 24, 26–28	13.5
9	13.6
30	13.7
29	13.8
25	13.9

Chapter Test 13

1. $3\overline{).63}$ 2. $4\overline{)8.4}$ 3. $3\overline{)1.2}$ 4. $7\overline{)35.7}$

5. $7\overline{).18}$ 6. $7\overline{)7.007}$ 7. $2\overline{).084}$ 8. $9\overline{).081}$

9. a. $.6 \div 100 =$ b. $4.27 \div 10 =$ c. $63.8 \div 1,000 =$

10. Change $\dfrac{3}{5}$ to a decimal fraction.

11. Change $\dfrac{2}{7}$ to a two-place decimal.

12. Wai Hung used 13 gal. of gasoline to drive a distance of 158.6 mi. On the average, how many miles did she drive with 1 gal. of gasoline?

13. If 45 gumdrops weigh 14 oz., find the weight of a gumdrop to the nearest hundredth of an ounce.

14. $.4\overline{).8}$ 15. $.8\overline{).72}$ 16. $.6\overline{)4.2}$ 17. $.5\overline{)25}$

18. $.7\overline{).84}$ 19. $.04\overline{)16.8}$ 20. $.6\overline{).003}$ 21. $.87\overline{)104.4}$

22. $6.3\overline{)15.12}$

23. Divide 48,334.5 by .69.

24. Potatoes sell for $.79 per lb. How many pounds did Mrs. Phillips buy if she paid $4.74 for her potatoes?

25. Diane paid $318.55 for $34\dfrac{5}{8}$ yd. of drapery material. What was the cost of 1 yd. of material?

26. If 3 lb. of onions cost $1.47, how many pounds can be bought for $4.41?

27. Mary worked a year, earning $20,982. How much did she earn weekly? (52 weeks = 1 year)

28. Find the average of 38, 41, 35, 45, 49, and 37 to the nearest tenth.

29. Change $.08\frac{1}{5}$ to a decimal.

30. Change $.62\frac{1}{2}$ to a fraction in its simplest form.

YOU ARE ALMOST THERE

This completes your study of decimals. You have seen how, where, and why decimals are used in everyday life. You have only scratched the surface in these chapters but we hope that this study will help you to use mathematics on your job, to achieve a high school diploma, to feel better about yourself as a person, or just to feel more secure when mathematics is used to explain a happening. Now you are ready for Chapter 14.

HOLD IT!

Are your arithemetic skills improving? This test will allow you to find out. Use your calculator wherever you can to check your results.

1. $\frac{5}{8} \times 2\frac{3}{10} =$

2. $3.16 \times .42 =$

3. $.036 \div 12 =$

4. Change $\frac{2}{9}$ to a two-place decimal.

5. $1.6 + 12 + .483 =$

6. $14\frac{5}{8}$
$+ 4\frac{1}{2}$

7. Change .025 to a common fraction in its simplest form.

8. The sales tax on an item is $4.292. What is this amount rounded off to the nearest cent?

9. If a runner sets a new world record of eight and eighty-seven hundredths seconds, how would this be written as a decimal numeral?

10. Find the quotient of $5.16 \div 1.2$.

11. The cost of 18,800 ft.^{2} of property is \$35,500. What is the cost per ft.^{2} correct to the nearest cent?

12. a. What is the cost of $2\frac{1}{2}$ lb. of rump roast and a $2\frac{1}{4}$ lb. pork tenderloin at the following prices?

rump roast	\$3.29 lb.
pork tenderloin	\$4.59 lb.

   b. If you give the butcher a \$20 bill, how much change do you get?

ANSWERS

1. $1\frac{7}{16}$ 2. 1.3272   3. .003   4. .22   5. 14.083

6. $19\frac{1}{8}$ 7. $\frac{1}{40}$ 8. \$4.29   9. 8.87 s.   10. 4.3

11. \$1.89 per ft.^{2} 12. a. $\begin{array}{r} \$\ 8.23 \\ +\ 10.33 \\ \hline \$18.56 \end{array}$ b. $\begin{array}{r} \$20.00 \\ -\ 18.56 \\ \hline \$\ 1.44 \end{array}$

Percents

Much time and energy can be wasted when studying percents. However, you can avoid this inefficiency if you recognize that percent is a way of using knowledge you already have.

Percent is probably the most common method we have of expressing comparisons between quantities. Every day we see in our newspapers a sale that will allow us 40% off on sweaters or up to 30% off on all furniture, interest rates of 5.47%, etc.

Because these uses of percent occur daily, it is important to know what percent means and how it is related to fractions and decimals, the study of which you have already completed. Chapter 14 will deal with percents and how they are used in practical situations.

Check your skill on this topic in the pretest which follows.

Pretest 14–See What You Know and Remember

Do as many of these problems as you can. Try hard for a satisfactory grade.

1. Write $\frac{5}{100}$ as a percent.

2. Change 43% to a fraction with a denominator of 100.

3. Sara gets 65 correct answers on an examination which consists of 100 questions. What percent of the problems are correct?

Change each of the following percents to decimals.

4. 45%

5. $6\frac{1}{4}\%$

6. 215%

Change each of the following decimals to percents.

7. .06   8. 1.25   9. $.00\frac{1}{2}$

Change each of the following percents to mixed numbers or fractions in their simplest form.

10. 25%   11. 5.5%   12. $3\frac{1}{2}\%$ 13. 125%

Change each of the following fractions or mixed numbers to percents.

14. $\frac{3}{5}$ 15. $\frac{1}{10}$ 16. $\frac{1}{3}$ 17. $4\frac{3}{4}$

18. Mary was able to do 6 problems correctly in a test consisting of 8 examples. What percent of the problems were correct?

19. Find 40% of $60.

20. What percent of 8 is 3?

21. If 30% of a number is 24, what is the number?

22. At a sale, a dress sold for $56. This represented 80% of the original price. What was the original price?

23. A sporting goods store sold a pair of skates for $102. The skates cost the store $68.
a. What was the profit?
b. What percent of increase was this of the original cost?

24. Jane bought a dress for $38.80 that originally cost $48.50. By what percent was the dress reduced from the original price?

25. Sue is reading a book that has 300 pages. She has already read 200 pages. What percent of the book is left for her to read?

Check your answers by turning to the back of the book. Add up all that you had correct.

A Score of   *Means That You*

23–25   Did very well. You can move on to Chapter 15.

20–22   Know the material except for a few points. Read the sections about the ones you missed.

17–19   Need to check carefully on the sections you missed.

0–16   Need to work with this chapter to refresh your memory and improve your skills.

Questions   *Are Covered in Section*

Questions	Are Covered in Section
1–3	14.1
4–6	14.2
7–9	14.3
10–13	14.4
14–18	14.5
19	14.6
20, 23–25	14.7
21, 22	14.8

14.1 Meaning of Percent

EXAMPLE

Stemkowski made 20 base hits in 50 times at bat, and Russo made 10 base hits in 20 times at bat. Who has the better record?

SOLUTION

In previous chapters ratios were expressed as fractions. To determine which batter has the better record write each one as a fraction.

$$
\begin{array}{cc}
\textit{Stemkowski} & \textit{Russo} \\[4pt]
\left.\begin{array}{l} \text{Hits} \;\;\;= 20 \\ \text{At Bats} = 50 \end{array}\right\} = \dfrac{20}{50} & \left.\begin{array}{l} \text{Hits} \;\;\;= 10 \\ \text{At Bats} = 20 \end{array}\right\} = \dfrac{10}{20}
\end{array}
$$

These ratios can be compared by writing them as fractions having the same denominator or as decimal fractions.

As fractions they become

$$\frac{20 \times 2}{50 \times 2} = \frac{40}{100}$$

and

$$\frac{10 \times 5}{20 \times 5} = \frac{50}{100}$$

Since $\frac{50}{100}$ is greater than $\frac{40}{100}$, Russo has the better record.

Stemkowski hit safely 40 out of 100 tries, and this can be expressed as $\frac{40}{100}$ or 40 hundredths. This is written as 40 percent.

The symbol % means percent. Thus, 40 percent is written 40%.

REMEMBER
Percent means
hundredths.

Russo hit safely 50 out of 100 tries and this is written as $\frac{50}{100}$, 50%, or .500. Baseball averages are usually written as three-place decimals for greater accuracy.

In the following diagram, consisting of 100 squares, 4 shaded areas are indicated by the letters *A*, *B*, *C*, and *D*.

Area A covers 25 squares, which is stated as $\frac{25}{100}$, .25, or 25%.

Area B covers one square or $\frac{1}{100}$, .01, or 1%.

Area C covers 10 squares or $\frac{10}{100}$, .10, or ?%. What percent is 10 out of 100? That's correct, 10%.

Area D covers ? squares or $\frac{?}{100}$, .?, or ?%. Fill in the missing values yourself.

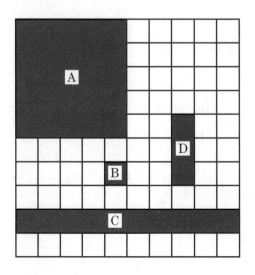

Notice that *A* covers 25% of the whole square while *B* covers 1%, *C* covers 10%, and *D* covers 3%. What percent is unshaded? We count 61 unshaded squares or 61% of the whole square. The same result could have been obtained this way.

The whole	−	The shaded part $(A + B + C + D)$	=	The unshaded part
$\dfrac{100}{100}$	−	$\dfrac{39}{100}$	=	$\dfrac{61}{100}$
100%		39%		61%

PRACTICE EXERCISE 136

Write each of the following fractions as percents.

1. $\dfrac{6}{100}$ 2. $\dfrac{50}{100}$ 3. $\dfrac{7}{10}$

Write these as percents.

4. 28 hundredths   5. .05   6. 2.15

7. .9   8. .725   9. $.04\dfrac{1}{2}$

Write these percents as common fractions with a denominator of 100.

10. 12%   11. 2%   12. 76%   13. $6\dfrac{1}{2}\%$

Supply the missing numbers or signs.

14. $\dfrac{3}{4} = \dfrac{?}{100} = .75 = 75$ percent = 75?

15. $\dfrac{2}{5} = \dfrac{?}{100} = .40 = ?$ percent = 40%

16. $\dfrac{5}{8} = \dfrac{62\frac{1}{2}}{?} = .62\dfrac{1}{2} = ?$ percent = $62\dfrac{1}{2}\%$

17. $\dfrac{1}{2} = \dfrac{?}{100} = .? = ?$ percent $= ?\%$

18. $\dfrac{100}{100} = ?\%$

19.   If John had 85% of all the questions correct on his last mathematics examination, what percent did he have *incorrect*?

14.2 Changing Percents to Decimals

EXAMPLE

At a linen sale, Mrs. Volpe saved 40% on a purchase of $80 worth of sheets and pillowcases. How much did she save? How much was the cost of the linens?

SOLUTION

Percent is a way of expressing a fraction or a decimal, so no new methods must be learned for finding 40% of $80. Percent or its symbol % means hundredths, so change 40% to .40 and multiply .40 times $80. Recall that the word *of* in 40% *of* $80 means to multiply. Here is the completed example:

Problem	Step 1	Step 2	Step 3
40% of $80 =	40% = .40	$80 × .40 —— $32.00	$80 − 32 —— $48
	Change 40% to a decimal: .40.% = .40	Multiply: $32 is the amount of money saved.	Subtract: $80 − 32 = $48 The cost of the linens is $48.

Percents can be written in many different ways, such as 4%, $2\frac{1}{4}$%, 3.2%, 410%, etc.

How can these percents be changed to decimals in the simplest and easiest way?

Let's look at a few examples first to see if you get a hint from these.

$$40\% \text{ means } \frac{40}{100} = .40$$

$$4\% \text{ means } \frac{4}{100} = .04$$

$$3.2\% \text{ means } \frac{3.2}{100} = \frac{3.2 \times 10}{100 \times 10} = \frac{32}{1,000} = .032$$

Notice that the digits in the percent and the decimal are the same. *The difference lies in the removal of the percent sign and the decimal point.*

Division by 100 was done previously and you recall it was simply accomplished by moving the decimal point in the original number two places to the left. Since $\frac{40}{100}$ means 40 is divided by 100, and the decimal point is on the right of 40, dividing by 100 moves it two places to the left (.40.).

For 4%, place the decimal point to the right of 4 and, when moving it two places to the left, annex a zero (.04.).

In 3.2%, moving the decimal point two places to the left results in the answer .03.2, which checks with the result above.

TO CHANGE A PERCENT TO A DECIMAL

1. Divide the number by 100 by moving the decimal point two places to the left in the original number.
2. Omit the percent sign.

EXAMPLE

Change 45% to a decimal fraction.

SOLUTION

Problem	Step 1	Step 2	Step 3
Change 45% to a decimal.	45% = 45.%	.45.%	.45
	Place the decimal point to the right of 45: 45.%	Divide by 100 by moving the decimal point 2 places to the left: .45. Omit the % sign.	

And a few other illustrations—

1. $12\frac{1}{3}\% = 12.\frac{1}{3}\% = .12.\frac{1}{3}\% = .12\frac{1}{3}$
2. $5.47\% = 5.47\% = .05.47\% = .0547$
3. $\frac{1}{4}\% = .25\% = .00.25\% = .0025$
4. $215\% = 215.\% = 2.15.\% = 2.15$

PRACTICE EXERCISE 137

Change each of the following percents to decimals.

1. 75%   2. $12\frac{1}{4}\%$ 3. 5%   4. 99%   5. 125%

6. $2\frac{1}{2}\%$ 7. $\frac{1}{2}\%$ 8. 80%   9. 1.5%   10. $66\frac{2}{3}\%$

14.3 Changing Decimals to Percents

EXAMPLE

The winning team in a baseball league won .558 of the games it played. What percent of the games played did the team win?

SOLUTION

Since percent means hundredths, find the number of hundredths to which the decimal .558 is equal.

$$.558 = \frac{558}{1,000}$$

Divide both terms of the fraction by 10. This results in a denominator of 100.

$$\frac{558 \div 10}{1,000 \div 10} = \frac{55.8}{100} = 55.8\%$$

Look at the following examples, so that you can discover a short cut for changing a decimal into a percent.

$$.05 = \frac{5}{100} = 5\%$$

$$.12 = \frac{12}{100} = 12\%$$

$$.125 = \frac{125 \div 10}{1,000 \div 10} = \frac{12.5}{100} = 12.5\% \text{ or } 12\frac{1}{2}\%$$

$$1.25 = 1\frac{25}{100} = 125\%$$

> Notice that the digits in the decimal and percent remain unchanged. *The difference lies in adding a percent sign or a decimal point.*

Checking the position of the decimal point in the decimal and in its new position in the percent, you notice it moved two places to the right in all four examples. This is equivalent to multiplying the decimal by 100. Recall that multiplying by 100 moves the decimal point two places to the right.

TO CHANGE A DECIMAL TO A PERCENT

1. Multiply by 100 by moving the decimal point two places to the right.
2. Affix the percent sign.

EXAMPLE

Change .62 to a percent.

SOLUTION

Problem	Step 1	Step 2
Change .62 to a percent.	.62 becomes 62.	62%
	Multiply the decimal by 100: .62. × 100 = 62.	Affix the % sign.

Look at these.

1. .09: .09. = 9%

2. .10: .10. = 10%

3. .125: .12.5 = 12.5% or $12\frac{1}{2}\%$

4. .0075: .00.75 = .75% = $\frac{3}{4}\%$

PRACTICE EXERCISE 138

Change each of the following decimals to percents.

1. .38

2. .06

3. .375

4. $.06\frac{1}{4}$

5. .005

6. .0325

7. 1.25

8. $.00\frac{1}{4}$

9. 3.12

10. .6

14.4 Changing Percents to Fractions

EXAMPLE

In a recent drought 20% of the corn crop was destroyed. What fractional part of the entire crop does this represent?

SOLUTION

To find the fractional part change the percent to a fraction. By the definition of a percent, 20% means $\dfrac{20}{100}$, which is equivalent to $\dfrac{1}{5}$. Therefore, $\dfrac{1}{5}$ of the entire crop of corn was destroyed. See the following problem.

Problem	Step 1	Step 2
Change 20% to a fraction.	$20\% = \dfrac{20}{100}$	$\dfrac{20 \div 20}{100 \div 20} = \dfrac{1}{5}$
	Write 20% as a fraction with a denominator of 100: $\dfrac{20}{100}$	Simplify the fraction (if possible): $\dfrac{20}{100} = \dfrac{1}{5}$

TO CHANGE A PERCENT TO A FRACTION

1. Write a fraction whose numerator is the given number and whose denominator is 100.
2. Simplify the fraction (if possible).

Here are more examples:

1. $45\% = \dfrac{45}{100} = \dfrac{9}{20}$

2. $5\dfrac{1}{2}\% = \dfrac{5\dfrac{1}{2}}{100} = 5\dfrac{1}{2} \div 100 = \dfrac{11}{2} \div 100 = \dfrac{11}{2} \times \dfrac{1}{100} = \dfrac{11}{200}$

3. $.5\% = \dfrac{.5}{100} = \dfrac{.5 \times 10}{100 \times 10} = \dfrac{5}{1,000} = \dfrac{1}{200}$

PRACTICE EXERCISE 139

Change each of the following percents to fractions. Write each answer in its simplest form.

1. 75%   2. 5%   3. 99%   4. 5.46%   5. 22.2%

6. $\frac{1}{4}$%   7. $2\frac{1}{2}$%   8. $16\frac{2}{3}$%   9. 7.5%   10. .025%

14.5 Changing Fractions to Percents

EXAMPLE

On a test that consisted of 125 questions, Mr. Colon got 75 correct answers. What percent of all the questions were correct?

SOLUTION

You know that he correctly answered 75 of the 125 questions. This is $\frac{75}{125}$ of all the questions. This fraction is equivalent to $\frac{3}{5}$. Since percent means hundredths, multiply the terms of the fraction $\frac{3}{5}$ by a number which will give a product of 100 in the denominator. Thus, you multiply both terms of the fraction $\frac{3}{5}$ by 20. You obtain

$$\frac{3 \times 20}{5 \times 20} = \frac{60}{100} = 60\%$$

This is one method you can use to change a fraction into a percent.

EXAMPLE

Find the percent equivalent to the fraction $\frac{3}{4}$.

SOLUTION

You must multiply the denominator by a number which will give a product of 100. In this case, multiply both terms of the fraction by 25.

Problem	Step 1	Step 2
$\dfrac{3}{4} = ?\%$	$\dfrac{3 \times 25}{4 \times 25} = \dfrac{75}{100}$	$\dfrac{75}{100} = 75\%$
	Multiply the terms of the fraction by 25. This results in a denominator of 100.	Change the fraction into a percent: $\dfrac{75}{100} = 75\%$

TO CHANGE A FRACTION TO A PERCENT

1. Change the given fraction to an equivalent fraction whose denominator is 100 and whose numerator is the number of percent.
2. Affix the percent sign.

This method is limited, for example, when you change a fraction like $\dfrac{7}{8}$ into a percent.

By what number should you multiply the terms of the fraction so that the denominator is 100? Since this is not a familiar number, let's look for another method to supplement the one just demonstrated.

Earlier, you learned that a fraction can be changed to a decimal. In section 16.3, decimals were changed to percents. Therefore, a fraction can be changed to a decimal and the decimal can be changed to a percent.

$$\text{FRACTION} \rightarrow \text{DECIMAL} \rightarrow \text{PERCENT}$$

EXAMPLE

At a sale, Cassandra bought a coat for $75 which usually sells for $125. What percent of the original price did she pay?

SOLUTION

Problem	Step 1	Step 2	Step 3
$\dfrac{\$75}{\$125} = ?\%$	$\dfrac{75}{125} = 75 \div 125$	$125\overline{)75.00}^{.60}$	$.60 = 60\%$
	$\dfrac{a}{b}$ means $a \div b$.	Divide the whole numbers: $\dfrac{75}{125}$. Carry the division to two decimal places.	Change the decimal to a %: $.60 = 60\%$

TO CHANGE A FRACTION TO A DECIMAL AND THEN TO A PERCENT

1. Divide the denominator of the fraction into the numerator and find the quotient.
2. Write the decimal as a percent.

Each method is illustrated below.

Method 1	Method 2
Fraction ⟶ Percent	Fraction ⟶ Decimal ⟶ Percent
1. $\dfrac{1}{2} = \dfrac{1 \times 50}{2 \times 50} = \dfrac{50}{100} \quad = \quad 50\%$	$\dfrac{1}{2} \quad = \quad 2\overline{)1.00}^{.50} \quad = \quad 50\%$
2. $\dfrac{7}{8} = \dfrac{7 \times 12\frac{1}{2}}{8 \times 12\frac{1}{2}} = \dfrac{87\frac{1}{2}}{100} \quad = \quad 87\frac{1}{2}\%$	$\dfrac{7}{8} \quad = \quad 8\overline{)7.000}^{.875} \quad = \quad 87.5\%$ $= \quad 87\frac{1}{2}\%$

PRACTICE EXERCISE 140

Change each of the following fractions into percents. Use whichever method is easier for you to handle. One method may be easier than the other.

1. $\dfrac{4}{5}$ 2. $\dfrac{1}{8}$ 3. $\dfrac{3}{10}$ 4. $\dfrac{1}{3}$ 5. $\dfrac{1}{20}$

6. $\dfrac{3}{7}$ 7. $\dfrac{5}{16}$ 8. $\dfrac{1}{4}$ 9. $\dfrac{5}{8}$ 10. $\dfrac{2}{9}$

Some of the following numbers in parentheses ( ) are not correct. To get an example right, you must copy all the correct numbers and no others. Can you find all of them?

11. 60% equals $\left(\dfrac{60}{100}, 0.06, \dfrac{6}{10}, \dfrac{3}{4}, \dfrac{3}{5} \right)$.

12. $12\dfrac{1}{2}\%$ equals $\left(1.25, .125, \dfrac{11}{8}, \dfrac{1}{8}, 1.12\dfrac{1}{2} \right)$.

13. $\dfrac{5}{8}$ equals $\left(0.62, 62\dfrac{1}{2}, 6.25, .625 \right)$.

14. 20% equals $\left(\dfrac{20}{100}, \dfrac{1}{5}, .020, .2 \right)$.

15. $\dfrac{2}{3}$ equals $\left(\dfrac{4}{6}, 66\dfrac{2}{3}, \text{about } .67 \right)$.

14.6 Finding a Percent of a Number

EXAMPLE

Mr. Woo works in a department store as a salesman. He earns a base rate of pay of $320 per week plus a 10% commission on his sales. At the end of his first week of work he had sold $2,200 worth of merchandise. How much money did he earn that week?

SOLUTION

Commission is computed by finding the percent of the sales. In this example:

$$\text{commission} = 10\% \text{ of } \$2,\!200 \text{ or}$$
$$\text{commission} = 10\% \times \$2,\!200$$

since the word *of* means *to multiply*. Ten percent is changed to either a decimal or a fraction before multiplying. See the following two examples:

Problem	Step 1	Step 2	Step 3
10% of $2,200 =	.10 × $2,200 =	$2,200 × ___10 $220.00	$320 + 220 $540
	Change 10% to a decimal, 10% = .10, and multiply.	This represents the commission.	Add his commission ($220) to his base rate of pay ($320): $220 + $320 = $540

OR

Problem	Step 1	Step 2	Step 3
10% of $2,200 =	$\dfrac{1}{10} \times \$2,\!200 =$	$\dfrac{1}{\cancel{10}} \times \dfrac{\overset{220}{\cancel{2,\!200}}}{1} =$ $\dfrac{\$220}{1} = \$220$	$320 + 220 $540
	Change 10% to a fraction $\dfrac{1}{10}$ and then multiply.	$220 represents his commission.	$540 is his total salary for the first week of work.

You notice that there is no one method for finding a percent of a number. You return to those numbers learned previously, such as fractions and decimals.

TO FIND A PERCENT OF A NUMBER

1. Change the percent to either a fraction or a decimal. (It is your choice to pick that one which makes the computation easier for you.)
2. Multiply the number by the fraction or decimal.

EXAMPLE

Find 25% of 80.

SOLUTION

25% = .25
Therefore: .25 × 80 =

$$
\begin{array}{r}
80 \\
\times\,.25 \\
\hline
4\,00 \\
16\,0 \\
\hline
20.00 = 20
\end{array}
$$

or $\dfrac{25}{100} = \dfrac{1}{4}$

$$\dfrac{1}{\overset{}{\underset{1}{4}}} \times \dfrac{\overset{20}{\cancel{80}}}{1} = \dfrac{1 \times 20}{1 \times 1} = \dfrac{20}{1} = 20$$

EXAMPLE

At a sale, a dress is marked 30% off. If the dress was originally priced at $60, how much money does Helen save if she buys the dress?

SOLUTION

You need to know what 30% of $60 is in order to find the answer to the problem. Changing 30% to its equivalent fraction $\dfrac{30}{100}$ or $\dfrac{3}{10}$ and multiplying by $60, you have:

$$\dfrac{3}{10} \times \$60 =$$

$$\dfrac{3}{\underset{1}{\cancel{10}}} \times \dfrac{\overset{6}{\cancel{60}}}{1} = \$18$$

Helen will save $18 if she buys the dress.

PRACTICE EXERCISE 141

Find:

1. 4% of 60

2. 3% of 200

3. 8% of 8

4. 125% of 20

5. 300% of 6

6. $\frac{1}{2}$% of 120

7. 40% of 65

8. $2\frac{1}{2}$% of 82

9. 5.7% of $490

10. $8\frac{1}{4}$% of 200

11. 6.5% of $2.74

12. .02% of 1,250

13. A worker paid a tax of 18.6% on $912.24 of his income. How much money did he pay?

14. Ordinary brass is made up of 61.6% copper, 2.9% lead, .2% tin, and 35.3% zinc. How many pounds of each metal would there be in 60 lb. of this brass?

15. The price of a new car is $20,450. Mr. Pepper made a down payment of 15% when he purchased the car. How much money did he pay at the time?

16. Mr. Harris earns a yearly salary of $22,500. He also earns a commission of 6% on his sales. If his sales for the past year totaled $230,000, what did Mr. Harris earn?

14.7 Finding the Percent That One Number Is of Another

EXAMPLE

Mr. Kline, working as a parts inspector, rejected 21 parts from a total of 420 parts. Of the parts inspected what percent was *accepted*?

SOLUTION
Since 21 of 420 were rejected, then 420 − 21 = 399 were accepted. The fraction $\frac{399}{420}$ represents the part of the whole which was accepted. Change this fraction into a decimal by dividing:

$$399 \div 420 =$$

$$\begin{array}{r} .95 \\ 420\overline{)399.000} \\ \underline{378\ 0} \\ 21\ 00 \\ \underline{21\ 00} \end{array} = \frac{95}{100} = 95\%$$

While this example seems simple enough, there are a few troublemakers that you will run into that may cause difficulty. Let's look at a few of these before writing the method for solving this type of problem.

Since in the first example we wrote 399 as part of 420 in the form of a fraction, let's begin there.

EXAMPLE

What part of 16 is 12?

SOLUTION

Since 12 is part of 16, it is written as $\dfrac{12}{16}$.

EXAMPLE

What part of 12 is 16?

SOLUTION

This is written as the fraction $\dfrac{16}{12}$ (notice the difference in the two examples).

Another wording of the problem may be: 16 is what part of 12? This is written as the fraction $\dfrac{16}{12}$. Writing the fraction correctly is most important or else the fraction may be reversed resulting in an incorrect answer. You know that

$$\frac{12}{16} \neq \frac{16}{12}$$

Looking again at the previous three examples, you notice that in each case the number which follows the word *of* becomes the denominator of the fraction.

Look at the two problems which follow.

To find what part one number is of another, you write a fraction whose denominator is the number following the word *of* and whose numerator is the other number.

EXAMPLE

What part *of* 24 is 9?

SOLUTION

Since 24 follows the word *of*, the fraction $\dfrac{9}{24}$ is the result.

EXAMPLE

Four hundred twelve is what part *of* 380?

SOLUTION

Since 380 follows the word *of*, the fraction is written as $\dfrac{412}{380}$.

PRACTICE EXERCISE 142

Write the correct fraction and solve:

1. Eighteen is what part of 48?

2. What part of 75 is 50?

3. What part of 8 is 100?

4. Nine is what part of 24?

5. Eight is what part of 20?

6. What part of 15 is 6?

Once the fraction is written, showing the relationship between the two numbers, you have almost completed the problem. All that remains is to change the fraction into a percent. This was illustrated in the previous chapter but let's review it again. The easiest method to use when changing a fraction to a percent is to rewrite the fraction with 100 as its denominator. This is then changed to a percent. While this is the simplest method to use it may not be easy to change the denominator into 100. In the original problem in this section you learned that you can change the fraction into a decimal first and then into a percent. Since this method can be used in all cases, take another look at it.

EXAMPLE

Our town baseball team won 48 games and lost 37. What percent of its games has it won?

SOLUTION

The team won 48 games out of 85 games played. (Remember to add 48 to 37 to find the total number of games played.) Write a fraction representing the games won out of the total number of games played.

> What part of *its games* has it *won*?
> What part of *85* is *48*?
> What percent of its games has it won?

The correct fraction is $\dfrac{48}{85}$, since 85 follows the word *of* in the question. Change the fraction to a decimal by dividing:

$$
\begin{array}{r}
.5647 \\
85\overline{)48.0000} \\
42\,5 \\
\overline{} \\
5\,50 \\
5\,10 \\
\overline{} \\
400 \\
340 \\
\overline{} \\
600 \\
595 \\
\overline{}
\end{array}
$$

.5647 = 56.47% or 56.5% to the nearest tenth

TO FIND WHAT PERCENT ONE NUMBER IS OF ANOTHER

1. Write a fraction whose denominator is the number which follows the word *of* and whose numerator is the other number.
2. Change the fraction into a percent.

PRACTICE EXERCISE 143

1. What percent of 16 is 12?

2. What percent of 15 is 6?

3. Ten is what percent of 30?

4. Fifty is what percent of 75?

5. What percent of 100 is 8?

6. Eight is what percent of 20?

7. James did 12 problems correctly in Practice Exercise 88. There are 16 problems in the exercise. What percent of his problems are correct?

8. Our community fund has set $105,600 as its goal. To date our citizens have collected $44,352. What percent of the goal has been raised?

9. The Durable Company has increased its output per month from 2,560 to 2,944 machines. How much is this increase expressed as a percent?

10. Three things, A, B, and C, are combined by taking 5 parts of A, 3 parts of B, and 1 part of C.
    a. What percent of the mixture is A?
    b. What percent of the mixture is B?
    c. What percent of the mixture is C?
    d. How many pounds of A are there in 80 lb. of the mixture?
    e. How many pounds of B are there in 80 lb. of the mixture?
    f. How many pounds of C are there in 80 lb. of the mixture?

14.8 Finding a Number When a Percent of It Is Known

EXAMPLE

The R & P Company increased all wages 5%. For Jesus Ortiz this meant an increase of $27 per week. What was his salary before the increase?

SOLUTION

Since a 5% increase in his salary is $27, then $\frac{1}{5}$ of the increase (1%) is equivalent to $\frac{\$27}{5}$ or $5.40. If a 1% increase is equal to $5.40, then 100% would be $540—his total salary. Thus, the problem can be written in the following manner:

$$5\% \text{ increase} = \$27$$

$$\frac{1}{5} \text{ of } 5\% \text{ increase} = \frac{1}{5} \text{ of } \$27$$

$$1\% \text{ increase} = \$5.40$$

and

$$100\% = \$540$$

To check this result find 5% of 540:

$$\begin{array}{r} \$540 \\ \times \quad .05 \\ \hline \end{array}$$

$27.00 increase (which verifies the correctness of our result)

A second method may also be used but it is very similar to the one shown above. The difference is in the representation of the percent. The percent in the first method was kept but it could also be shown as a decimal or as a fraction. As a fraction is would look like this:

$$\frac{5}{100} \text{ increase} = \$27$$

$$\frac{1}{\cancel{5}} \times \frac{\overset{1}{\cancel{5}}}{100} \text{ increase} = \frac{1}{5} \text{ of } \$27$$

$$\frac{1}{100} \text{ increase} = \$5.40$$

$$\frac{100}{100} = \$540$$

TO FIND A NUMBER WHEN A PERCENT OF IT IS KNOWN

1. Find 1% of the number.
2. Then find 100% of that number.

EXAMPLE

Nine percent of what number is 18?

SOLUTION

Problem	Step 1	Step 2	Step 3
Nine percent of what number is 18?	9% of the number = 18 1% of the number = 2	100% of the number = 200	9% of 200 = ? 200 × .09 ‾‾‾‾‾‾ 18.00
	Find 1% of the number by dividing by 9: 1% of the number = 2.	Find 100% by multiplying by 100: 100% of the number = 200.	Check the result.

EXAMPLE

In one day, the Gowanus Club raised $75, which was 125% of its quota. What was the organization's quota?

SOLUTION

Problem	Step 1	Step 2	Step 3
125% of its quota = $75	125% of its quota = $75 1% of its quota = $\dfrac{\$75}{\$125}$ = $.60	100% of the quota = $60	125% of $60 = 1.25 × 60 ‾‾‾‾‾‾ $75.00
	Find 1% of the quota by dividing by 125: 1% of the quota = $.60.	Find 100% by multiplying by 100: 100% of the quota = $60.	Check the result.

EXAMPLE

Alec, the car salesman, sold a pre-owned car for $3,840. This represented a gain of 20% figured on the original cost. How much did the car cost the dealer?

SOLUTION

This problem is one of those troublemarkers, because you cannot say that a 20% gain = $3,840. Looking at the problem again you see that 20% represents a gain figured on the original cost. Thus, the original cost, 100% + the 20% gain, represents the selling price of $3,840. Stating it in another way you have:

$$\text{Cost} + \text{Profit} = \text{Selling price}$$
$$(100\% + 20\%) \ = \$3,840$$
$$120\% = \$3,840$$
$$1\% = \frac{\$3,840}{120} = \$32$$
$$100\% = \$3,200,$$
$$\text{the original cost}$$

Checking the result, you have an original cost of $3,200 plus a gain of 20%, which together represents the selling price.

$$20\% \text{ of } \$3,200 = \frac{1}{5} \text{ of } 3,200 \qquad \text{and} \qquad \begin{array}{r} \$3,200 \\ + \quad 640 \\ \hline \$3,840 \end{array}$$

$$= \frac{1}{\cancel{5}} \times \frac{\overset{640}{\cancel{3,200}}}{1}$$
$$\phantom{=\frac{1}{5}}_{1}$$

$$= \$640 \text{ gain}$$

PRACTICE EXERCISE 144

1. What is the number if you know that 5% of the number is 20?

2. Eight is 2% of what number?

3. Twelve is 100% of what number?

4. What is the number if you know that 20% of the number is 8?

5. Twenty-two percent of what number is 44?

6   Sixty-six is 75% of what number?

7. One hundred is 125% of what number?

8. Twenty-two and five-tenths is 7.5% of what number?

9. An engine delivers 102 horsepower at 85% efficiency. What is the rated horsepower based on 100% efficiency?

10. On a mathematics test Bill received a grade of 75% correct. If each problem counted an equal amount, and Bill had 9 problems correct, then how many problems were on the test?

11. The payroll of the T.T.R. Company increased 4% over last month. If the present payroll is $130,000, how much was the payroll last month?

REVIEW OF IMPORTANT IDEAS

Some of the most important ideas in Chapter 14 were:

Percent means hundredths.

Percents, decimals, and fractions are intertwined with each other.

Fraction ⟷ Decimal

Percent

To express a percent, compare the number with one hundred and affix a percent sign (%).

To find the percent of a number—
1. Change the percent to its fraction or a decimal equivalent.
2. Multiply the number by the fraction or decimal.

To find the percent one number is another—
1. Write the numbers as a ratio with the number following *of* as the denominator.
2. Change the fraction into a percent.

To find a number when a percent of it is known—
1. Find 1% of the number.
2. Find 100% of that number.

CHECK WHAT YOU HAVE LEARNED

You were introduced to percents in this chapter. Percents were related to other ideas you learned previously. The problems covered in this chapter are a small sample of examples which involve percents. Probably more areas of mathematics use percents than use fractions or decimals. Percents are easier to understand since they are related to the denominator of 100. The chapter test which follows will provide you with enough examples to see if you are still having difficulty. If you are, review the portion of the chapter that applies to that problem.

A Score of	Means That You
22–26	Did very well. You can move to Chapter 15.
20–22	Know this material except for a few points. Reread the sections about the ones you missed.
17–19	Need to check carefully on the sections you missed.
0–16	Need to review the chapter again to refresh your memory and improve your skills.

Questions	Are Covered in Section
1–3	14.1
4–6	14.2
7–18	14.3
10–13	14.4
14–18	14.5
19	14.6
20, 23–25	14.7
21, 22	14.7

Chapter Test 14

Use a calculator to check your answers after you have completed the entire chapter test.

1. Write $\dfrac{8}{100}$ as a percent.

2. Change 39% to a fraction with a denominator of 100.

3. On an examination of 100 questions, Mike gets 20 incorrect answers. What percent of the problems are incorrect?

Change each of the following percents to decimals.

4.   55%   5. $5\frac{1}{2}\%$ 6.   185%

Change each of the following decimals to percents.

7.   .08   8.   1.35   9. $.00\frac{1}{4}$

Change each of the following percents to mixed numbers or fractions in their simplest forms.

10.   15%   11.   4.5%   12. $4\frac{1}{4}\%$ 13.   150%

Change each of the following fractions or mixed numbers into percents.

14. $\frac{2}{5}$ 15. $\frac{3}{10}$ 16. $\frac{2}{3}$ 17. $3\frac{1}{4}$

18.   Yvette was able to complete 9 problems correctly on an exam consisting of 12 examples. What percent of the problems are correct?

19.   Find 3% of $80.

20.   What percent of 30 is 12?

21.   What is the number if 16% of it is 24?

22.   Mr. Kline saved $48 on a coat which he purchased at a clearance sale. This represented 40% of the original price. What was the original price of the coat?

23.   A recent homecoming dance was held by the local high school. There were 320 graduates in the class but only 112 graduates returned for the homecoming.
a. How many graduates did not attend?
b. What percent of the whole graduating class did attend the homecoming dance?

24.   Julio bought a suit for $227.40 that originally cost $379. By what percent was the suit reduced from the original price?

25.   A baseball player made 19 hits out of 57 times at bat. Express his batting average as a three-place decimal.

Measurement

Which weighs more, 16 oz. of feathers or a pound of steel? Which is larger, 4 ft. or $1\frac{1}{3}$ yd.? Which is the greater amount of time, $\frac{2}{3}$ h. or 40 min.?

To answer these questions you are required to know the measurements associated with weight, length, and time. The four fundamental operations of mathematics can be applied to measurements, too. Check your skill on this topic on the pretest which follows.

Pretest 15–See What You Know and Remember

Work these exercises carefully, doing as many problems as you can. Write your answers in simplified form below each question.

1. 4 da. 6 h.
    +3 da. 10 h.
    _____

2. 2 h. 40 min.
    +1 h. 50 min.
    _____

3. 3 lb. 6 oz.
    +2 lb. 5 oz.
    _____

4. 5 wk. 4 da.
    +3 wk. 5 da.
    _____

5. 5 gal. 2 qt.
    −2 gal. 1 qt.
    _____

6. 52 min. 40 s.
    −38 min. 50 s.
    _____

7. 2 qt.
    −1 qt. 1 pt.
    _____

8. 8 yr. 13 wk. 4 da.
    −2 yr. 40 wk. 5 da.
    _____

9. 5 hr. 30 min.
         ×3
    _____

10. 4 da. 10 min.
          ×4
     _____

11. 5 yd. 2 ft.
          ×4
     _____

12. 3)12 ft. 9 in.
     _____

13. $6\overline{)13 \text{ ft. 6 in.}}$ 14. $8\overline{)17 \text{ yd. 4 in.}}$

15. Find the number of minutes in a day.

16. Find the number of pounds in 2 tons.

17. Mr. Jimenez works 7 h. 15 min. each day for 6 days a week. Find the amount of time he works each week.

18. Mrs. Olaf wishes to divide a piece of lumber 12 ft. 1 in. long into 5 equal parts. What is the length of each part?

19. Mrs. Villaneueva wishes to triple a recipe calling for 1 pt. 2 oz. of milk. How much milk does she require?

20. The radiator in Mr. Richardson's automobile contains 8 gal. of water. He removes 1 gal. 1 pt. of water to make room for anti-freeze. How much water is in the tank?

21. 42 m = _____ cm

22. 265 mm = _____ m

23. 3 km = _____ m

24. 8 m = _____ km

25. 576 cm = _____ mm

Check your answers by turning to the back of the book. Add up all that you had correct.

A Score of	Means That You
23–25	Did very well. You have successfully completed *Arithmetic The Easy Way*.
20–22	Know this material except for a few points. Read the sections about the ones you missed.
17–19	Need to check carefully on the sections you missed.
0–16	Need to work with this chapter to refresh your memory and improve our skills.

15.1 Units of Measurement

Some useful units of measurement for length, time, and weight are listed below. Most tables of measurement are not based on 10 or the decimal system. Most units of measure are arbitrary multiples of others. There are 7 days in a week, 24 hours in a day, and 16 ounces in a pound. There is no logical way of remembering the number of feet in a yard or the number of pounds in a ton. The tables below state the relationship between some of these measures. Notice the abbreviated form associated with each measure.

Measures of Length

1 foot = 12 inches (in.)
1 yard (yd.) = 3 feet (ft.)
1 mile (mi.) = 5,280 feet

Measures of Weight

1 pound = 16 ounces (oz.)
1 ton (T.) = 2000 pounds (lb.)

Liquid Measures

1 pint = 16 fluid ounces (fl. oz.)
1 quart = 2 pints (pt.)
1 gallon (gal.) = 4 quarts (qt.)

Measures of Time

1 minute = 60 seconds (s.)
1 hour = 60 minutes (min.)
1 day = 24 hours (h.)
1 week = 7 days (da.)
1 month = 4 weeks (wk.)
1 year (yr.) = 12 months (mo.)
1 year ≈ 52 weeks (≈ means approximately equal to, since there are 365 days in a year, except leap year, which is 366 days.)

If you know and remember these common measurements and their relationships, you can compute easily with them. If you don't remember them, refer to the tables. You are now ready to compute with measures.

15.2 Adding Measures

EXAMPLE

Cliff found that he needed two pieces of felt to cover a jewelry box. One piece must measure 2 ft. 3 in. long and the other 1 ft. 8 in. long. How much felt did he need in all?

SOLUTION
You must add the two measures to find the total amount.

Problem	Step 1	Step 2
2 ft. 3 in. +1 ft. 8 in.	2 ft. 3 in. +1 ft. 8 in. ——— 11 in.	2 ft. 3 in. +1 ft. 8 in. ——— 3 ft. 11 in.
	Add the number of inches together: 8 + 3 = 11 in.	Add the number of feet together: 2 + 1 = 3 ft.

Thus: 2 ft. 3 in.
       +1 ft. 8 in.
       —————
        3 ft. 11 in.

EXAMPLE

Mrs. Eden bought two chickens; one weighed 4 lb. 8 oz. and the other weighed 3 lb. 12 oz. What was the total weight of her purchase?

SOLUTION

Problem	Step 1	Step 2
4 lb. 8 oz. +3 lb. 12 oz.	4 lb. 8 oz. +3 lb. 12 oz. ——— 20 oz. 1 lb. 4 oz.	4 lb. 8 oz. +3 lb. 12 oz. ——— 20 oz. 1 lb. 4 oz. ——— 8 lb. 4 oz.
	Add the number of ounces together: 8 + 12 = 20 oz. Since 1 lb. = 16 oz., 20 oz. = 1 lb. 4 oz. You have *converted* smaller units to larger ones.	Add the number of pounds together: 4 + 3 + 1 = 8 lb.

Thus: 4 lb. 8 oz.
   +3 lb. 12 oz.
   ———————
   8 lb. 4 oz.

PRACTICE EXERCISE 145

Add. Be careful to convert smaller units to larger ones when necessary.

1. 12 da. 4 h.
   + 6 da. 10 h.

2. 4 ft. 6 in.
   +3 ft. 5 in.

3. 7 wk. 3 da.
   +5 wk. 6 da.

4. 5 yd. 2 ft.
   +2 yd. 1 ft.

5. 2 hr. 43 min. 35 s.
   +   12 min. 46 s.

6. 4 yd. 1 ft. 10 in.
   +2 yd. 1 ft. 4 in.

7. 4 lb. 8 oz.
   2 lb. 1 oz.
   +5 lb. 13 oz.

8. 6 gal. 2 qt. 1 pt.
   +5 gal. 3 qt. 1 pt.

15.3 Subtracting Measures

EXAMPLE

From a board measuring 14 ft. 11 in. long, Juan cuts a piece measuring 6 ft. 3 in. long. How much of the board still remains?

SOLUTION
You must subtract the two measurements to find the remainder.

Problem	Step 1	Step 2
14 ft. 11 in. − 6 ft. 3 in.	14 ft. 11 in. − 6 ft. 3 in. ———— 8 in.	14 ft. 11 in. − 6 ft. 3 in. ———— 8 ft. 8 in.
	Subtract the number of inches: 11 − 3 = 8 in.	Subtract the number of feet: 14 − 6 = 8 ft.

Thus: 14 ft. 11 in.
       − 6 ft. 3 in.
       ————————
        8 ft. 8 in.

EXAMPLE

Carmela's automobile has a gas tank which holds 16 gal. of gasoline. How much gasoline was in the tank if it takes 6 gal. 1 pt. to fill it?

SOLUTION

Problem	Step 1	Step 2
16 gal. – 6 gal. 1 pt.	$\overset{15}{\cancel{16}}$ gal. $\overset{3}{\cancel{4}}$ qt. 2 pt. – 6 gal. 1 pt.	15 gal. 3 qt. 2 pt. – 6 gal. 1 pt. 9 gal. 3 qt. 1 pt.
	Since you can't subtract 1 pt. from 0, you exchange: 1 gal. = 4 qt. and 4 qt. = 3 qt. 2 pt. Thus: 16 gal. = 15 gal. 3 qt. 2 pt.	Subtract like measures.

Thus: 16 gal.
     – 6 gal.   1 pt.
      9 gal. 3 qt. 1 pt.

PRACTICE EXERCISE 146

Subtract. Exchange larger units for smaller units when necessary.

1.   12 da. 10 hr.
   – 6 da.   4 hr.

2.   4 ft. 6 in.
   –3 ft. 5 in.

3.   7 wk. 3 da.
   –5 wk. 6 da.

4.   5 yd. 1 ft.
   –2 yd. 1 ft.

5.   4 yd. 1 ft. 10 in.
   –2 yd. 1 ft.   4 in.

6.   2 hr. 43 min. 35 sec.
   –   12 min. 46 sec.

7.   4 lb.   8 oz.
   –1 lb. 12 oz.

8.   6 gal. 2 qt.
   –5 gal. 3 qt. 1 pt.

15.4 Multiplying Measures

EXAMPLE

Mr. Alvarez wants to enclose a rectangular field with fencing. How much fencing is required if the field's dimensions are 32 yd. 1 ft. by 23 yd. 2 ft.?

SOLUTION

In this example you wish to find the perimeter of a rectangle. Recall that the formula is: $P = 2L + 2W$. You must multiply the length by 2 and the width by 2. Here is how to solve the problem:

Problem	Step 1	Step 2
32 yd. 1 ft. ×2	32 yd. 1 ft. ×2 ———— 2 ft.	32 yd. 1 ft. ×2 ———— 64 yd. 2 ft.
	Multiply the number of feet by 2: $1 \times 2 = 2$ ft.	Multiply the number of yards by 2: $32 \times 2 = 64$ yd.

Thus: 32 yd. 1 ft.
            ×2
      ————————
      64 yd. 2 ft.

and

Problem	Step 1	Step 2	Step 3
23 yd. 2 ft. ×2	23 yd. 2 ft. ×2 ———— 4 ft.	23 yd. 2 ft. ×2 ———— 46 yd. 4 ft.	46 yd 1 yd. 1 ft. ———— 47 yd. 1 ft.
	Multiply the number of feet by 2: $2 \times 2 = 4$ ft.	Multiply the number of yards by 2: $23 \times 2 = 46$ yd.	Convert the smaller units into larger ones: 4 ft. = 1 yd. 1 ft. Add.

Thus: 23 yd. 2 ft.

$\underline{\qquad \times 2 \qquad}$

47 yd. 1 ft.

Therefore: $P = 2L + 2W$

$P = 64$ yd. 2 ft. $+ 47$ yd. 1 ft.

$P = 111$ yd. 3 ft.

$P = 112$ yd. (Remember: 3 ft. = 1 yd.)

PRACTICE EXERCISE 147

Multiply. Be sure to convert smaller units to larger ones when necessary.

1. 4 da. 6 h.
     $\underline{\times 3 \qquad}$

2. 2 h. 40 min.
     $\underline{\times 5 \qquad}$

3. 3 lb. 6 oz.
     $\underline{\times 4 \qquad}$

4. 5 wk. 4 da.
     $\underline{\times 9 \qquad}$

5. 5 gal. 2 qt.
     $\underline{\times 7 \qquad}$

6. 52 min. 40 s.
     $\underline{\times 2 \qquad}$

7. 1 qt. 1 pt.
     $\underline{\times 3 \qquad}$

8. 2 T. 1,500 lb. 10 oz.
     $\underline{\times 2 \qquad}$

15.5 Dividing Measures

EXAMPLE

Mrs. Kelly attends school 3 h. 45 min. each evening. If her 3 classes are of equal time, how long is each class?

SOLUTION

Divide 3 h. 45 min. by 3.

Problem	Step 1	Step 2
$3\overline{)3\text{ h. }45\text{ min.}}$	$\dfrac{1\text{ h.}}{3\overline{)3\text{ h. }45\text{ min.}}}$	$\dfrac{1\text{ h. }15\text{ min.}}{3\overline{)3\text{ h. }45\text{ min.}}}$
	Divide: 3 h. ÷ 3 = 1 h.	Divide: 45 min. ÷ 3 = 15 min.

Thus: $3\overline{)3\text{ h. }45\text{ min.}}$ = 1 h. 15 min.

EXAMPLE

Mr. Svenson cuts a piece of lumber 14 ft. 4 in. into 4 equal pieces. What is the length of each piece?

SOLUTION

Problem	Step 1	Step 2
$4\overline{)14\text{ ft. }\quad 4\text{ in.}}$	3 ft. $4\overline{)14\text{ ft.}\quad\quad 4\text{ in.}}$ $\underline{12\text{ ft.}}$ 2 ft. = $\underline{24\text{ in.}}$ 28 in.	3 ft. 7 in. $4\overline{)14\text{ ft.}\quad\quad 4\text{ in.}}$ $\underline{12\text{ ft.}}$ 28 in. $\underline{28\text{ in.}}$
	Divide: 14 ft. ÷ 4 = 3 ft. R 2 ft. Convert 2 ft. = 24 in. and add to the 4 in.: 24 + 4 = 28 in.	Divide: 28 in. ÷ 4 = 7 in.

Thus: $4\overline{)14\text{ ft. }\quad 4\text{ in.}}$ = 3 ft. 7 in.

PRACTICE EXERCISE 148

Divide. Exchange larger units for smaller units when necessary.

1. 2)‾4 da. 6 h.

2. 3)‾4 h. 15 min.

3. 2)‾3 lb. 2 oz.

4. 4)‾9 wk. 1 da

5. 6)‾14 yd. 2 ft.

6. 9)‾52 min. 3 s.

7. 5)‾3 qt. 1 pt. 3 fl. oz.

8. 7)‾9 gal. 2 qt. 1 pt.

15.6 Word Problems Requiring Measures

Read Chapter 2, section 2.7 to review the general instructions for solving word problems.

PRACTICE EXERCISE 149

1. If 2 lb. 4 oz. of candy are divided among 3 people, how much does each person receive?

2. Apples weighed 3 lb. 2 oz. and pears weighed 1 lb. 6 oz. What was the total weight of Mrs. Wong's purchase?

3. An entire rack of lamb weighed 6 lb. 12 oz. After the butcher cut off some lamb chops weighing 2 lb. 3 oz., how many pounds still remain?

4. A shelf measures 3 ft. 2 in. How long a piece of lumber would be required to make 4 of these shelves?

5. Mrs. Katonah wishes to fill individual containers which hold 1 pt. 2 fl. oz. How large a quantity does she need if she wants to fill 20 containers?

6. The train for Philadelphia leaves New York City at 8 h. 30 min. (8:30 A.M.) and arrives at 10 h. 15 min. (10:15 A.M.). How long is the trip?

7. Mrs. Danzig purchased 4 pieces of ribbon that measured 2 yd. 2 ft. 6 in.; 1 yd. 7 in.; 1 yd. 1 ft. 3 in.; and 2 ft. 11 in. What is the total length of ribbon she purchasd?

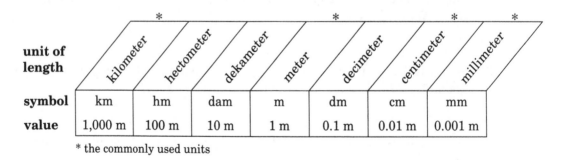

8. Mrs. Haines works 23 hours a week at a part-time job. If she works an equal number of hours for 4 days a week, how many hours and minutes does she work each day?

9. The Cotten family drove their automobile on a three-day trip. How long did they travel if they drove 6 h. 10 min.; 5 h. 30 min.; and 5 h. 40 min. each of the three days?

10. Djerassi won the 35-lb. hammer throw with a record toss of 67 ft. 4 in. By how much did he surpass the previous mark of 66 ft. 10^1/$_2$ in.?

15.7 Converting Metric Measures

The basic unit of length in the metric system is the meter. The metric system is based on 10. Each unit is 10 times the nearest unit to the right and one-tenth of the nearest unit to the left.

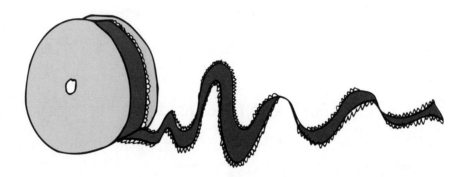

unit of length	kilometer	hectometer	dekameter	meter	decimeter	centimeter	millimeter
symbol	km	hm	dam	m	dm	cm	mm
value	1,000 m	100 m	10 m	1 m	0.1 m	0.01 m	0.001 m

* the commonly used units

To convert from a larger to a smaller metric unit move the decimal point to the right. (Refer to section 12.3 on page 333).

EXAMPLE

7 m = __?__ cm

SOLUTION

Think   1 m = 100 cm

so,   7 m = 7 × 100

= 700 cm

 To convert from a smaller to a larger metric unit move the decimal point to the left. (Refer to section 13.7 on page 356.)

EXAMPLE

123 cm = __?__ m

SOLUTION

Think   1 cm = 0.01 m

so,   123 cm = 123 × 0.001 m

= 1.23 cm

PRACTICE EXERCISE 150

Complete.

1.   5 m = _____ cm

2.   21 cm = _____ mm

3.   9 m = _____ mm

4.   13 km = _____ m

5.   0.03 m = _____ cm

6.   0.6 cm = _____ mm

7.   0.004 m = _____ mm

8.   0.039 km = _____ m

9.   342 cm = _____ m

10.   796 mm = _____ cm

11.   2,500 mm = _____ m   12.   3,800 m = _____ km

13.   82 cm = _____ m   14.   400 mm = _____ cm

15.   53 mm = _____ m   16.   68 m = _____ km

The same prefixes: kilo, hecto, deka, deci, centi, and milli are also used with the basic metric unit of weight, which is gram, and the metric unit of liquid measure, which is the liter.

TERMS YOU SHOULD REMEMBER

Measure   A unit specified by a scale, as an inch.
Convert   To change into another form.

REVIEW OF IMPORTANT IDEAS

Some of the most important ideas in Chapter 15 were:

To add measures—
1. Arrange the same measures in vertical columns and add.
2. Convert smaller units to larger ones if necessary.

To subtract measures—
1. Arrange the same measures in vertical columns and subtract.
2. Exchange larger units for smaller units when you can't subtract.

To multiply measures—
1. Multiply each measure by the multiplier.
2. Convert smaller units to larger ones if necessary.

To divide measures—
1. Divide each measure by the divisor.
2. Convert larger units to smaller units if a remainder is obtained after dividing. Continue the division.

 To convert metric measures—
1. Decide whether going from a smaller to a larger unit of measure, or a larger to smaller unit.
2. Move the decimal point to the left or right.

CHECK WHAT YOU HAVE LEARNED

Good luck on the chapter test. See if you can achieve a satisfactory grade.

A Score of	Means That You
23–25	Did very well. You have successfully completed *E-Z Arithmetic*.
20–22	Know this material except for a few points. Reread the sections about the ones you missed.
17–19	Need to check carefully on the sections you missed.
0–16	Need to review the chapter again to refresh your memory and improve your skills.

Questions	Are Covered in Section
15, 16	15.1
1–4	15.2
5–8	15.3
9–11	15.4
12–14	15.5
17–20	15.6
21–25	15.7

Chapter Test 15

Solve the following measurement problems. Convert answers to larger terms when necessary. Write each answer in the space provided.

1. 5 da. 3 h.
    +4 da. 11 h.
    _____

2. 2 h. 40 min.
    +1 h. 40 min.
    _____

3. 2 T. 150 lb.
    +1 T. 400 lb.
    _____

4. 7 wk. 3 da.
    +2 wk. 5 da.
    _____

5. 6 gal. 3 qt.
    −2 gal. 2 qt.
    _____

6. 4 h. 38 min.
    −2 h. 52 min.
    _____

7. 5 qt.
    −2 qt. 1 pt.
    _____

8. 2 yd. 1 ft. 3 in.
    −1 yd. 2 ft. 9 in.
    _____

9. 11 min. 13 s.
            ×5
    _____

10. 2 pt. 1 fl. oz.
            ×5
    _____

11. 6 yd. 1 ft.
            ×4
    _____

12. $5\overline{)15 \text{ ft. } 10 \text{ in.}}$

13. $7\overline{)15 \text{ ft. } 9 \text{ in.}}$

14. $4\overline{)5 \text{ yd. } 4 \text{ in.}}$

15. Find the number of seconds in an hour.

16. Find the number of fluid ounces in 4 qt.

17. Mr. Alvarado works 8 h. 15 min. each day for 5 days a week. Find the amount of time he works each week.

18. Mr. Penner divided a 12 ft. 11 in. board into 5 equal parts. What was the length of each part?

19. The chef at a dairy restaurant wishes to triple a recipe which calls for 1 pt. 3 oz. of milk. How much milk does he require?

20. After removing 3 qt. 1 pt. of water from a 10-gal. automobile radiator, how much water still remains?

21. 36 m = _____ cm

22. 5,000 m = _____ km

23. 122 cm = _____ mm

24. 6,800 mm = _____ m

25. 14 km = _____ m

Answers to Tests and Exercises

Chapter 1

PRETEST 1, pages 2–3

1. 0, 1, 2, 3, 4, 5, 6, 7, 8, 9

2. ten or 10

3. 卌 I

4. 32

5. a. 1   b. 2   c. 1   d. 3

6. a. ones   b. ones

7. a. hundreds   b. hundred-thousands   c. thousands   d. ten-thousands
   e. millions

8. a. 423   b. 2,035   c. 5,091   d. 18,367   e. 1,002,590

9. a. 10,000 100,000   b. 300,000 3,000,000   c. 60 6   d. 150 15

10. a. VII   d. DCXX   g. 18
    b. XXVIII   e. MCLXIII   h. 256
    c. LXI   f. 2   i. 1,776

PRACTICE EXERCISE 1, page 7

1. 37 = 3 tens and 7 ones

2. 73 = 7 tens and 3 ones

3. 239 = 2 hundreds, 3 tens, and 9 ones

4. 932 = 9 hundreds, 3 tens, and 2 ones

5. 444 = 4 hundreds, 4 tens, and 4 ones

PRACTICE EXERCISE 2, pages 8–9

1. a. 8 ones   b. 7 tens   c. 5 hundreds

2. 3,721 = 3 thousands, 7 hundreds, 2 tens and 1 one

3. 15,526 = 1 ten-thousand, 5 thousands, 5 hundreds, 2 tens, and 6 ones

4. 737,985 = 7 hundred-thousands, 3 ten-thousands, 7 thousands, 9 hundreds, 8 tens, and 5 ones

5. 46   6. 398

7. 4,219   8. 21,437

9. 623,121   10. 53

11. 346   12. three thousand, six hundred ninety-three

13. seven hundred twenty-six

14. four hundred fifty-nine thousand, eight hundred twenty-two

15. eight million, three hundred thirty-three thousand, three hundred thirty-three

16. a. 9,865   b. 5,689

17. a. 10,000 100,000   b. 300,000 3,000,000   c. 10 1   d. 60 6

PRACTICE EXERCISE 3, pages 10–11

1. a. 40   b. 980   c. 601   d. 7,304   e. 5,060

2. a. 2,032   b. 3,060   c. 690   d. 40,310   e. 10,000

PRACTICE EXERCISE 4, page 13

1. a. LXVII   e. CI   h. MXXXI
    b. CV   f. DCCCLXXIII   i. MDXLI
    c. CCCXLV   g. MCMLXXIV   j. MMDCCCLXV
    d. DVIII

2. a. 50 + 10 + 2 = 62   f. 1,000 + 100 + 50 = 1,150
    b. 50 − 10 + 5 − 1 = 44   g. 1,000 − 100 + 50 + 10 + 1 = 961
    c. 100 + 50 − 10 + 10 − 1 = 149 h. 500 − 100 + 10 = 410
    d. 200 + 50 + 10 + 5 + 2 = 267   i. 1,000 + 1,000 − 100 + 50 + 20 + 3 = 1,973
    e. 500 + 100 + 50 − 10 + 1 = 641 j. 2,000 + 1,000 − 100 + 10 − 1 = 2,909

3. MCMXII = 1,000 + 1,000 − 100 + 10 + 2 = 1912

CHAPTER TEST 1, pages 16–17

1. 0, 1, 2, 3, 4, 5, 6, 7, 8, 9   3. IIII   5. a. 2   b. 1

2. ten or 10   4. 22   c. 3   d. 2

6. a. tens   b. thousands   c. ten-thousands   d. ones   e. hundreds

7. a. 38 = *3* tens and *8* ones   b. 3,822 = *3* thousands, *8* hundreds, *2* tens, and *2* ones
    c. 83,213 = *8* ten-thousands, *3* thousands, *2* hundreds, *1* ten, and *3* ones
    d. 8 = *8* ones   e. 991 = *9* hundreds, *9* tens, and *1* one

8. a. 532
    b. 3,035
    c. 5,901
    d. 18,325
    e. 2,002,509

9. a. 16 = XVI
    b. 97 = XCVII
    c. 495 = CDXCV
    d. 1,974 = MCMLXXIV
    e. 10 + 5 + 3 = 18
    f. 50 − 10 = 40
    g. 1,000 + 500 + 100 + 50 + 10 + 5 + 2 = 1,667
    h. 1,000 + 1,000 − 100 + 50 + 20 + 5 − 1 = 1,974

MATHEMATICAL MAZE SOLUTION, page 18

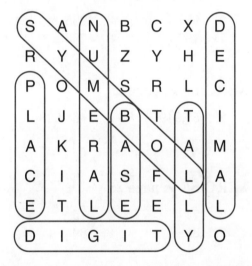

Chapter 2

PRETEST 2A, pages 19–20

1.	7	2.	6	3.	38	4.	67	5.	9
6.	567	7.	289	8.	9	9.	369	10.	66
11.	897	12.	57	13.	895	14.	29,178		

15. a. 4 < 7   b. 6 > 4   c. 7 = 7   d. 15 > 13   e. 1 < 3

PRACTICE EXERCISE 5, page 22

1. 9	2. 2	3. 5	4. 9	5. 4	6. 6
7. 7	8. 7	9. 4	10. 6	11. 8	12. 8
13. 5	14. 6	15. 8	16. 6	17. 9	18. 9
19. 8	20. 5	21. 8	22. 9	23. 7	24. 3
25. 9	26. 9	27. 8	28. 5	29. 6	30. 9
31. 7	32. 3	33. 7	34. 8	35. 4	36. 7

PRACTICE EXERCISE 6, page 23

1. 58	2. 59	3. 49	4. 78	5. 82
6. 98	7. 58	8. 99	9. 89	10. 87

PRACTICE EXERCISE 7, page 24

1. 577	2. 967	3. 985	4. 989	5. 779
6. 856	7. 799	8. 763	9. 629	10. 892

PRACTICE EXERCISE 8, page 25

1. 5,958	2. 58,949	3. 468,668	4. 58,794

5. 8,495,695

PRACTICE EXERCISE 9, page 25

1. 86	2. 74	3. 75	4. 96	5. 98
6. 52	7. 83	8. 89	9. 76	10. 94

PRACTICE EXERCISE 10, page 27

1. $7 > 6$	2. $5 > 2$	3. $3 < 8$	4. $6 = 6$	5. $1 < 2$
6. $10 = 10$	7. $5 < 7$	8. $9 > 3$	9. $8 = 8$	10. $4 < 5$

PRACTICE EXERCISE 11, pages 27–28

1.

2. 6	3. 9	4. 5	5. 6	
6. 12	7. $6, 6 > 5$	8. $7°$	9. $9	

PRACTICE EXERCISE 12, page 29

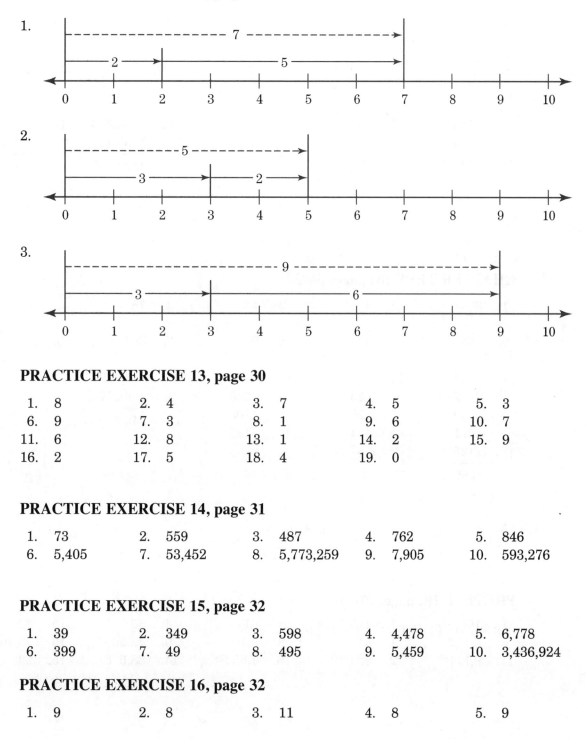

PRACTICE EXERCISE 13, page 30

1. 8	2. 4	3. 7	4. 5	5. 3
6. 9	7. 3	8. 1	9. 6	10. 7
11. 6	12. 8	13. 1	14. 2	15. 9
16. 2	17. 5	18. 4	19. 0	

PRACTICE EXERCISE 14, page 31

1. 73	2. 559	3. 487	4. 762	5. 846
6. 5,405	7. 53,452	8. 5,773,259	9. 7,905	10. 593,276

PRACTICE EXERCISE 15, page 32

1. 39	2. 349	3. 598	4. 4,478	5. 6,778
6. 399	7. 49	8. 495	9. 5,459	10. 3,436,924

PRACTICE EXERCISE 16, page 32

1. 9	2. 8	3. 11	4. 8	5. 9

PRACTICE EXERCISE 17, page 34

1. 2
    3
    +1
    ───
    6

2. 43
    +22
    ───
    65

3. $65
    + 10
    ───
    $75

4. 4
    +3
    ──
    7

5. 4
    3
    +2
    ──
    9

6. 103
    + 46
    ────
    149

7. No. 3
         1
         2
         1
        +2
        ───
         9

8. 41
    23
    +14
    ───
    78

9. $34,324
    + 35,225
    ───────
    $69,549

10. $5,035
    + 1,703
    ───────
    $6,738

CHAPTER TEST 2A, pages 36–37

1. 8   2. 4   3. 46   4. 78   5. 8

6. 768   7. 498   8. 9   9. 389   10. 77

11. 143
       2
      30
     401
    +123
    ────
     699

12. 12
     15
     20
    +21
    ───
     68

13. 222
     233
     403
    +120
    ────
     978

14. 43,622
    + 6,257
    ───────
     49,879

15. a. 6 > 4   b. 8 > 3   c. 1 < 10
     d. 9 = 9   e. 13 < 15

PRETEST 2B, pages 37–38

1. 16   2. 115   3. 61   4. 80   5. 626
6. 6,179   7. 41   8. 9,132   9. 4,000   10. 13,463
11. 9,174   12. $2,079   13. 7,834   14. $8,673   15. 233 ft.

PRACTICE EXERCISE 18, pages 39–40

1. 14	2. 14	3. 11	4. 11	5. 15
6. 15	7. 11	8. 11	9. 13	10. 13
11. 12	12. 18	13. 13	14. 12	15. 14
16. 13	17. 14	18. 15	19. 17	20. 11
21. 16	22. 16	23. 10	24. 10	25. 12
26. 15	27. 12	28. 12	29. 10	30. 16
31. 11	32. 17	33. 13	34. 10	35. 12
36. 10	37. 13	38. 10	39. 11	40. 10
41. 14	42. 10	43. 12	44. 11	45. 10

PRACTICE EXERCISE 19, page 41

1. 124	2. 147	3. 1,259	4. 1,195	5. 1,127
6. 11,949	7. 157,964	8. 1,373,895	9. 12,959,000	10. 1,899

PRACTICE EXERCISE 20, page 42

1. 84	2. 97	3. 84	4. 81	5. 94
6. 77	7. 65	8. 94	9. 55	10. 46
11. 31	12. 62	13. 53	14. 37	15. 62
16. 53	17. 22	18. 88	19. 60	20. 36

PRACTICE EXERCISE 21, page 43

1. 81	2. 1,029	3. 9,139	4. 58,873	5. 649,326
6. 125	7. 426	8. 9,659	9. $41,770	10. 6,170,000

PRACTICE EXERCISE 22, page 44

1. 154	2. 1,213	3. 5,519	4. 91,000	5. 847,110
6. 143	7. 1,231	8. 10,022	9. 60,419	10. 13,821,000

PRACTICE EXERCISE 23, page 45

1. 17	2. 24	3. 15	4. 12	5. 31
6. 25	7. 22	8. 11	9. 21	10. 28

PRACTICE EXERCISE 24, page 47

1. 8,971	2. 1,778	3. 10,481	4. 3,618	5. 14,626
6. 4,085	7. 7,799	8. 11,585	9. 10,945	10. 25,221

PRACTICE EXERCISE 25, pages 48–50

1. 1	2. 6	3. 6	4. 7	5. 13
6. 2	7. 8	8. 11	9. 18	10. 4
11. 6	12. 8	13. 3	14. 7	15. 9
16. 5	17. 9	18. 14	19. 9	20. 13
21. 7	22. 2	23. 13	24. 3	25. 10
26. 9	27. 11	28. 3	29. 12	30. 17
31. 12	32. 6	33. 4	34. 7	35. 14
36. 10	37. 0	38. 15	39. 13	40. 10
41. 10	42. 16	43. 11	44. 12	45. 8
46. 4	47. 8	48. 7	49. 7	50. 15
51. 10	52. 5	53. 11	54. 16	55. 5
56. 16	57. 8	58. 9	59. 15	60. 6
61. 10	62. 3	63. 15	64. 2	65. 10
66. 6	67. 14	68. 11	69. 14	70. 11
71. 12	72. 4	73. 7	74. 12	75. 10
76. 9	77. 9	78. 17	79. 7	80. 12
81. 8	82. 11	83. 7	84. 9	85. 10
86. 13	87. 14	88. 12	89. 5	90. 8
91. 6	92. 5	93. 4	94. 13	95. 5
96. 11	97. 8	98. 9	99. 1	100. 8

PRACTICE EXERCISE 26, pages 52-53

1. 6
   +7
   ―――
   13

2. 2
    6
    9
    1
   +5
   ―――
   23

3. $615
    643
    626
    619
    637
   + 625
   ―――
   $3,765

4. 4,325
    2,786
   +1,938
   ―――
   9,049

5. $2 + $5 + $5 + $3
   + $2 + $1 + $5 +
   $7 + $10 + $5 +
   $3 + $2 + $2 +
   $5 + $1 = $58

6. 178
   + 35
   ―――
   213

7. 3,423
   + 789
   ―――
   4,212

8. 232
    246
    207
    198
   + 163
   ―――
   1,046

9. 3
    2
    4
   + 5
   ―――
   14 h.

10. 162
     146
     87
     243
    +109
    ―――
    747

PRACTICE EXERCISE 27, pages 55–56

1. a. $P = 142'$

    b. $P = 27' + 35' + 53'$
       $P = 115'$

    c. $P = 67' + 67' + 34' + 34'$
       $P = 202'$

    d. $P = 63'' + 63'' + 48'' + 48''$
       $P = 222''$

2. a.

    b. $P = 150'$

3. $P = 164'$

CHAPTER TEST 2B, page 58

1.	13	2.	114	3.	75	4.	70	5.	528
6.	7,438	7.	41	8.	8,643	9.	7,000	10.	15,666

11. 16,991

12.
```
    6
    8
   12
    4
    8
    5
  +18
  ———
   61
```

13.
```
  1,089
      2
    674
     80
 +6,872
 ——————
  8,717
```

14.
```
    286
    340
    197
    432
    263
    307
  + 169
  ——————
  1,994
```

15. $P = 13$ ft. $+ 8$ ft. $+ 13$ ft. $+ 8$ ft.
     $P = 42$ feet

CROSS-NUMBER PUZZLE SOLUTION, page 59

Chapter 3

PRETEST 3, pages 61–62

1. 3	2. 22	3. 222	4. 201	5. 216
6. 93	7. 6	8. 65	9. 154	10. 245
11. 915	12. 18	13. 3,792	14. 11¢	15. $36,489
16. 5 dollars 18 cents		17. 420	18. $108,195	19. 59 ft.

20.

```
      ◄- - - - 3 - - - - ►◄━━━━━━━━ 5 ━━━━━━━━►
      │                  │                         │
      │◄━━━━━━━━━━━━━ 8 ━━━━━━━━━━━━━►│
◄─────┼────┼────┼────┼────┼────┼────┼────┼────┼────┼────►
      0    1    2    3    4    5    6    7    8    9    10
```

PRACTICE EXERCISE 28, pages 66–67

1. 5	2. 0	3. 5	4. 6	5. 5
6. 0	7. 2	8. 1	9. 1	10. 0
11. 1	12. 7	13. 1	14. 3	15. 1
16. 6	17. 0	18. 6	19. 1	20. 0
21. 1	22. 2	23. 4	24. 2	25. 7
26. 2	27. 3	28. 2	29. 2	30. 0
31. 1	32. 5	33. 0	34. 3	35. 3
36. 4	37. 8	38. 3	39. 5	40. 3
41. 6	42. 2	43. 3	44. 7	45. 0

46. 9   47. 4   48. 8   49. 0   50. 4
51. 2   52. 4   53. 1   54. 4   55. 0

PRACTICE EXERCISE 29, pages 67–68

1. 12   2. 35   3. 11   4. 32   5. 63
6. 73   7. 23   8. 57   9. 41   10. 25

PRACTICE EXERCISE 30, pages 68–69

1. 25   2. 325   3. 1,602   4. 2,110   5. 411,080
6. 14   7. 202   8. 397   9. 22,235   10. $1,233,000
11. 225   12. 6,420   13. 227   14. $245   15. 232,055

PRACTICE EXERCISE 31, page 70

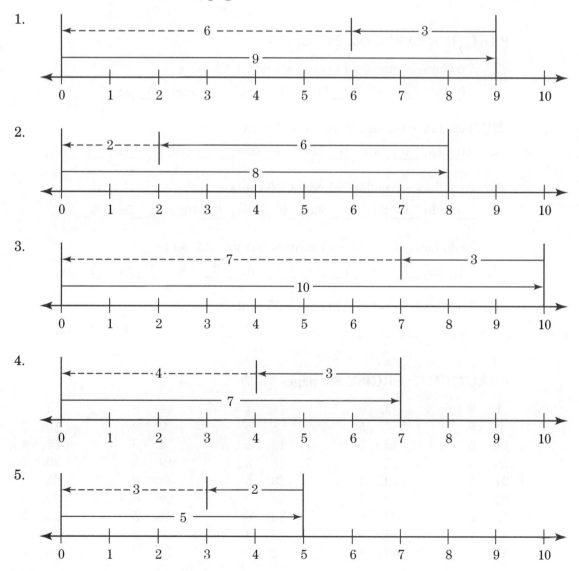

PRACTICE EXERCISE 32, page 71

1.	31	2.	327	3.	514	4.	4,212	5.	6,704
6.	471	7.	5,412	8.	375	9.	45	10.	$420,000

PRACTICE EXERCISE 33, pages 72–73

1.
$$\begin{array}{r} \$48 \\ -\ 45 \\ \hline \$\ 3 \end{array}$$
2.
$$\begin{array}{r} \$173 \\ -\ 151 \\ \hline \$\ 22 \end{array}$$
3.
$$\begin{array}{r} \$29 \\ -\ 21 \\ \hline \$\ 8 \end{array}$$
4.
$$\begin{array}{r} 338{,}400\ \text{lb.} \\ -231{,}200\ \text{lb.} \\ \hline 107{,}200\ \text{lb.} \end{array}$$
5.
$$\begin{array}{r} \$48{,}764 \\ -\ 46{,}560 \\ \hline \$\ 2{,}204 \end{array}$$

6.
$$\begin{array}{r} \$259{,}900 \\ -\ 256{,}000 \\ \hline \$\ 3{,}900 \end{array}$$
7.
$$\begin{array}{r} 45 \\ -31 \\ \hline 14 \end{array}$$
8.
$$\begin{array}{r} \$20{,}995 \\ -\ 20{,}075 \\ \hline \$\ 920 \end{array}$$
9.
$$\begin{array}{r} \$53 \\ -\ 42 \\ \hline \$11 \end{array}$$
10.
$$\begin{array}{r} 43 \\ -22 \\ \hline 21 \end{array}$$

PRACTICE EXERCISE 34, page 74

1. 5 dollars – 3 dollars 74 cents = 1 dollar 26 cents

   __0__ $5; __1__ $1; __0__ 50¢; __1__ 25¢; __0__ 10¢; __0__ 5¢; __1__ 1¢;

2. 2 dollars – 1 dollar 59 cents = 41 cents

   __0__ $5; __0__ $1; __0__ 50¢; __1__ 25¢; __1__ 10¢; __1__ 5¢; __1__ 1¢;

3. 10 dollars – 9 dollars 95 cents = 5 cents

   __0__ $5; __0__ $1; __0__ 50¢; __0__ 25¢; __0__ 10¢; __1__ 5¢; __0__ 1¢;

4. 20 dollars – 11 dollars 43 cents = 8 dollars 57 cents

   __1__ $5; __3__ $1; __1__ 50¢; __0__ 25¢; __0__ 10¢; __1__ 5¢; __2__ 1¢;

5. 7 dollars – 6 dollars 67 cents = 33 cents

   __0__ $5; __0__ $1; __0__ 50¢; __1__ 25¢; __0__ 10¢; __1__ 5¢; __3__ 1¢;

PRACTICE EXERCISE 35, pages 75–76

1. 3	2. 9	3. 4	4. 8	5. 9
6. 5	7. 5	8. 8	9. 9	10. 3
11. 6	12. 7	13. 5	14. 8	15. 4
16. 9	17. 7	18. 2	19. 3	20. 8
21. 1	22. 4	23. 6	24. 9	25. 7
26. 7	27. 5	28. 9	29. 8	30. 2
31. 6	32. 8	33. 6	34. 5	35. 6
36. 7	37. 8	38. 7	39. 9	40. 6
41. 9	42. 8	43. 9	44. 7	45. 4

PRACTICE EXERCISE 36, page 77

1. 58	2. 17	3. 23	4. 16	5. 16
6. 18	7. 25	8. 35	9. 35	10. 39

PRACTICE EXERCISE 37, page 81

1. 13 ones	2. 10 ones	3. 14 ones	4. 14 tens	5. 12 tens
6. 327	7. 948	8. 249	9. 229	10. 111
11. 102	12. 63	13. 734	14. 3,154	15. 629
16. 574	17. 174	18. 4,296	19. 6,961	20. 823

PRACTICE EXERCISE 38, pages 83–84

1.
$$\begin{array}{r} \$50 \\ -\ 38 \\ \hline \$12 \end{array}$$

2.
$$\begin{array}{r} 218 \text{ lb.} \\ -\ 39 \text{ lb.} \\ \hline 179 \text{ lb.} \end{array}$$

3.
$$\begin{array}{r} \$498 \\ -\ 449 \\ \hline \$\ 49 \end{array}$$

4.
$$\begin{array}{r} 7,500 \text{ lb.} \\ -3,600 \text{ lb.} \\ \hline 3,900 \text{ lb.} \end{array}$$

5.
$$\begin{array}{r} \$155,125 \\ -\ 147,750 \\ \hline \$\ \ 7,375 \end{array}$$

6.
$$\begin{array}{r} \$1,804 \\ -\ \ \ \ 84 \\ \hline \$1,719 \end{array}$$

7.
$$\begin{array}{r} 27 \text{ ft.} \\ +32 \text{ ft.} \\ \hline 59 \text{ ft.} \end{array}$$

$$\begin{array}{r} 100 \text{ ft.} \\ -\ 59 \text{ ft.} \\ \hline 41 \text{ ft.} \end{array}$$

8.
$$\begin{array}{r} 510 \text{ mi.} \\ -385 \text{ mi.} \\ \hline 125 \text{ mi.} \end{array}$$

9.
$$\begin{array}{r} 8,634 \\ -\ 796 \\ \hline 7,838 \end{array}$$

10.
$$\begin{array}{r} 240 \\ -229 \\ \hline 11 \end{array}$$

CHAPTER TEST 3, pages 85–86

1. 7	2. 21	3. 202	4. 358	5. 73
6. 322	7. 6	8. 56	9. 154	10. 258
11. 5,735	12. 127	13. 4,691	14.	15.

14.
$$\begin{array}{r} 96¢ \\ -63¢ \\ \hline 33¢ \end{array}$$

15.
$$\begin{array}{r} 3,265 \\ -2,390 \\ \hline 875 \end{array}$$

16. 6 dollars and 44 cents

17.
$$\begin{array}{r} 1,812 \\ -\ 345 \\ \hline 1,467 \end{array}$$

18.
$$\begin{array}{r} 101 \text{ ft.} \\ -\ 66 \text{ ft.} \\ \hline 35 \text{ ft.} \end{array}$$

19.
$$\begin{array}{r} \$254,000 \\ -\ 249,995 \\ \hline \$4,005 \end{array}$$

20.

CROSS-NUMBER PUZZLE SOLUTION, page 87

¹2	3	5			²5	³9		
0	■		⁴2	5	⁵4	■	⁶2	7	⁷5
⁸1	7	⁹6	5	■	¹⁰1	¹¹3	8	■	2
	¹²1	8	■	¹³1	7	■	¹⁴9		
¹⁵3	■	0	■	3	■	¹⁶1	1	¹⁷5
¹⁸8	¹⁹1	8	■	²⁰1	■	²¹8	■	8
	4	■	²²1	6	■	²³5	²⁴2	
²⁵3	■	²⁶1	8	9	■	²⁷7	2	1	²⁸3
²⁹6	3	3	■	³⁰2	5	0	■	9
	³¹7	7				³²4	0	5

Chapter 4

PRETEST 4, pages 89–90

1.	35	2.	68	3.	848	4.	219	5.	2,412
6.	660	7.	420	8.	1,462	9.	2,632	10.	28,458
11.	35,700	12.	21,090	13.	100,000	14.	37,044	15.	558¢

16.

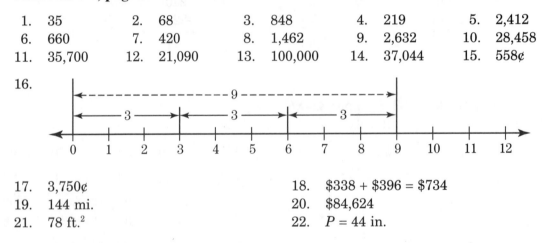

17. 3,750¢   18. $338 + $396 = $734
19. 144 mi.   20. $84,624
21. 78 ft.²   22. $P = 44$ in.

PRACTICE EXERCISE 39, page 94

1.

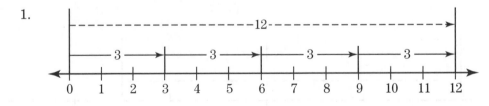

2.

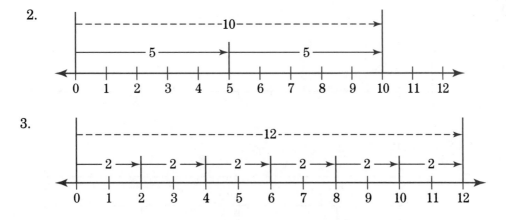

3.

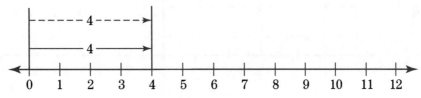

PRACTICE EXERCISE 40, pages 95–96

1. (1×4)

$3 \times (1 \times 4) = 12$

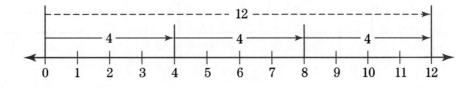

and (3×1)

$(3 \times 1) \times 4 = 12$

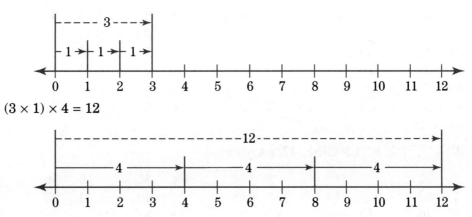

2. (2×2)

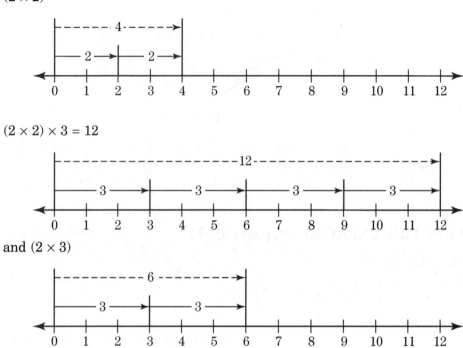

$(2 \times 2) \times 3 = 12$

and (2×3)

$2 \times (2 \times 3) = 12$

PRACTICE EXERCISE 41, pages 97–98

1. 0	2. 0	3. 0	4. 0	5. 0	6. 0
7. 0	8. 0	9. 0	10. 0	11. 2	12. 1
13. 0	14. 4	15. 9	16. 5	17. 0	18. 0
19. 0	20. 0	21. 0	22. 0	23. 3	24. 8
25. 2	26. 9	27. 0	28. 3	29. 5	30. 0
31. 6	32. 4	33. 8	34. 6	35. 7	36. 7

PRACTICE EXERCISE 42, pages 98–99

1. 30	2. 18	3. 14	4. 45	5. 56	6. 24
7. 42	8. 54	9. 48	10. 49	11. 63	12. 63
13. 40	14. 16	15. 30	16. 21	17. 36	18. 20
19. 24	20. 42	21. 21	22. 28	23. 81	24. 64
25. 27	26. 35	27. 40	28. 24	29. 36	30. 27
31. 16	32. 32	33. 36	34. 10	35. 15	36. 56
37. 28	38. 54	39. 25	40. 18	41. 72	42. 15

43. 72	44. 14	45. 48	46. 4	47. 32	48. 6
49. 6	50. 45	51. 8	52. 9	53. 20	54. 10
55. 18	56. 24	57. 16	58. 35	59. 12	60. 12
61. 12	62. 12	63. 18	64. 8		

PRACTICE EXERCISE 43, page 101

| 1. 86 | 2. 48 | 3. 60 | 4. 93 | 5. 48 |
| 6. 88 | 7. 66 | 8. 78 | 9. 70 | 10. 65 |

PRACTICE EXERCISE 44, page 104

1. 495	2. 308	3. 228	4. 174	5. 378
6. 846	7. 387	8. 1,715	9. 2,456	10. 5,411
11. 5,190	12. 5,856	13. 3,780	14. 825	15. 19,112

PRACTICE EXERCISE 45, page 106

1.
$$\begin{array}{r} \$25 \\ \times\ 6 \\ \hline \$150 \end{array}$$

2.
$$\begin{array}{r} 8 \\ \times 6 \\ \hline 48 \end{array}$$

3.
$$\begin{array}{r} \$8 \\ \times 4 \\ \hline \$32 \end{array}$$

4.
$$\begin{array}{r} 18 \\ \times 8 \\ \hline 144 \text{ mi.} \end{array}$$

5.
$$\begin{array}{r} 46 \\ \times\ 9 \\ \hline 414 \end{array}$$

6.
$$\begin{array}{r} 35 \\ \times \$9 \\ \hline \$315 \end{array}$$

7.
$$\begin{array}{r} 1,063 \\ \times\ \ \ 7 \\ \hline 7,441 \end{array}$$

8.
$$\begin{array}{r} 246 \\ \times\ 4 \\ \hline 984 \end{array}$$

9.
$$\begin{array}{r} 163 \\ \times\ 3 \\ \hline 489 \end{array}$$

10.
$$\begin{array}{r} 4,873 \\ \times\ \ \ 4 \\ \hline 19,492 \end{array}$$

PRACTICE EXERCISE 46, page 108

| 1. 140 | 2. 5,600 | 3. 460 | 4. 12,400 | 5. 1,080 |
| 6. 3,570 | 7. 27,540 | 8. 6,000 | 9. 1,230 | 10. 100,000 |

PRACTICE EXERCISE 47, page 110

1. 2,494	2. 1,092	3. 1,960	4. 8,004	5. 2,867
6. 28,220	7. 7,260	8. 42,619	9. 11,180	10. 80,109
11. 39,234	12. 29,250	13. 100,920	14. 676,236	15. 879,296

PRACTICE EXERCISE 48, page 113

| 1. 324 ft.^{2} | 2. 30 yd.^{2} | 3. 252 in.^{2} | 4. $360 |

PRACTICE EXERCISE 49, page 116

1. $P = 3s$
    $P = 3 \times 34$ ft.
    $P = 102$ ft.

2. $P = 2L + 2W$
    $P = 2 \times 15$ ft. $+ 2 \times 12$ ft.
    $P = 30$ ft. $+ 24$ ft.
    $P = 54$ ft.

3. $P = 4s$
    $P = 4 \times s$
    $P = 4 \times 16$ ft.
    $P = 64$ ft.

4. $P = 2L + 2W$
    $P = 2 \times 32$ ft. $+ 2 \times 10$ ft.
    $P = 64$ ft. $+ 20$ ft.
    $P = 84$ ft.

5. $P = 12x$
    $P = 12 \times 17$ in.
    $P = 204$ in.

6. $P = a + a + b$
    $P = 2 \times a + b$
    $P = 2a + b$

7. $P = 2a + b$
    62 ft. $= 2 \times 23$ ft. $+ b$
    62 ft. $= 46$ ft. $+ b$
    16 ft. $= b$

PRACTICE EXERCISE 50, pages 117–118

1. 625
2. 29,340¢
3. 11,100
4. $7,490
5. 17,220
6. $1,021,440
7. 1,584¢
8. 231,800 lb.
9. $1,000
10. $26,100

CHAPTER TEST 4, pages 119–120

1. 48
2. 86
3. 939
4. 128
5. 1,520
6. 880
7. 312
8. 1,638
9. 4,872
10. 34,572
11. 48,800
12. 180,000
13. 43,343

14.

15.
$25
$\times\ \ 50$
$1,250

16.
$168
$\times\ \ \ \ 6$
$1,008

17.
$99
$\times\ \ 4$
$396

$169
$\times\ \ 2$
$338

$396
+338
$734

18.
12
$\times 12$
24
12
144 mi.

19.
54,260
$\times 23$
162 780
1 085 20
1,247,980

20.

26'

45'

$A = L \times W$
$A = 45$ ft. $\times 26$ ft.
$A = 1{,}170$ ft.^{2}

21.
```
      1,723
   ×    391
   ─────────
      1 723
    155 07
    516 9
   ─────────
    673,693
```

22. $P = 6s$
       $P = 6 \times 39$ ft.
       $P = 234$ ft.

CROSS-NUMBER PUZZLE SOLUTION, page 121

Chapter 5

PRETEST 5, pages 123–124

1. 7	2. 23	3. 213	4. 101
5. 52	6. 110	7. 7 R 2	8. 171
9. 206	10. 128	11, $7	12. $6 \div 2 = 3$

13. $8,007   14. 5   15. 21   16. 52
17. 114 R 18   18. mean = 63 in   19. 6 ft.   20. $750
            median = 62 in.
            mode = 62 in.

PRACTICE EXERCISE 51, pages 128–129

1.	1	2.	8	3.	3	4.	3	5.	4
6.	2	7.	1	8.	7	9.	0	10.	1
11.	4	12.	5	13.	2	14.	5	15.	2
16.	0	17.	6	18.	9	19.	6	20.	2
21.	2	22.	6	23.	6	24.	3	25.	3
26.	7	27.	1	28.	8	29.	8	30.	0
31.	7	32.	7	33.	7	34.	4	35.	5
36.	0	37.	7	38.	1	39.	7	40.	4
41.	3	42.	9	43.	0	44.	2	45.	8
46.	8	47.	8	48.	6	49.	4	50.	5
51.	6	52.	2	53.	3	54.	4	55.	0
56.	7	57.	3	58.	8	59.	5	60.	1
61.	9	62.	0	63.	5	64.	6	65.	9
66.	6	67.	9	68.	3	69.	2	70.	5
71.	4	72.	4	73.	9	74.	2	75.	9
76.	0	77.	4	78.	9	79.	9	80.	1
81.	7	82.	1	83.	1	84.	6	85.	3
86.	8	87.	5	88.	8	89.	0	90.	5

PRACTICE EXERCISE 52, pages 130–131

1. $8 \div 2 = 4$

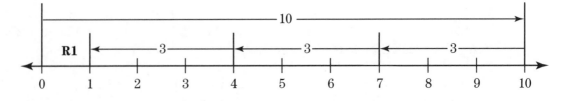

2. $10 \div 3 = 3 \text{ R } 1$

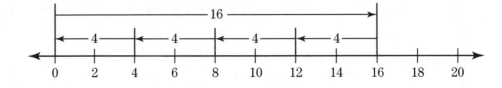

3. $16 \div 4 = 4$

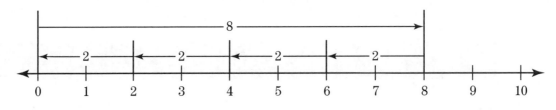

4. $11 \div 5 = 2$ R 1

PRACTICE EXERCISE 53, page 132

1. 31	2. 21	3. 11	4. 43	5. 23
6. 12	7. 11	8. 31	9. 11	10. 12

PRACTICE EXERCISE 54, page 134

1. 223	2. 201	3. 4,321	4. 101	5. 3,112
6. 101	7. 421	8. 2,011	9. 13,101	10. 1,111
11. 10,012	12. 111	13. 1,012	14. 322	15. 434

PRACTICE EXERCISE 55, page 135

1. 61	2. 31	3. 72	4. 51	5. 81
6. 81	7. 31	8. 31	9. 64	10. 81

PRACTICE EXERCISE 56, page 138

1. 104	2. 40	3. 105	4. 106	5. 105
6. 90	7. 90	8. 102	9. 1,305	10. 1,031
11. 1,200	12. 1,060			

PRACTICE EXERCISE 57, page 139

1. 8 R 1	2. 7 R 4	3. 7 R 1	4. 5 R 1	5. 4 R 3
6. 8 R 1	7. 4 R 3	8. 6 R 2	9. 7 R 5	10. 9 R 2
11. 7 R 3	12. 9 R 1	13. 8 R 3	14. 3 R 2	15. 3 R 5

PRACTICE EXERCISE 58, page 142

1. 13	2. 83	3. 19	4. 71	5. 77
6. 29 R 1	7. 651	8. 791	9. 679 R 1	10. 813
11. 778	12. 3,402 R 3			

PRACTICE EXERCISE 59, page 143

1. $\dfrac{12}{6} = 2$

2. $2\overline{)86¢}$ → $43¢$

3. $7\overline{)350}$ → 50 lb.

4. $9\overline{)144}$ → 16

5. $\$12 \div 4 = \$3$

6. $\$210 \div 3 = \$70$

7. $\begin{array}{r} \$12,400 \\ -\quad 400 \\ \hline \$12,000 \end{array}$ $8\overline{)\$12,000}$ → $\$ 1,500$

8. $6\overline{)2,070}$ → 345 lb.

9. $8\overline{)424}$ → 53 gal.

10. $6\overline{)\$312}$ → $\$ 52$

PRACTICE EXERCISE 60, page 147

1. 6
2. 13
3. 39
4. 46
5. 3
6. 3 R 6
7. 26
8. 1,142
9. 91
10. 88
11. 102
12. 72 R 18
13. 18 R 4
14. 57
15. 128 R 64
16. 3

PRACTICE EXERCISE 61, pages 148–149

1. $8\overline{)216}$ → 27

2. $12\overline{)\$37,776}$ → $\$ 3,148$

3. $24\overline{)\$11,280}$ → $\$ 470$

4. $108\overline{)6,804}$ → 63 lb.

5. $16\overline{)192}$ → 12

6. $47\overline{)\$21,855}$ → $\$ 465$

7. $12\overline{)16,464}$ → $1,372$

8. $12\overline{)420}$ → 35 gal.

9. $54\overline{)432}$ → 8

10. $375\overline{)13,875}$ → 37

PRACTICE EXERCISE 62, page 150

1. $\dfrac{132+117+94+73}{4} = \dfrac{416}{4} = 104$ lb.

2. a. $\dfrac{85\%+94\%+64\%+88\%+89\%}{5} = \dfrac{420}{5} = 84\%$

   b. 88%

   c. No

3. a. $\dfrac{13+11+5+11+10+9+4}{7} = \dfrac{63}{7} = 9$ gal.

   b. 10 gal.

   c. Yes. 11 gal.

4. $11\overline{)2{,}552}$ $\overset{232}{}$ lb.

5. $5\overline{)\$365}$ $\overset{\$\,73}{}$

6. $\dfrac{63+61+57+54+43+40}{6}=\dfrac{318}{6}=53$

PRACTICE EXERCISE 63, page 151

1. $L \times W = A$
    $43 \times ? = 1{,}462$
    $? = 34$ ft.

2. $L \times W = A$
    $34 \times ? = 612$
    $? = 18$ ft.

3. $L \times W = A$
    $12 \times ? = 108$
    $? = 9$ in.

CHAPTER TEST 5, page 153

1. $7\overline{)35}$ $\overset{5}{}$

2. $4\overline{)84}$ $\overset{21}{}$

3. $2\overline{)628}$ $\overset{314}{}$

4. $6\overline{)606}$ $\overset{101}{}$

5. $3\overline{)186}$ $\overset{62}{}$

6. $9\overline{)990}$ $\overset{110}{}$

7. $\dfrac{51}{9} = 5 \text{ R } 6$

8. $2\overline{)348}$ $\overset{174}{}$

9. $3\overline{)618}$ $\overset{206}{}$

10. $4\overline{)512}$ $\overset{128}{}$

11. $\$16 \div 8 = \$2$

12. $9 \div 3 = 3$

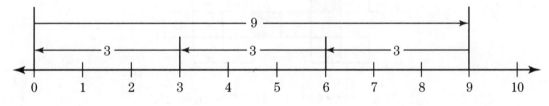

13. $6\overline{)\$54{,}048}$ $\overset{\$\,9{,}008}{}$

14. $19\overline{)76}$ $\overset{4}{}$

17. $32\overline{)4{,}626}$ $\overset{144 \text{ R } 18}{}$
$$\begin{array}{r} 3\,2\downarrow \\ \hline 1\,42 \\ 1\,28\downarrow \\ \hline 146 \\ 128 \\ \hline 18 \end{array}$$

15. $26\overline{)546}$ $\overset{21}{}$
$$\begin{array}{r} 52\downarrow \\ \hline 26 \\ 26 \\ \hline \end{array}$$

16. $42\overline{)2{,}184}$ $\overset{52}{}$
$$\begin{array}{r} 210 \\ \hline 84 \\ 84 \\ \hline \end{array}$$

18.   30 ft., 30 ft., 32 ft., 32 ft., 32 ft., 35 ft., 36 ft., 38 ft., 41 ft.

mean $= \dfrac{30+30+32+32+32+35+36+38+41}{9} = \dfrac{306}{9} = 34$ ft.

median = 30, 30, 32, 32, (32), 35, 36, 38, 41 = 32 ft.

mode = 32 ft.

19. $L \times W = A$

$23 \times ? = 138$

$? = 6$ yd.

20.

$$\begin{array}{r} 3{,}140 \\ \$25{\overline{\smash{\big)}\,\$78{,}500}} \\ \underline{75{\downarrow}} \\ 35 \\ 25{\downarrow} \\ \overline{100} \\ \underline{100{\downarrow}} \\ 0 \\ \underline{0} \end{array}$$

CROSS-NUMBER PUZZLE SOLUTION, page 154
Division of Whole Numbers

CROSS-NUMBER PUZZLE SOLUTION, page 155

All Operations

Chapter 6

PRETEST 6, pages 159–161

1. $\dfrac{4}{6} = \dfrac{2}{3}$

2. $\dfrac{2}{8} = \dfrac{1}{4}$

3. 3 is the numerator
    16 is the denominator

4. a. improper
    b. proper
    c. improper

5. $\dfrac{1}{12}$

6. b. $\begin{array}{r} 5 \\ -2 \\ \hline 3 \end{array}$ $\dfrac{3}{5}$

7. a. $\dfrac{17}{27}$ men

    b. $\begin{array}{r} 27 \\ -17 \\ \hline 10 \end{array}$ $\dfrac{10}{27}$ women

8. $\dfrac{3}{8}$

9. 4

10. $2\dfrac{4}{5}$

11. $\dfrac{10}{3}$

12. a. $\dfrac{10 \div 2}{24 \div 2} = \dfrac{5}{12}$

     b. $\begin{array}{r} 24 \\ -10 \\ \hline 14 \end{array}$ $\dfrac{14 \div 2}{24 \div 2} = \dfrac{7}{12}$

13. L.C.D. = 12

14. $\dfrac{11}{12} > \dfrac{7}{8}$

15. $\dfrac{5}{12}, \dfrac{1}{2}, \dfrac{2}{3}$

PRACTICE EXERCISE 64, pages 164–165

1. a. 3
    b. $\dfrac{1}{3}$
    c. $\dfrac{2}{3}$

2. a. 4
    b. $\dfrac{1}{4}$
    c. $\dfrac{2}{4}$

3. a. 14
    b. $\dfrac{1}{14}$
    c. $\dfrac{6}{14}$

4. a. 9
    b. $\dfrac{1}{9}$
    c. $\dfrac{5}{9}$

5. a. 18
    b. $\dfrac{1}{18}$
    c. $\dfrac{10}{18}$

6. c. $\dfrac{1}{3}$

7. $\dfrac{5}{12}$

8. $\dfrac{4}{7}$

9. b. $8\overline{)5}$

10. b. $\dfrac{6}{9}$

PRACTICE EXERCISE 65, pages 167–168

1. a. improper
   b. proper
   c. proper
   d. improper
   e. improper

2. a. $\dfrac{5}{4}$
   b. improper

3. a. $\dfrac{8}{3}$
   b. improper

4. a. $\dfrac{12}{3}$
   b. improper

5. a. $8\left(\dfrac{8}{8} = \text{entire pie}\right)$
   b. improper

6. a. $\dfrac{7}{12}$
   b. proper

PRACTICE EXERCISE 66, page 171

1. $\dfrac{1}{4}$ 2. $\dfrac{1}{2}$ 3. $\dfrac{1}{5}$ 4. $\dfrac{1}{4}$ 5. $\dfrac{1}{2}$

6. $\dfrac{3}{13}$ 7. $\dfrac{1}{2}$ 8. $\dfrac{4}{9}$ 9. $\dfrac{12}{17}$ 10. $\dfrac{7}{8}$

11. $\dfrac{2}{5}$ 12. $\dfrac{3}{4}$ 13. $\dfrac{1}{2}$ 14. $\dfrac{2}{3}$ 15. $\dfrac{7}{8}$

PRACTICE EXERCISE 67, page 173

1. $\dfrac{1}{2}$ 2. $\dfrac{4}{7}$ 3. $\dfrac{35}{47}$ 4. $\dfrac{5}{7}$ 5. $\dfrac{5}{7}$

6. $\dfrac{1}{2}$ 7. $\dfrac{5}{19}$ 8. $\dfrac{19}{34}$ 9. $\dfrac{3}{5}$ 10. $\dfrac{7}{8}$

PRACTICE EXERCISE 68, page 175

1. $\dfrac{4}{6}$ 2. $\dfrac{14}{35}$ 3. $\dfrac{15}{25}$ 4. $\dfrac{6}{8}$ 5. $\dfrac{3}{24}$

6. $\dfrac{4}{8}$ 7. $\dfrac{8}{36}$ 8. $\dfrac{15}{40}$ 9. $\dfrac{36}{40}$ 10. $\dfrac{15}{18}$

11. $\dfrac{6}{15}$ 12. $\dfrac{8}{12}$ 13. $\dfrac{4}{24}$ 14. $\dfrac{14}{16}$ 15. $\dfrac{12}{16}$

PRACTICE EXERCISE 69, page 178

1. $\dfrac{9}{12}$ 2. $\dfrac{5}{6}$ 3. $\dfrac{2}{3}$ 4. $\dfrac{3}{5}$ 5. $\dfrac{8}{10}$

6. $\dfrac{11}{16}$ 7. $\dfrac{5}{24}$ 8. $\dfrac{7}{18}$

PRACTICE EXERCISE 70, page 180

1. 9   2. 12   3. 10   4. 15   5. 30

6. 15   7. $\frac{1}{2}, \frac{3}{5}, \frac{5}{6}$ 8. $\frac{5}{12}, \frac{4}{5}, \frac{5}{6}$ 9. $\frac{4}{5}, \frac{2}{3}, \frac{1}{2}$ 10. $\frac{3}{10}, \frac{4}{15}, \frac{7}{30}$

PRACTICE EXERCISE 71, page 183

1. $\frac{11}{5}$ 2. $\frac{8}{5}$ 3. $\frac{11}{3}$ 4. $1\frac{3}{4}$ 5. $2\frac{1}{3}$

6. $5\frac{1}{3}$ 7. $2\frac{3}{5}$ 8. 1   9. $6\frac{1}{2}$ 10. $1\frac{1}{3}$

11. $5\frac{1}{8}$ 12. $10\frac{2}{3}$ 13. $4\frac{3}{4}$ 14. $1\frac{2}{3}$ 15. $2\frac{1}{8}$

16. 3   17. $8\frac{8}{9}$ 18. $7\frac{2}{5}$ 19. $21\frac{1}{2}$ 20. $6\frac{2}{5}$

21. $4\frac{1}{2}$ 22. 1   23. $6\frac{1}{3}$ 24. 3   25. $3\frac{2}{3}$

PRACTICE EXERCISE 72, page 186

1. $\frac{37}{9}$ 2. $\frac{7}{2}$ 3. $\frac{33}{5}$ 4. $\frac{30}{5}$ 5. $\frac{35}{4}$

6. $\frac{73}{7}$ 7. $\frac{71}{10}$ 8. $\frac{58}{5}$ 9. $\frac{101}{8}$ 10. $\frac{33}{7}$

11. $\frac{37}{6}$ 12. $\frac{20}{4}$ 13. $\frac{24}{3}$ 14. $\frac{44}{3}$ 15. $\frac{381}{16}$

CHAPTER TEST 6, pages 188–189

1. $\frac{5}{7}$

2. $\frac{6}{8} = \frac{3}{4}$

3. 15 is the numerator
   16 is the denominator

4. a. proper
   b. improper
   c. improper

5. $\frac{1}{3}$

6. $\frac{40}{60} = \frac{40 \div 20}{60 \div 20} = \frac{2}{3}$

7. a. $\frac{15}{25} = \frac{3}{5}$ women
   b. $\begin{array}{r} 25 \\ -15 \\ \hline 10 \end{array}$ $\frac{10}{25} = \frac{2}{5}$ men

8. $\frac{24 \div 8}{32 \div 8} = \frac{3}{4}$

9. 5

10. $\dfrac{23}{8} = 8\overline{)23}\,^{2}\ R\ 7 = 2\dfrac{7}{8}$

11. $4\dfrac{2}{7} = \dfrac{4 \times 7 + 2}{7} = \dfrac{28 + 2}{7}$
$= \dfrac{30}{7}$

12. a. $\dfrac{2 \div 2}{8 \div 2} = \dfrac{1}{4}$

   b. $\begin{array}{r} 8 \\ -2 \\ \hline 6 \end{array}$ $\dfrac{6 \div 2}{8 \div 2} = \dfrac{3}{4}$

13. L.C.D. = 24

14. $\dfrac{3 \times 6}{4 \times 6} = \dfrac{18}{24}$
$\dfrac{19}{24} > \dfrac{18}{24}$ $\dfrac{19}{24} > \dfrac{3}{4}$

15. $\dfrac{7 \times 2}{8 \times 2} = \dfrac{14}{16}$
$\dfrac{3 \times 4}{4 \times 4} = \dfrac{12}{16}$
Thus:
$\dfrac{12}{16} < \dfrac{13}{16} < \dfrac{14}{16}$ or
$\dfrac{3}{4} < \dfrac{13}{16} < \dfrac{7}{8}$

Chapter 7

PRETEST 7, pages 191–193

1. $\dfrac{7}{8}$

2. $\dfrac{2}{6} = \dfrac{1}{3}$

3. $\dfrac{5}{5} = 1$

4. $\dfrac{11}{9} = 1\dfrac{2}{9}$

5. $\dfrac{12}{10} = 1\dfrac{2}{10} = 1\dfrac{1}{5}$

6. $7\dfrac{2}{3}$

7. $11\dfrac{1}{2}$

8. $8\dfrac{2}{3}$

9. $9\dfrac{4}{8} = 9\dfrac{1}{2}$

10. $19\dfrac{6}{6} = 20$

11. $11\dfrac{9}{7} = 12\dfrac{2}{7}$

12. $7\dfrac{24}{16} = 8\dfrac{1}{2}$

13. $\dfrac{3}{8} + \dfrac{1}{8} = \dfrac{4}{8} = \dfrac{1}{2}$

14. $\begin{array}{r} 3\dfrac{1}{4} \\ 3\dfrac{3}{4} \\ +5\dfrac{1}{4} \\ \hline 11\dfrac{5}{4} = 12\dfrac{1}{4}\ \text{lb.} \end{array}$

15. L.C.D. = 12

16. $\dfrac{7}{8}$

17. $\dfrac{21}{24} = \dfrac{7}{8}$

18. $1\dfrac{1}{4}$

19. $1\dfrac{23}{30}$

20. $6\dfrac{1}{2}$

21. $16\dfrac{7}{16}$

22. $P = 23\dfrac{1}{2} + 16\dfrac{3}{4} + 32\dfrac{1}{4} = 20\dfrac{5}{16}$

$P = 92\dfrac{13}{16}$ ft.

23. $4\dfrac{5}{6} + 10\dfrac{1}{2} + 7\dfrac{1}{12} = 22\dfrac{5}{12}$ yd.

PRACTICE EXERCISE 73, page 195

1. $\dfrac{3}{5}$

2. $\dfrac{2}{3}$

3. $\dfrac{3}{4}$

4. $\dfrac{5}{7}$

5. $\dfrac{7}{9}$

6. $\dfrac{9}{11}$

7. $\dfrac{5}{6}$

8. $\dfrac{7}{8}$

9. $\dfrac{11}{12}$

10. $\dfrac{7}{10}$

PRACTICE EXERCISE 74, pages 196–197

1. $\dfrac{3}{9} = \dfrac{1}{3}$

2. $\dfrac{4}{6} = \dfrac{2}{3}$

3. $\dfrac{4}{8} = \dfrac{1}{2}$

4. $\dfrac{5}{10} = \dfrac{1}{2}$

5. $\dfrac{4}{12} = \dfrac{1}{3}$

6. $\dfrac{3}{6} = \dfrac{1}{2}$

7. $\dfrac{8}{10} = \dfrac{4}{5}$

8. $\dfrac{9}{12} = \dfrac{3}{4}$

9. $\dfrac{6}{8} = \dfrac{3}{4}$

10. $\dfrac{6}{9} = \dfrac{2}{3}$

PRACTICE EXERCISE 75, page 198

1. $\dfrac{5}{5} = 1$

2. $\dfrac{2}{2} = 1$

3. $\dfrac{7}{8}$

4. $\dfrac{3}{3} = 1$

5. $\dfrac{6}{6} = 1$

6. $\dfrac{7}{7} = 1$

7. $\dfrac{8}{8} = 1$

8. $\dfrac{9}{10}$

9. $\dfrac{12}{12} = 1$

10. $\dfrac{6}{6} = 1$

PRACTICE EXERCISE 76, page 199

1. $\dfrac{5}{4} = 1\dfrac{1}{4}$

2. $\dfrac{4}{3} = 1\dfrac{1}{3}$

3. $\dfrac{6}{5} = 1\dfrac{1}{5}$

4. $\dfrac{11}{9} = 1\dfrac{2}{9}$

5. $\dfrac{12}{10} = 1\dfrac{2}{10} = 1\dfrac{1}{5}$

6. $\dfrac{12}{8} = 1\dfrac{4}{8} = 1\dfrac{1}{2}$

7. $\dfrac{14}{12} = 1\dfrac{2}{12} = 1\dfrac{1}{6}$

8. $\dfrac{15}{6} = 2\dfrac{3}{6} = 2\dfrac{1}{2}$

9. $\dfrac{9}{7} = 1\dfrac{2}{7}$

10. $\dfrac{6}{3} = 2$

PRACTICE EXERCISE 77, pages 200–201

1. $4\dfrac{3}{5}$

2. $11\dfrac{2}{3}$

3. $6\dfrac{3}{4}$

4. $7\dfrac{5}{7}$

5. $8\dfrac{7}{9}$

6. $13\dfrac{9}{11}$

7. $7\dfrac{2}{3}$

8. $7\dfrac{1}{2}$

PRACTICE EXERCISE 78, page 202

1. $6\dfrac{4}{8} = 6\dfrac{1}{2}$

2. $7\dfrac{4}{6} = 7\dfrac{2}{3}$

3. $23\dfrac{4}{12} = 23\dfrac{1}{3}$

4. $22\dfrac{5}{10} = 22\dfrac{1}{2}$

5. $26\dfrac{8}{10} = 26\dfrac{4}{5}$

6. $12\dfrac{9}{12} = 12\dfrac{3}{4}$

7. $11\dfrac{6}{8} = 11\dfrac{3}{4}$

8. $7\dfrac{6}{9} = 7\dfrac{2}{3}$

PRACTICE EXERCISE 79, page 203

1. $7\dfrac{2}{2} = 8$

2. $6\dfrac{8}{8} = 7$

3. $18\dfrac{3}{3} = 19$

4. $41\dfrac{6}{6} = 42$

5. $9\dfrac{7}{7} = 10$

6. $7\dfrac{8}{8} = 8$

7. $12\dfrac{12}{12} = 13$

8. $11\dfrac{6}{6} = 12$

PRACTICE EXERCISE 80, page 205

1. $7\dfrac{4}{3} = 8\dfrac{1}{3}$

2. $10\dfrac{6}{5} = 11\dfrac{1}{5}$

3. $11\dfrac{11}{9} = 12\dfrac{2}{9}$

4. $12\dfrac{12}{10} = 13\dfrac{1}{5}$

5. $11\dfrac{12}{8} = 12\dfrac{1}{2}$

6. $14\dfrac{14}{12} = 15\dfrac{1}{6}$

7. $7\dfrac{9}{7} = 8\dfrac{2}{7}$

8. $9\dfrac{6}{3} = 11$

PRACTICE EXERCISE 81, pages 206–207

1. $\begin{aligned}\dfrac{1}{4}\\ +\dfrac{3}{4}\\ \hline \dfrac{4}{4} = 1\end{aligned}$

2. $\begin{aligned}\dfrac{2}{7}\\ +\dfrac{3}{7}\\ \hline \dfrac{5}{7}\end{aligned}$

3. $\begin{aligned}4\dfrac{1}{8}\\ +6\dfrac{5}{8}\\ \hline 10\dfrac{6}{8} = 10\dfrac{3}{4}\text{ yd.}\end{aligned}$

4. $\begin{aligned}6\dfrac{1}{4}\\ +3\dfrac{1}{4}\\ \hline 9\dfrac{2}{4} = 9\dfrac{1}{2}\text{ h.}\end{aligned}$

5. $\begin{aligned}4\dfrac{5}{6}\\ 10\dfrac{1}{6}\\ + 7\\ \hline 21\dfrac{6}{6} = 22\text{ yd.}\end{aligned}$

6. $\begin{aligned}\dfrac{6}{20}\\ \dfrac{1}{20}\\ +\dfrac{1}{20}\\ \hline \dfrac{8}{20} = \dfrac{2}{5}\end{aligned}$

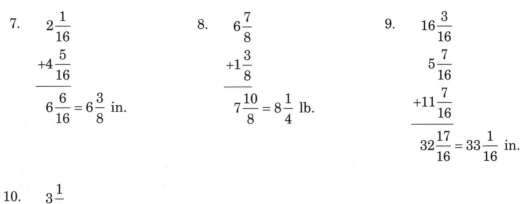

7. $2\dfrac{1}{16}$

    $+4\dfrac{5}{16}$

    $6\dfrac{6}{16} = 6\dfrac{3}{8}$ in.

8. $6\dfrac{7}{8}$

    $+1\dfrac{3}{8}$

    $7\dfrac{10}{8} = 8\dfrac{1}{4}$ lb.

9. $16\dfrac{3}{16}$

    $5\dfrac{7}{16}$

    $+11\dfrac{7}{16}$

    $32\dfrac{17}{16} = 33\dfrac{1}{16}$ in.

10. $3\dfrac{1}{4}$

    $5\dfrac{3}{4}$

    $+1\dfrac{3}{4}$

    $9\dfrac{7}{4} = 10\dfrac{3}{4}$ mi.

PRACTICE EXERCISE 82, page 209

1. $\dfrac{7}{8}$

2. $\dfrac{1}{2}$

3. $\dfrac{17}{21}$

4. $1\dfrac{1}{40}$

5. $1\dfrac{4}{15}$

6. $1\dfrac{1}{8}$

7. $1\dfrac{1}{12}$

8. $\dfrac{11}{12}$

9. $1\dfrac{3}{20}$

10. $1\dfrac{13}{60}$

11. $2\dfrac{1}{8}$

12. $1\dfrac{23}{30}$

PRACTICE EXERCISE 83, page 212

1. $5\dfrac{5}{6}$

2. $11\dfrac{1}{4}$

3. $11\dfrac{3}{20}$

4. $18\dfrac{5}{8}$

5. $8\dfrac{25}{28}$

6. $13\dfrac{3}{16}$

7. $5\dfrac{13}{16}$

8. $4\dfrac{1}{10}$

9. $12\dfrac{7}{24}$

10. $6\dfrac{23}{30}$

11. 15

PRACTICE EXERCISE 84, pages 212–214

1. $3\dfrac{1}{3} = 3\dfrac{2}{6}$

    $+4\dfrac{5}{6} = 4\dfrac{5}{6}$

    $7\dfrac{7}{6} = 8\dfrac{1}{6}$ h.

2. $13\dfrac{3}{10} = 13\dfrac{3}{10}$

    $+ 15\dfrac{1}{2} = 15\dfrac{5}{10}$

    $28\dfrac{8}{10} = 28\dfrac{4}{5}$ mi.

3. $1\dfrac{1}{2} = 1\dfrac{4}{8}$

    $+2\dfrac{5}{8} = 2\dfrac{5}{8}$

    $3\dfrac{9}{8} = 4\dfrac{1}{8}$ yd.

4.
$$8\frac{7}{8} = 8\frac{7}{8}$$
$$6\frac{1}{4} = 6\frac{2}{8}$$
$$+7\frac{1}{2} = 7\frac{4}{8}$$
$$\rule{2cm}{0.4pt}$$
$$21\frac{13}{8} = 22\frac{5}{8} \text{ in.}$$

5.
$$3\frac{2}{3} = 3\frac{8}{12}$$
$$7\frac{3}{4} = 7\frac{9}{12}$$
$$3 = 3$$
$$+6\frac{1}{2} = 6\frac{6}{12}$$
$$\rule{2cm}{0.4pt}$$
$$19\frac{23}{12} = 20\frac{11}{12}$$

6.
$$8\frac{5}{8} = 8\frac{10}{16}$$
$$10\frac{1}{2} = 10\frac{8}{16}$$
$$7\frac{3}{16} = 7\frac{3}{16}$$
$$+\quad 9 = 9$$
$$\rule{2cm}{0.4pt}$$
$$34\frac{21}{16} = 35\frac{5}{16} \text{ in.}$$

7.
$$\frac{1}{2} = \frac{8}{16}$$
$$+1\frac{9}{16} = 1\frac{9}{16}$$
$$\rule{2cm}{0.4pt}$$
$$1\frac{17}{16} = 2\frac{1}{16} \text{ in.}$$

8.
$$3\frac{3}{8} = 3\frac{6}{16}$$
$$5\frac{1}{2} = 5\frac{8}{16}$$
$$+2\frac{9}{16} = 2\frac{9}{16}$$
$$\rule{2cm}{0.4pt}$$
$$10\frac{23}{16} = 11\frac{7}{16} \text{ in.}$$

9.
$$6\frac{1}{2} = 6\frac{4}{8}$$
$$8\frac{5}{8} = 8\frac{5}{8}$$
$$+\quad 2 = 2$$
$$\rule{2cm}{0.4pt}$$
$$16\frac{9}{8} = 17\frac{1}{8} \text{ lb.}$$

10.
$$73\frac{5}{6} = 73\frac{10}{12}$$
$$63\frac{1}{2} = 63\frac{6}{12}$$
$$+69\frac{1}{12} = 69\frac{1}{12}$$
$$\rule{2cm}{0.4pt}$$
$$205\frac{17}{12} = 206\frac{5}{12} \text{ ft.}$$

CHAPTER TEST 7, pages 215–217

1.
$$\frac{3}{5}$$
$$+\frac{1}{5}$$
$$\rule{1.5cm}{0.4pt}$$
$$\frac{4}{5}$$

2.
$$\frac{1}{8}$$
$$+\frac{1}{8}$$
$$\rule{1.5cm}{0.4pt}$$
$$\frac{2}{8} = \frac{1}{4}$$

3.
$$\frac{4}{7}$$
$$+\frac{3}{7}$$
$$\rule{1.5cm}{0.4pt}$$
$$\frac{7}{7} = 1$$

4.
$$\frac{7}{11}$$
$$+\frac{5}{11}$$
$$\rule{1.5cm}{0.4pt}$$
$$\frac{12}{11} = 1\frac{1}{11}$$

5.
$$\frac{3}{4}$$
$$+\frac{3}{4}$$
$$\rule{1.5cm}{0.4pt}$$
$$\frac{6}{4} = 1\frac{2}{4} = 1\frac{1}{2}$$

6.
$$5\frac{1}{6}$$
$$+2$$
$$\rule{1.5cm}{0.4pt}$$
$$7\frac{1}{6}$$

10.
$$11\frac{1}{8}$$
$$+\ 6\frac{7}{8}$$
$$17\frac{8}{8} = 17 + 1$$

11.
$$7\frac{3}{5}$$
$$+3\frac{3}{5}$$
$$10\frac{6}{5} = 10 + 1\frac{1}{5}$$
$$= 11\frac{1}{5}$$

12.
$$9\frac{7}{12}$$
$$+\ \frac{11}{12}$$
$$9\frac{18}{12} = 9 + 1\frac{6}{12}$$
$$= 9 + 1\frac{1}{2}$$
$$= 10\frac{1}{2}$$

13.
$$\frac{5}{16}$$
$$+\frac{3}{16}$$
$$\frac{8}{16} = \frac{1}{2}\ \text{h.}$$

14.
$$14\frac{3}{4}$$
$$9\frac{1}{4}$$
$$+12\frac{3}{4}$$
$$35\frac{7}{4} = 35 + 1\frac{3}{4}$$
$$= 36\frac{3}{4}$$

15. L.C.D. = 24

16.
$$\frac{1}{8} = \frac{2}{16}$$
$$+\frac{5}{16} = \frac{5}{16}$$
$$\frac{7}{16}$$

17.
$$\frac{2}{3} = \frac{8}{12}$$
$$+\frac{7}{12} = \frac{7}{12}$$
$$\frac{15}{12} = 1\frac{3}{12}$$
$$= 1\frac{1}{4}$$

18.
$$\frac{3}{4} = \frac{6}{8}$$
$$+\frac{5}{8} = \frac{5}{8}$$
$$\frac{11}{8} = 1\frac{3}{8}$$

19.
$$\frac{2}{5} = \frac{12}{30}$$
$$\frac{1}{2} = \frac{15}{30}$$
$$+\frac{1}{3} = \frac{10}{30}$$
$$\frac{37}{30} = 1\frac{7}{30}$$

20.
$$4\frac{5}{6} = \frac{5}{6}$$
$$+3\frac{2}{3} = \frac{4}{6}$$
$$7\quad \frac{9}{6} = 7 + 1\frac{3}{6}$$
$$= 7 + 1\frac{1}{2}$$
$$= 8\frac{1}{2}$$

21.
$$6\frac{5}{12} = \frac{5}{12}$$
$$7\frac{1}{2} = \frac{6}{12}$$
$$+5\frac{2}{3} = \frac{8}{12}$$
$$18\quad \frac{19}{12} = 18 + 1\frac{7}{12}$$
$$= 19\frac{7}{12}$$

22. $P = 14\dfrac{3}{8} + 15\dfrac{3}{4} + 21\dfrac{1}{2} + 9\dfrac{1}{16}$

$$14\dfrac{3}{8} = \dfrac{6}{16}$$

$$15\dfrac{3}{4} = \dfrac{12}{16}$$

$$21\dfrac{1}{2} = \dfrac{8}{16}$$

$$+\,9\dfrac{1}{16} = \dfrac{1}{16}$$

$$59 \quad \dfrac{27}{16} = 59 + 1\dfrac{1}{16}$$

$$= 60\dfrac{11}{16} \text{ yd.}$$

14. $3\dfrac{3}{4} = \dfrac{9}{12}$

$$5\dfrac{1}{2} = \dfrac{6}{12}$$

$$+\,1\dfrac{5}{12} = \dfrac{5}{12}$$

$$9 \qquad \dfrac{20}{12} = 9 + 1\dfrac{8}{12}$$

$$= 9 + 1\dfrac{2}{3}$$

$$= 10\dfrac{2}{3} \text{ ft.}$$

Chapter 8

PRETEST 8, pages 219–220

1. $\dfrac{1}{3}$

2. $\dfrac{2}{4} = \dfrac{1}{2}$

3. $\dfrac{8}{8} = 1$

4. $\dfrac{20}{16} = 1\dfrac{4}{16} = 1\dfrac{1}{4}$

5. $1\dfrac{2}{5}$

6. $1\dfrac{2}{8} = 1\dfrac{1}{4}$

7. $8\dfrac{1}{7}$

8. $\dfrac{3}{5}$

9. $4\dfrac{4}{5}$

10. $1\dfrac{2}{6} = 1\dfrac{1}{3}$

11. $9\dfrac{1}{2}$

12. $4\dfrac{2}{3}$

13. $2\dfrac{1}{4} = 1\dfrac{5}{4}$

$$-1\dfrac{3}{4} = 1\dfrac{3}{4}$$

$$\dfrac{2}{4} = \dfrac{1}{2} \text{ lb.}$$

14. $\dfrac{3}{4}$

$$-\dfrac{1}{4}$$

$$\dfrac{2}{4} = \dfrac{1}{2} \text{ h.}$$

15. $\dfrac{3}{6} = \dfrac{1}{2}$

16. $\dfrac{3}{10}$

17. $1\dfrac{4}{15}$

18. $2\dfrac{3}{4}$

19. $8\dfrac{21}{40}$

20. $31\dfrac{1}{4} = 31\dfrac{3}{12} = 30\dfrac{15}{12}$

$\quad -\ 6\dfrac{2}{3} =\ 6\dfrac{8}{12} =\ 6\dfrac{8}{12}$

$\qquad\qquad\qquad\qquad 24\dfrac{7}{12}$ h.

21. $2\dfrac{1}{2} = 2\dfrac{5}{10}$

$\ +3\dfrac{3}{5} = 3\dfrac{6}{10}$

$\qquad\qquad 5\dfrac{11}{10}$

$\quad 5\dfrac{11}{10} = 5\dfrac{22}{20}$

$\ -1\dfrac{1}{4}\ = 1\dfrac{5}{20}$

$\qquad\qquad 4\dfrac{17}{20}$

22. $2\dfrac{1}{2} = 2\dfrac{2}{4}$

$\ +6\dfrac{3}{4} = 6\dfrac{3}{4}$

$\qquad\qquad 8\dfrac{5}{4}$

$\quad 8\dfrac{5}{4} = 8\dfrac{10}{8}$

$\ -3\dfrac{5}{8} =\ 3\dfrac{5}{8}$

$\qquad\qquad 5\dfrac{5}{8}$ yd.

23. $24\dfrac{7}{8} = 24\dfrac{7}{8}$

$\ -12\dfrac{3}{4} = 12\dfrac{6}{8}$

$\qquad\qquad 12\dfrac{1}{8}$ yd.

PRACTICE EXERCISE 85, page 223

1. $\dfrac{1}{5}$ 2. $\dfrac{1}{7}$ 3. $\dfrac{3}{11}$ 4. $\dfrac{1}{3}$ 5. $\dfrac{3}{8}$

6. $\dfrac{1}{4}$ 7. $\dfrac{7}{12}$ 8. $\dfrac{7}{10}$ 9. $\dfrac{5}{6}$ 10. $\dfrac{0}{9} = 0$

PRACTICE EXERCISE 86, page 224

1. $\dfrac{4}{6} = \dfrac{2}{3}$ 2. $\dfrac{6}{8} = \dfrac{3}{4}$ 3. $\dfrac{4}{12} = \dfrac{1}{3}$

4. $\dfrac{8}{20} = \dfrac{2}{5}$ 5. $\dfrac{2}{4} = \dfrac{1}{2}$ 6. $\dfrac{6}{10} = \dfrac{3}{5}$

7. $\dfrac{6}{16} = \dfrac{3}{8}$ 8. $\dfrac{4}{32} = \dfrac{1}{8}$ 9. $\dfrac{2}{14} = \dfrac{1}{7}$

PRACTICE EXERCISE 87, page 226

1. $\dfrac{5}{5} = 1$ 2. $\dfrac{4}{3} = 1\dfrac{1}{3}$ 3. $\dfrac{6}{4} = 1\dfrac{2}{4} = 1\dfrac{1}{2}$

4. $\dfrac{11}{10} = 1\dfrac{1}{10}$ 5. $\dfrac{16}{8} = 2$ 6. $\dfrac{9}{7} = 1\dfrac{2}{7}$

7. $\dfrac{14}{12} = 1\dfrac{2}{12} = 1\dfrac{1}{6}$ 8. $\dfrac{20}{16} = 1\dfrac{4}{16} = 1\dfrac{1}{4}$ 9. $\dfrac{32}{32} = 1$

PRACTICE EXERCISE 88, page 228

1. $4\dfrac{2}{7}$ 2. $5\dfrac{2}{4} = 5\dfrac{1}{2}$ 3. $9\dfrac{4}{9}$ 4. $5\dfrac{1}{2}$

5. $3\dfrac{4}{6} = 3\dfrac{2}{3}$ 6. $5\dfrac{4}{16} = 5\dfrac{1}{4}$ 7. $7\dfrac{1}{3}$ 8. $11\dfrac{6}{12} = 11\dfrac{1}{2}$

9. $12\dfrac{5}{6}$ 10. $4\dfrac{8}{16} = 4\dfrac{1}{2}$

PRACTICE EXERCISE 89, page 230

1. $\dfrac{1}{8}$ 2. $\dfrac{3}{5}$ 3. $2\dfrac{1}{4}$ 4. $3\dfrac{1}{2}$

5. $5\dfrac{5}{12}$ 6. $\dfrac{13}{16}$ 7. $1\dfrac{3}{8}$ 8. $7\dfrac{5}{6}$

9. $14\dfrac{4}{7}$ 10. $14\dfrac{1}{3}$ 11. $\dfrac{27}{32}$ 12. $27\dfrac{5}{12}$

PRACTICE EXERCISE 90, page 233

1. $\dfrac{2}{3}$ 2. $5\dfrac{3}{5}$ 3. $3\dfrac{4}{8} = 3\dfrac{1}{2}$

4. $8\dfrac{2}{4} = 8\dfrac{1}{2}$ 5. $5\dfrac{4}{7}$ 6. $23\dfrac{2}{10} = 23\dfrac{1}{5}$

7. $7\dfrac{6}{16} = 7\dfrac{3}{8}$ 8. $4\dfrac{12}{15} = 4\dfrac{4}{5}$ 9. $44\dfrac{18}{32} = 44\dfrac{9}{16}$

PRACTICE EXERCISE 91, pages 234–235

1. $\begin{array}{r} 1 \ \ = \dfrac{8}{8} \\ -\dfrac{3}{8} = \dfrac{3}{8} \\ \hline \dfrac{5}{8} \end{array}$

2. $\begin{array}{r} 4 \ \ = 3\dfrac{2}{2} \\ -1\dfrac{1}{2} = 1\dfrac{1}{2} \\ \hline 2\dfrac{1}{2}\ \text{h.} \end{array}$

3. $\begin{array}{r} 114\dfrac{3}{4} \\ -106\dfrac{1}{4} \\ \hline 8\dfrac{2}{4} = 8\dfrac{1}{2}\ \text{lb.} \end{array}$

4. $\begin{array}{r} 2\dfrac{3}{8} = 1\dfrac{11}{8} \\ -1\dfrac{7}{8} = 1\dfrac{7}{8} \\ \hline \dfrac{4}{8} = \dfrac{1}{2}\ \text{in.} \end{array}$

5. $\begin{array}{r} 3\dfrac{5}{6} \\ -2 \\ \hline 1\dfrac{5}{6}\ \text{h.} \end{array}$

6. $\begin{array}{r} 12\dfrac{3}{10} \\ -\ 7\dfrac{1}{10} \\ \hline 5\dfrac{2}{10} = 5\dfrac{1}{5}\ \text{mi.} \end{array}$

7.
$$\frac{2}{8}$$
$$+\frac{1}{8}$$
$$\frac{3}{8}$$

$$1 = \frac{8}{8}$$
$$-\frac{3}{8} = \frac{3}{8}$$
$$\frac{5}{8}$$

8.
$$4 = 3\frac{6}{6}$$
$$-1\frac{5}{6} = 1\frac{5}{6}$$
$$2\frac{1}{6} \text{ h.}$$

9.
$$58 = 57\frac{2}{2}$$
$$-39\frac{1}{2} = 39\frac{1}{2}$$
$$18\frac{1}{2} \text{ in.}$$

10.
$$4\frac{11}{16}$$
$$5\frac{3}{16}$$
$$+ 3\frac{5}{16}$$
$$12\frac{19}{16} = 13\frac{3}{16}$$

$$16 = 15\frac{16}{16}$$
$$-13\frac{3}{16} = 13\frac{3}{16}$$
$$2\frac{13}{16} \text{ ft.}$$

PRACTICE EXERCISE 92, pages 237–238

1. $\frac{1}{10}$ 2. $\frac{1}{4}$ 3. $\frac{7}{12}$ 4. $\frac{1}{12}$ 5. $\frac{1}{8}$

6. $\frac{3}{20}$ 7. $\frac{1}{6}$ 8. $\frac{3}{10}$ 9. $\frac{5}{16}$ 10. $\frac{11}{16}$

PRACTICE EXERCISE 93, page 239

1. $2\frac{7}{16}$ 2. $5\frac{11}{24}$ 3. $6\frac{9}{20}$

4. $4\frac{1}{6}$ 5. $2\frac{1}{8}$ 6. $3\frac{11}{40}$

7. $1\frac{3}{12} = 1\frac{1}{4}$ 8. $14\frac{1}{16}$ 9. $8\frac{1}{20}$

PRACTICE EXERCISE 94, pages 241–242

1. $2\frac{3}{4}$ 2. $4\frac{5}{6}$ 3. $5\frac{7}{10}$

4. $\frac{23}{30}$ 5. $\frac{13}{21}$ 6. $3\frac{7}{16}$

7. $1\frac{9}{10}$ 8. $4\frac{5}{6}$ 9. $\frac{17}{24}$

Answers to Tests and Exercises

PRACTICE EXERCISE 95, pages 242–243

1. $\dfrac{7}{8} = \dfrac{7}{8}$

   $-\dfrac{1}{2} = \dfrac{4}{8}$

   $\dfrac{3}{8}$ lb.

2. $24\dfrac{1}{8} = 24\dfrac{1}{8} = 23\dfrac{9}{8}$

   $-12\dfrac{3}{4} = 12\dfrac{6}{8} = 12\dfrac{6}{8}$

   $11\dfrac{3}{8}$ yd.

3. $47\dfrac{7}{8} = 47\dfrac{7}{8}$

   $-12\dfrac{3}{4} = 12\dfrac{6}{8}$

   $35\dfrac{1}{8}$ tons

4. $\dfrac{5}{8} = \dfrac{25}{40}$

   $-\dfrac{3}{5} = \dfrac{24}{40}$

   $\dfrac{1}{40}$

5. $\dfrac{2}{3} = \dfrac{32}{48}$

   $-\dfrac{9}{16} = \dfrac{27}{48}$

   $\dfrac{5}{48}$ mi.

6. $3\dfrac{1}{3} = 3\dfrac{2}{6} = 2\dfrac{8}{6}$

   $-1\dfrac{5}{6} \qquad = 1\dfrac{5}{6}$

   $1\dfrac{3}{6} = 1\dfrac{1}{2}$ ft.

7. $\quad 6 \quad = \quad 6$

   $5\dfrac{5}{8} = \quad 5\dfrac{10}{16}$

   $2\dfrac{1}{2} = \quad 2\dfrac{8}{16}$

   $+1\dfrac{11}{16} = \quad 1\dfrac{11}{16}$

   $14\dfrac{29}{16} = 14 + 1\dfrac{13}{16} = 15\dfrac{13}{16}$

   $16 \quad = 15\dfrac{16}{16}$

   $-15\dfrac{13}{16} = 15\dfrac{13}{16}$

   $\dfrac{3}{16}$ ft.

8. $19\dfrac{3}{16} = 19\dfrac{6}{32}$

   $+20\dfrac{11}{32} = 20\dfrac{11}{32}$

   $39\dfrac{17}{32}$

   $58\dfrac{1}{8} = 58\dfrac{4}{32} = 57\dfrac{36}{32}$

   $-39\dfrac{17}{32} \qquad = 39\dfrac{17}{32}$

   $18\dfrac{19}{32}$ ft.

9. $13\dfrac{1}{2} = 13\dfrac{4}{8}$

   $13\dfrac{1}{2} = 13\dfrac{4}{8}$

   $14\dfrac{7}{8} = 14\dfrac{7}{8}$

   $+14\dfrac{7}{8} = 14\dfrac{7}{8}$

   $54\dfrac{22}{8} = 54 + 2\dfrac{6}{8} = 56\dfrac{6}{8} = 56\dfrac{3}{4}$

   $64 \quad = 63\dfrac{4}{4}$

   $-56\dfrac{3}{4} = 56\dfrac{3}{4}$

   $7\dfrac{1}{4}$ ft.

10. $45\dfrac{1}{8} = 45\dfrac{1}{8}$ $\qquad\qquad\qquad$ $231 \ = 230\dfrac{4}{4}$

$\quad\ \ 49\dfrac{1}{2} = 49\dfrac{4}{8}$ $\qquad\qquad\qquad$ $\underline{-203\dfrac{1}{4} = 203\dfrac{1}{4}}$

$\quad\ \ 32\dfrac{7}{8} = 32\dfrac{7}{8}$ $\qquad\qquad\qquad\qquad\qquad 27\dfrac{3}{4}$ lb.

$\underline{+75\dfrac{3}{4} = \ 75\dfrac{6}{8}}$

$\qquad 201\dfrac{18}{8} = 201 + 2\dfrac{2}{8} = 203\dfrac{2}{8} = 203\dfrac{1}{4}$

CHAPTER TEST 8, pages 245–246

1. $\dfrac{3}{5}$ $\qquad\qquad$ 2. $\dfrac{7}{10}$ $\qquad\qquad$ 3. $\dfrac{4}{3}$

$\underline{-\dfrac{1}{5}}$ $\qquad\qquad\qquad$ $\underline{-\dfrac{1}{10}}$ $\qquad\qquad\qquad$ $\underline{-\dfrac{1}{3}}$

$\ \ \dfrac{2}{5}$ $\qquad\qquad\qquad$ $\ \ \dfrac{6}{10} = \dfrac{3}{5}$ $\qquad\qquad$ $\ \ \dfrac{3}{3} = 1$

4. $\dfrac{13}{8}$ $\qquad\qquad$ 5. $5\dfrac{2}{3}$ $\qquad\qquad$ 6. $7\dfrac{3}{4}$

$\underline{-\dfrac{3}{8}}$ $\qquad\qquad\qquad$ $\underline{-3\dfrac{1}{3}}$ $\qquad\qquad\qquad$ $\underline{-6\dfrac{1}{4}}$

$\ \ \dfrac{10}{8} = 1\dfrac{2}{8} = 1\dfrac{1}{4}$ $\quad$ $\ \ 2\dfrac{1}{3}$ $\qquad\qquad$ $\ \ 1\dfrac{2}{4} = 1\dfrac{1}{2}$

7. $9\dfrac{5}{16}$ $\qquad\qquad$ 8. $1 \ = \dfrac{8}{8}$ $\qquad\qquad$ 9. $6\dfrac{1}{3} = 5 + 1 + \dfrac{1}{3} = 5\dfrac{4}{3}$

$\underline{-\ \dfrac{4}{16}}$ $\qquad\qquad$ $\underline{-\ \dfrac{3}{8} = \dfrac{3}{8}}$ $\qquad\qquad$ $\underline{-\ \dfrac{2}{3} \qquad\qquad = \ \dfrac{2}{3}}$

$\ \ 9\dfrac{1}{16}$ $\qquad\qquad\qquad$ $\ \ \dfrac{5}{8}$ $\qquad\qquad\qquad\qquad\qquad\qquad 5\dfrac{2}{3}$

10. $7\dfrac{1}{8} = 6 + 1 + \dfrac{1}{8} = 6\dfrac{9}{8}$ $\quad$ 11. $14\dfrac{4}{7}$ $\qquad$ 12. $8 \ = 7 + 1 = 7\dfrac{5}{5}$

$\underline{-3\dfrac{5}{8} \qquad\qquad\quad = 3\dfrac{5}{8}}$ $\qquad$ $\underline{-\ 3}$ $\qquad\qquad$ $\underline{-5\dfrac{3}{5} \qquad\quad = 5\dfrac{3}{5}}$

$\qquad\qquad\qquad 3\dfrac{4}{8} = 3\dfrac{1}{2}$ $\qquad$ $11\dfrac{4}{7}$ $\qquad\qquad\qquad\qquad 2\dfrac{2}{5}$

13.
$$2\frac{1}{4} = 1 + 1 + \frac{1}{4} = 1\frac{5}{4}$$
$$-1\frac{3}{4} \qquad\quad = 1\frac{3}{4}$$
$$\overline{\qquad\qquad\qquad \frac{2}{4} = \frac{1}{2} \text{ lb.}}$$

14.
$$\frac{7}{8}$$
$$-\frac{1}{8}$$
$$\overline{\frac{6}{8} = \frac{3}{4} \text{ in.}}$$

15.
$$\frac{11}{12} = \frac{11}{12}$$
$$-\frac{3}{4} = \frac{9}{12}$$
$$\overline{\frac{2}{12} = \frac{1}{6}}$$

16.
$$\frac{7}{9} = \frac{14}{18}$$
$$-\frac{1}{2} = \frac{9}{18}$$
$$\overline{\frac{5}{18}}$$

17.
$$8\frac{5}{8} = 8\frac{25}{40}$$
$$-7\frac{2}{5} - 7\frac{16}{40}$$
$$\overline{1\frac{9}{40}}$$

18.
$$7\frac{1}{2} = 7\frac{2}{4} = 6 + 1 + \frac{2}{4} = 6\frac{6}{4}$$
$$-5\frac{3}{4} \qquad\qquad\qquad\quad = 5\frac{3}{4}$$
$$\overline{\qquad\qquad\qquad\qquad\qquad 1\frac{3}{4}}$$

19.
$$19\frac{1}{4} = 19\frac{7}{28} = 18 + 1 + \frac{7}{28} = 18\frac{35}{28}$$
$$- \ 6\frac{3}{7} = \ 6\frac{12}{28} \qquad\qquad\qquad = \ 6\frac{12}{28}$$
$$\overline{\qquad\qquad\qquad\qquad\qquad\qquad 12\frac{23}{28}}$$

20.
$$36\frac{1}{4} = 36\frac{3}{12} = 35 + 1 + \frac{3}{12} = 35\frac{15}{12}$$
$$-19\frac{2}{3} = 19\frac{8}{12} \qquad\qquad\qquad = 19\frac{8}{12}$$
$$\overline{\qquad\qquad\qquad\qquad\qquad\qquad 6\frac{7}{12} \text{ ft.}}$$

21.
$$4\frac{3}{8} = 4\frac{9}{24}$$
$$+1\frac{1}{3} = 1\frac{8}{24}$$
$$\overline{5\frac{17}{24}}$$

$$5\frac{17}{24} - 5\frac{34}{48}$$
$$-3\frac{1}{16} = 3\frac{3}{48}$$
$$\overline{2\frac{31}{48}}$$

22.
$$2\frac{1}{2} = 2\frac{2}{4}$$
$$+6\frac{3}{4} = 6\frac{3}{4}$$
$$\overline{8\frac{5}{4}}$$

$$8\frac{5}{4} = 8\frac{10}{8}$$
$$-3\frac{5}{8} = \ 3\frac{5}{8}$$
$$\overline{5\frac{5}{8} \text{ lb.}}$$

23.
$$24\frac{7}{8} = 24\frac{7}{8}$$
$$-12\frac{3}{4} = 12\frac{6}{8}$$
$$\overline{12\frac{1}{8} \text{ yd.}}$$

Chapter 9

PRETEST 9A, pages 251–252

1. $\frac{2}{15}$

2. $\frac{2}{5}$

3. $\frac{10}{21}$

4. $\frac{2}{3}$

5. 4

6. $1\frac{1}{2}$

7. 9

8. $\frac{1}{10}$

9. $\frac{1}{4}$ cup   10. $24\frac{1}{2}$ 11. 45   12. $1\frac{3}{32}$

13. $1\frac{19}{32}$ 14. $\frac{1}{3}$ 15. 2   16. $2\frac{11}{20}$

17. $10\frac{5}{6}$ 18. $\frac{1}{16}$ 19. $\frac{1}{2}$ 20. 14 mi.

21. $7\frac{1}{2}$ cups   22. $11\frac{1}{2}$ yd.

PRACTICE EXERCISE 96, page 254

1. $\frac{1}{24}$ 2. $\frac{18}{35}$ 3. $\frac{1}{4}$ 4. $\frac{1}{6}$

5. $\frac{1}{3}$ 6. $\frac{9}{14}$ 7. $\frac{2}{5}$ 8. $\frac{1}{9}$

9. $\frac{1}{3}$ 10. $\frac{3}{32}$ 11. $\frac{7}{12}$ 12. $\frac{1}{2}$

PRACTICE EXERCISE 97, page 256

1. $\frac{2}{5}$ 2. $\frac{1}{2}$ 3. $\frac{2}{5}$ 4. $\frac{5}{28}$ 5. $\frac{2}{5}$

6. $\frac{7}{15}$ 7. $\frac{9}{14}$ 8. $\frac{1}{3}$ 9. $\frac{1}{9}$

PRACTICE EXERCISE 98, page 258

1. 5   2. $\frac{75}{4} = 18\frac{3}{4}$ 3. 40   4. $\frac{40}{3} = 13\frac{1}{3}$ 5. $\frac{1}{4}$

6. $\frac{48}{5} = 9\frac{3}{5}$ 7. 9   8. $\frac{16}{3} = 5\frac{1}{3}$ 9. 8

PRACTICE EXERCISE 99, page 259

1. $\frac{1}{5}$ 2. $\frac{1}{3}$ 3. $\frac{5}{4} = 1\frac{1}{4}$ 4. $\frac{21}{10} = 2\frac{1}{10}$

PRACTICE EXERCISE 100, page 261

1. 7   2. 13   3. $17\frac{1}{3}$ 4. $10\frac{5}{8}$

5. $1\frac{3}{5}$ 6. $1\frac{1}{20}$ 7. 32   8. $16\frac{1}{3}$

9. $48\frac{3}{5}$ 10. 7   11. $19\frac{1}{4}$ 12. $13\frac{13}{20}$

PRACTICE EXERCISE 101, pages 263–264

1. $\dfrac{2}{3} \times 24 =$

$$\dfrac{2}{\cancel{3}_{1}} \times \dfrac{\cancel{24}^{8}}{1} = 16$$

2. $5 \times 8\dfrac{7}{8} =$

$$\dfrac{5}{1} \times \dfrac{71}{8} = \dfrac{355}{8} = 44\dfrac{3}{8} \text{ mi.}$$

3. $24 \times 7\dfrac{1}{16} =$

$$\dfrac{\cancel{24}^{3}}{1} \times \dfrac{113}{\cancel{16}_{2}} = \dfrac{339}{2} = 169\dfrac{1}{2} \text{ in.}$$

16 ft. = 192 in.

$$\dfrac{-169\dfrac{1}{2}}{22\dfrac{1}{2} \text{ in.}}$$

4. a. $\dfrac{1}{\cancel{10}_{2}} \times \dfrac{\cancel{925}^{185}}{1} = \dfrac{185}{2} = \$92\dfrac{1}{2}$

b. $52 \times \$92\dfrac{1}{2} =$

$$\dfrac{\cancel{52}^{26}}{1} \times \dfrac{185}{\cancel{2}_{1}} = \$4,810$$

5. $300 \times 3\dfrac{7}{8} =$

$$\dfrac{\cancel{300}^{75}}{1} \times \dfrac{31}{\cancel{8}_{2}} = \dfrac{2325}{2} = 1,162\dfrac{1}{2} \text{ yd.}$$

6. $A = L \times W$

$$A = 6\dfrac{3}{5} \times 8\dfrac{4}{11}$$

$$= \dfrac{\cancel{33}^{3}}{5} \times \dfrac{92}{\cancel{11}_{2}} = \dfrac{276}{5} = 55\dfrac{1}{5} \text{ ft.}^2$$

7. $\dfrac{1}{3} \times 3\dfrac{3}{4} =$

$$\dfrac{1}{\cancel{3}_{1}} \times \dfrac{\cancel{15}^{3}}{4} = \dfrac{5}{4} = 1\dfrac{1}{4} \text{ h.}$$

8. $A = \dfrac{1}{2}b \times h$

$$= \dfrac{1}{2} \times 4\dfrac{5}{8} \times 9\dfrac{1}{2}$$

$$= \dfrac{1}{2} \times \dfrac{37}{8} \times \dfrac{19}{2}$$

$$A = \dfrac{703}{32} = 21\dfrac{31}{32} \text{ in.}^2$$

9. $\dfrac{2}{\cancel{3}_{1}} \times \dfrac{\cancel{36}^{12}}{1} = 24 \text{ vanilla}$

$$\dfrac{36}{-24}$$
$$\overline{12 \text{ chocolate}}$$

10. $2 \times 2\frac{7}{8} =$ $\qquad$ $4 \times 2\frac{1}{16} =$ $\qquad$ $5\frac{3}{4}$

$\dfrac{\cancel{2}^{1}}{1} \times \dfrac{23}{\cancel{8}_{4}} = \dfrac{23}{4} = 5\frac{3}{4}$ $\dfrac{\cancel{4}^{1}}{1} \times \dfrac{33}{\cancel{16}_{4}} = \dfrac{33}{4} = 8\frac{1}{4}$ $\begin{aligned} &+8\frac{1}{4} \\ \hline &13\frac{4}{4} = 14 \text{ yd.} \end{aligned}$

CHAPTER TEST 9A, pages 266–267

1. $\dfrac{2}{5} \times \dfrac{1}{3} = \dfrac{2}{15}$

2. $\dfrac{1}{\cancel{3}_{1}} \times \dfrac{\cancel{6}^{2}}{7} = \dfrac{2}{7}$

3. $\dfrac{5}{\cancel{8}_{4}} \times \dfrac{\cancel{6}^{3}}{7} = \dfrac{15}{28}$

4. $\dfrac{\cancel{3}^{1}}{\cancel{5}_{1}} \times \dfrac{\cancel{10}^{2}}{\cancel{21}_{7}} = \dfrac{2}{7}$

5. $\dfrac{3}{\cancel{4}_{1}} \times \dfrac{\cancel{8}^{2}}{1} = \dfrac{6}{1} = 6$

6. $\dfrac{\cancel{3}^{1}}{1} \times \dfrac{5}{\cancel{6}_{2}} = \dfrac{5}{2} = 2\frac{1}{2}$

7. $\dfrac{\cancel{12}^{1}}{1} \times \dfrac{7}{\cancel{12}_{1}} = \dfrac{7}{1} = 7$

8. $\dfrac{\cancel{4}^{1}}{5} \times \dfrac{1}{\cancel{12}_{3}} = \dfrac{1}{15}$

9. $\dfrac{1}{3} \times \dfrac{1}{3} = \dfrac{1}{9}$ cup

10. $\dfrac{9}{2} \times \dfrac{9}{1} = \dfrac{81}{2} = 40\frac{1}{2}$

11. $\dfrac{10}{1} \times \dfrac{13}{\cancel{2}_{1}}^{5} = \dfrac{65}{1} = 65$

12. $\dfrac{5}{6} \times \dfrac{7}{6} = \dfrac{35}{36}$

13. $\dfrac{\cancel{2}^{1}}{3} \times \dfrac{17}{\cancel{8}_{4}} = 1\frac{5}{12}$

14. $\dfrac{\cancel{2}^{1}}{\cancel{21}_{3}} \times \dfrac{\cancel{7}^{1}}{\cancel{2}_{1}} = \dfrac{1}{3}$

15. $\dfrac{\cancel{9}^{1}}{\cancel{4}_{1}} \times \dfrac{\cancel{16}^{4}}{\cancel{9}_{1}} = 4$

16. $\dfrac{13}{\cancel{4}_{2}} \times \dfrac{\cancel{6}^{3}}{5} = 3\frac{9}{10}$

17. $\dfrac{10}{3} \times \dfrac{7}{3} = \dfrac{70}{9} = 7\frac{7}{9}$

18. $\dfrac{\cancel{2}^{1}}{3} \times \dfrac{\cancel{5}^{1}}{\cancel{6}_{3}} \times \dfrac{1}{\cancel{15}_{3}} = \dfrac{1}{27}$

19. $\dfrac{\cancel{2}^{1}}{\cancel{5}_{1}} \times \dfrac{\cancel{10}^{\cancel{2}^{1}}}{1} \times \dfrac{1}{\cancel{8}_{\cancel{4}_{2}}} = \dfrac{1}{2}$

20. $6 \times 1\frac{1}{2} = \dfrac{\cancel{6}^{3}}{1} \times \dfrac{3}{\cancel{2}_{1}} = 9$ mi.

21. $3 \times 3\frac{1}{2} = \dfrac{3}{1} \times \dfrac{7}{2} = \dfrac{21}{2} = 10\frac{1}{2}$ lb.

22. $4 \times 1\frac{1}{2} = \frac{\overset{2}{\cancel{4}}}{1} \times \frac{3}{\cancel{2}} = 6$ $5 \times 2\frac{1}{4} = \frac{5}{1} \times \frac{9}{4} = \frac{45}{4} = 11\frac{1}{4}$ $\begin{array}{r} \overset{6}{+11\frac{1}{4}} \\ \hline 17\frac{1}{4} \text{ ft.} \end{array}$

PRETEST 9B, pages 267–268

1. 27   2. 25   3. $13\frac{1}{3}$ 4. 8

5. $1\frac{11}{21}$ 6. $\frac{8}{9}$ 7. $9\frac{3}{5}$ 8. $1\frac{1}{8}$

9. $\frac{2}{3}$ 10. 4   11. $1\frac{3}{5}$ 12. $\frac{4}{35}$

13. $\frac{3}{40}$ 14. 60 pkges.   15. $6   16. $\frac{11}{12}$

17. $\frac{4}{5}$ 18. $1\frac{3}{4}$ 19. 4   20. $2\frac{2}{9}$

21. $5\frac{3}{4}$ 22. $\frac{15}{128}$ 23. $\frac{1}{8}$ 24. 6

25. $\frac{20}{21}$ 26. $2\frac{8}{25}$ 27. $1\frac{7}{8}$ 28. $\frac{3}{8}$ lb.

29. 3 lengths   30. $23\frac{3}{8}$ ft.

PRACTICE EXERCISE 102, page 272

1. 6   2. $2\frac{2}{3}$ 3. $\frac{3}{4}$ 4. 2

5. $1\frac{1}{2}$ 6. $1\frac{1}{2}$ 7. $1\frac{1}{4}$ 8. $1\frac{1}{3}$

9. $1\frac{1}{2}$ 10. 2   11. $\frac{1}{4}$ 12. $1\frac{1}{3}$

PRACTICE EXERCISE 103, page 275

1. 9   2. 12   3. 8   4. 2

5. $\frac{8}{17}$ 6. 3   7. $\frac{4}{5}$ 8. $\frac{25}{39}$

9. 2   10. $1\frac{11}{18}$ 11. $1\frac{2}{3}$ 12. $5\frac{3}{8}$

PRACTICE EXERCISE 104, pages 275–276

1. $27 \div 3\frac{3}{8} = \frac{27}{1} \div \frac{8}{27} = \frac{\overset{1}{27}}{1} \times \frac{8}{\underset{1}{27}} = 8$ pieces

2. $60 \div 2\frac{1}{2} = \frac{60}{1} \div \frac{5}{2} = \frac{\overset{12}{60}}{1} \times \frac{2}{\underset{1}{5}} = 24$ bags

3. $8\frac{3}{4} \div 7 = \frac{35}{4} \div \frac{7}{1} = \frac{\overset{5}{35}}{4} \times \frac{1}{\underset{1}{7}} = \frac{5}{4} = 1\frac{1}{4}$ ft.

4. $8\frac{1}{2} \div 17 = \frac{17}{2} \div \frac{17}{1} = \frac{\overset{1}{17}}{2} \times \frac{1}{\underset{1}{17}} = \frac{1}{2}$ acre

5. $182\frac{1}{2} \div 5 = \frac{365}{2} \div \frac{5}{1} = \frac{\overset{73}{365}}{2} \times \frac{1}{\underset{1}{5}} = \frac{73}{2} = 36\frac{1}{2}$ lb.   a. No   b. $1\frac{1}{2}$ lb. above weight

6. $\dfrac{12\frac{1}{2} + 14\frac{7}{8} \times 14\frac{3}{4}}{3} = \dfrac{42\frac{1}{8}}{3} = 42\frac{1}{8} \div 3 = \frac{337}{8} \times \frac{1}{3} = \frac{337}{24} = 14\frac{1}{24}$ ft.

7. $82 \div 2\frac{9}{16} = \frac{82}{1} \div \frac{41}{16} = \frac{\overset{2}{82}}{1} \times \frac{16}{\underset{1}{41}} = 32$ dresses

8. $12\frac{1}{4} \div 1\frac{3}{4} = \frac{49}{4} \div \frac{7}{4} = \frac{\overset{7}{\underset{1}{49}}}{4} \times \frac{\overset{1}{4}}{\underset{1}{7}} = 7$ rods

9. $W = \frac{A}{L}$ $W = \dfrac{45}{5\frac{5}{8}} = \frac{45}{1} \div \frac{45}{8} = \frac{\overset{1}{45}}{1} \times \frac{8}{\underset{1}{45}} = 8$ ft.

10. $1\frac{1}{3} \div 6 = \frac{4}{3} \div \frac{6}{1} = \frac{\overset{2}{4}}{3} \times \frac{1}{\underset{3}{6}} = \frac{2}{9}$ tablespoons

CHAPTER TEST 9B, pages 278–279

1. $8 \div \frac{1}{2} = \frac{8}{1} \times \frac{2}{1} = 16$

2. $24 \div \frac{3}{4} = \frac{24}{\underset{1}{\overset{8}{3}}} \times \frac{4}{3} = 32$

3. $7 \div \frac{4}{5} = \frac{7}{1} \times \frac{5}{4} = 8\frac{3}{4}$

4. $12 \div 1\frac{1}{5} = \frac{12}{1} \div \frac{6}{5} = \frac{12}{\underset{1}{\overset{2}{6}}} \times \frac{5}{6} = 10$

5. $3 \div 2\dfrac{5}{6} = \dfrac{3}{1} \div \dfrac{17}{6} = \dfrac{3}{1} \times \dfrac{6}{17} = 1\dfrac{1}{17}$

6. $6 \div 3\dfrac{3}{4} = \dfrac{6}{1} \div \dfrac{15}{4} = \dfrac{\cancel{6}^{2}}{1} \times \dfrac{4}{\cancel{15}_{5}} = 1\dfrac{3}{5}$

7. $18 \div 1\dfrac{5}{16} = \dfrac{18}{1} \div \dfrac{21}{16} = \dfrac{\cancel{18}^{6}}{1} \times \dfrac{16}{\cancel{21}_{7}} = 13\dfrac{5}{7}$

8. $\dfrac{2}{3} \div \dfrac{3}{4} = \dfrac{2}{3} \times \dfrac{4}{3} = \dfrac{8}{9}$

9. $\dfrac{3}{4} \div \dfrac{15}{16} = \dfrac{\cancel{3}^{1}}{\cancel{4}_{1}} \times \dfrac{\cancel{16}^{4}}{\cancel{15}_{5}} = \dfrac{4}{5}$

10. $\dfrac{5}{9} \div \dfrac{5}{27} = \dfrac{\cancel{5}^{1}}{\cancel{9}_{1}} \times \dfrac{\cancel{27}^{3}}{\cancel{5}_{1}} = 3$

11. $\dfrac{5}{6} \div \dfrac{7}{18} = \dfrac{5}{\cancel{6}_{1}} \times \dfrac{\cancel{18}^{3}}{7} = 2\dfrac{1}{7}$

12. $\dfrac{2}{5} \div 9 = \dfrac{2}{5} \div \dfrac{9}{1} = \dfrac{2}{5} \times \dfrac{1}{9} = \dfrac{2}{45}$

13. $\dfrac{9}{10} \div \dfrac{15}{1} = \dfrac{\cancel{9}^{3}}{10} \times \dfrac{1}{\cancel{15}_{5}} = \dfrac{3}{50}$

14. $40 \div \dfrac{5}{6} = \dfrac{\cancel{40}^{8}}{1} \times \dfrac{6}{\cancel{5}_{1}} = 48 \text{ cans}$

15. $\$60 \div 7\dfrac{1}{2} = \dfrac{60}{1} \div \dfrac{15}{2} = \dfrac{\cancel{60}^{4}}{1} \times \dfrac{2}{\cancel{15}_{1}} = \$8$

16. $3\dfrac{2}{5} \div 4 = \dfrac{17}{5} \div \dfrac{4}{1} = \dfrac{17}{5} \times \dfrac{1}{4} = \dfrac{17}{20}$

17. $2\dfrac{1}{4} \div 3 = \dfrac{9}{4} \div \dfrac{3}{1} = \dfrac{\cancel{9}^{3}}{4} \times \dfrac{1}{\cancel{3}_{1}} = \dfrac{3}{4}$

18. $12\dfrac{1}{2} \div 20 = \dfrac{25}{2} \div \dfrac{20}{1} = \dfrac{\cancel{25}^{5}}{2} \times \dfrac{1}{\cancel{20}_{4}} = \dfrac{5}{8}$

19. $1\dfrac{5}{8} \div \dfrac{13}{16} = \dfrac{13}{8} \div \dfrac{13}{16} = \dfrac{\cancel{13}^{1}}{\cancel{8}_{1}} \times \dfrac{\cancel{16}^{2}}{\cancel{13}_{1}} = 2$

20. $1\dfrac{3}{4} \div \dfrac{4}{5} = \dfrac{7}{4} \div \dfrac{4}{5} = \dfrac{7}{4} \times \dfrac{5}{4} = 2\dfrac{3}{16}$

21. $3\dfrac{5}{8} \div \dfrac{3}{4} = \dfrac{29}{8} \div \dfrac{3}{4} = \dfrac{29}{\cancel{8}_{2}} \times \dfrac{\cancel{4}^{1}}{3} = 4\dfrac{5}{6}$

22. $\dfrac{7}{20} \div 3\dfrac{1}{3} = \dfrac{7}{20} \div \dfrac{10}{3} = \dfrac{7}{20} \times \dfrac{3}{10} = \dfrac{21}{200}$

23. $\dfrac{3}{10} \div 5\dfrac{2}{5} = \dfrac{3}{10} \div \dfrac{27}{5} = \dfrac{\cancel{3}^{1}}{\cancel{10}_{2}} \times \dfrac{\cancel{5}^{1}}{\cancel{27}_{9}} = \dfrac{1}{18}$

24. $6\dfrac{2}{3} \div 1\dfrac{1}{9} = \dfrac{20}{3} \div \dfrac{10}{9} = \dfrac{\cancel{20}^{2}}{\cancel{3}_{1}} \times \dfrac{\cancel{9}^{3}}{\cancel{10}_{1}} = 6$

25. $3\dfrac{3}{4} \div 4\dfrac{1}{6} = \dfrac{15}{4} \div \dfrac{25}{6} = \dfrac{\cancel{15}^{3}}{\cancel{4}_{2}} \times \dfrac{\cancel{6}^{3}}{\cancel{25}_{5}} = \dfrac{9}{10}$

26. $4\dfrac{1}{5} \div 1\dfrac{2}{3} = \dfrac{21}{5} \div \dfrac{5}{3} = \dfrac{21}{5} \times \dfrac{3}{5} = 2\dfrac{13}{25}$

27. $2\frac{17}{32} \div 2\frac{1}{4} = \frac{81}{32} \div \frac{9}{4} = \frac{\overset{9}{\cancel{81}}}{\underset{8}{\cancel{32}}} \times \frac{\overset{1}{\cancel{4}}}{\cancel{9}} = 1\frac{1}{8}$ 28. $5\frac{1}{4} \div 9 = \frac{21}{4} \div \frac{9}{1} = \frac{\overset{7}{\cancel{21}}}{4} \times \frac{1}{\underset{3}{\cancel{9}}} = \frac{7}{12}$ lb.

29. $17\frac{1}{2} \div 3\frac{1}{2} = \frac{35}{2} \div \frac{7}{2} = \frac{\overset{5}{\cancel{35}}}{\underset{1}{\cancel{2}}} \times \frac{\overset{1}{\cancel{2}}}{\underset{1}{\cancel{7}}} = 5$

30. $\dfrac{39\frac{1}{2} + 38\frac{3}{4} + 39\frac{1}{8}}{3} = \dfrac{117\frac{3}{8}}{2} = 117\frac{3}{8} \div 3 = \dfrac{\overset{313}{\cancel{939}}}{8} \times \dfrac{1}{\underset{1}{\cancel{3}}} = 39\frac{1}{8}$ s.

Chapter 10

PRETEST 10, pages 281–283

1. .3

2. a. .6   b. .48   c. .162   d. .05   e. .022

3. a. $\frac{39}{100}$ b. $\frac{8}{10} = \frac{4}{5}$ c. $\frac{625}{1,000} = \frac{5}{8}$ d. $\frac{4}{100} = \frac{1}{25}$ e. $\frac{2}{1,000} = \frac{1}{500}$

4. 112.5 lb.   5. $\begin{array}{r} \$100 \\ -\ 86 \\ \hline \$\ 14 \end{array}$ $\frac{14}{100} = .14$ savings

6. (b) .008   7. a. two-tenths   b. three and seven hundredths

8. a. .44   b. 3.2   c. 3,004.063

9. a. >   b. =   c. <   d. >   e. <

10. a. .06, .6, .65   11. a. .9
    b. 7.037, 7.307, 7.37   b. 11.43
    c. 2.5, 2.505, 2.55   c. 3.142

PRACTICE EXERCISE 105, page 286

1. .6   2. .3   3. .1   4. .9   5. .7

6. $\frac{5}{10}$ 7. $\frac{2}{10}$ 8. $\frac{8}{10}$ 9. $\frac{1}{10}$ 10. $\frac{4}{10}$

PRACTICE EXERCISE 106, page 287

1. .37	2. .463	3. .99	4. .4
5. .036	6. .7834	7. .0834	8. .001

9. $\dfrac{53}{100}$
10. $\dfrac{25}{100}$
11. $\dfrac{8}{100}$
12. $\dfrac{32}{1,000}$

13. $\dfrac{205}{1,000}$
14. $\dfrac{1,034}{10,000}$
15. $\dfrac{78}{10,000}$
16. $\dfrac{5}{1,000}$

PRACTICE EXERCISE 107, pages 288–289

1. nine-tenths
2. twenty-seven hundredths
3. five-hundredths
4. six and one-tenth
5. three and one hundred forty-one thousandths
6. fifty-one ten-thousandths
7. five and one-quarter hundredths
8. seven and two-hundredths

9. 6.09
10. .003
11. 3.5
12. 312.04

13. $.04\dfrac{3}{4}$
14. .093
15. 7.0005
16. 40.43

17. .9
18. .25
19. .207
20. 3.014

21. 37.03
22. 900.009

PRACTICE EXERCISE 108, page 290

1. $\dfrac{5}{10} = \dfrac{1}{2}$
2. $\dfrac{25}{100} = \dfrac{1}{4}$
3. $\dfrac{625}{1,000} = \dfrac{5}{8}$

4. $\dfrac{75}{100} = \dfrac{3}{4}$
5. $\dfrac{34}{100} = \dfrac{17}{50}$
6. $\dfrac{125}{1,000} = \dfrac{1}{8}$

7. $\dfrac{9}{10}$
8. $\dfrac{12}{100} = \dfrac{3}{25}$
9. $\dfrac{8}{100} = \dfrac{2}{25}$

10. $3\dfrac{7}{10}$
11. $5\dfrac{3}{4}$
12. $12\dfrac{1}{10}$

13. $6\dfrac{3}{8}$
14. $1\dfrac{7}{8}$
15. $123\dfrac{1}{2}$

PRACTICE EXERCISE 109, page 293

1. .430

2. .040

3. .100

4. .56 and .560

5. .05, .05000, .050

6. .82

7. .95

8. .71

9. .04

10. .1

11. 4.01

12. .07 < .077 < .7

13. 3.001 < 3.01 < 3.1

14. 1.03 < 1.3 < 1.31

15. .0001 < .001 < .1

PRACTICE EXERCISE 110, page 295

1. .3

2. 2.4

3. 3.1

4. .7

5. 4.35

6. .77

7. 2.80

8. .87

9. 3.142

10. .678

11. 5.556

CHAPTER TEST 10, pages 298–299

1. $\dfrac{4}{10} = .4$

2. a. .5   b. .34   c. .612   d. .04   e. .087

3. a. $\dfrac{93}{100}$ b. $\dfrac{6}{10} = \dfrac{3}{5}$ c. $\dfrac{875}{1,000} = \dfrac{7}{8}$ d. $\dfrac{6}{100} = \dfrac{3}{50}$ e. $\dfrac{4}{1,000} = \dfrac{1}{250}$

4. 214.6

5. $\begin{array}{r} \$10,000 \\ -\ \ 9,600 \\ \hline \$\ \ \ \ 400 \end{array}$ $\dfrac{40}{1,000} = .040 \text{ or } .04$

6. (b) $\dfrac{6}{1,000} = .006$

7. a. four-tenths   b. seven and three-hundredths

8. a. .36   b. 2.3   c. 4,003.036

9. a. .7 > .5   b. .60 = .6   c. .090 < .90   d. .4 > .3925   e. 4.0 < 4.001

10. a. .05 < .50 < .56 or .05 < .5 < .56
     b. 8.034 < 8.304 < 8.340 or 8.034 < 8.304 < 8.34
     c. 5.200 < 5.202 < 5.220 or 5.2 < 5.202 < 5.22

11. a. .7̲8 ≈ .8   b. 14.21̲2 ≈ 14.21   c. 4.151̲8 ≈ 4.152

Chapter 11

PRETEST 11A, pages 301–302

1. .9   2. .19   3. .99   4. .09
5. 1.4   6. 1.5   7. 5.7   8. 2.484
9. 20.721   10. 23.5   11. 25.027   12. .019
13. $2.43   14. $3.17   15. 46.35 mi.   16. $467.94

PRACTICE EXERCISE 111, page 304

1. .9   2. .38   3. .90 = .9   4. .51
5. .618   6. .91   7. .812   8. .76

PRACTICE EXERCISE 112, page 305

1. .08   2. .082   3. .051   4. .0638
5. .083   6. .11

PRACTICE EXERCISE 113, page 306

1. .3494   2. .479   3. .724   4. .985
5. .912   6. .963

PRACTICE EXERCISE 114, page 309

1. 1.2   2. 8.29   3. 35.835   4. 53.045
5. 2.695   6. 16.232   7. $78.81   8. 131.078

PRACTICE EXERCISE 115, pages 310–311

1.   73.4
     55.8
   113.9
    93.7
 + 27.5
   364.3 mi.

2.   138.4
    258.
   212.9
   371.6
   157.9
   197.2
 + 164.
 1,500.0 lb.

    a. No
    b. 1,500 lb.

3.   383.7
   412.4
   139.9
 + 273.8
 1,209.8 mi.

4. $ 1.78
   2.58
   1.18
   1.38
    .58
 + 8.86
 $16.36

5. $P = 43.87 + 43.87 + 31.9 + 31.9$
   $P = 151.54$ ft.

6.   .12
  3.4
+5.375
 8.895 in.

7. $P = 17.5 + 6.8 + 21.92$
   $P = 46.22$ in.

8. $288.66

9.   2.73
  0.67
  1.2
+3.259
 7.859 in.

10. 8 min. 4.69 s.
          18.9 s.
 8 min. 23.59 s.

CHAPTER TEST 11A, page 313

1.   .6
+.3
 .9

2.   .12
+.06
 .18

3.   .67
+.32
 .99

4.   .03
+.06
 .09

5.   .9
+.8
1.7

6.   .7
 .6
+.4
1.7

7. 6.3
+1.2
7.5

8. 2.431
+ .35
2.781

9. 14.017
    + 7.704
    ─────
    21.721

10.   1.4
       .86
    + 14.
    ─────
      16.26

11. 8.37
     .03
     4.9
    + .002
    ─────
     13.302

12. .0301
    +.0006
    ─────
     .0307

13. $2.59
    + .48
    ─────
     $3.07

14. $2.59
    + 1.87
    ─────
     $4.46

15.   8.3
      4.1
     12.7
    + 2.46
    ─────
     27.56 mi.

16. $249.50
      39.99
       9.95
    + 47.90
    ─────
     $347.34

PRETEST 11B, pages 314–315

1. .6   2. .28   3. 4.6   4. .07   5. 2.13
6. 3.06   7. 5.9   8. 54.3   9. 4.3   10. 6.26
11. 17.4   12. .113   13. 4.04   14. .031   15. 1.090
16. .28   17. 34.59   18. 1.8   19. 19.48   20. .082
21. 5.5 ft.   22. 2.1 sec.   23 3.85 lb.   24. 3.7 gal.   25. $15.06

PRACTICE EXERCISE 116, page 317

1. .6   2. .21   3. .024
4. .153   5. .0255   6. .075
7. .014   8. .005   9. .5

PRACTICE EXERCISE 117, page 318

1. .257   2. .152   3. .457
4. .327   5. .088   6. .0009
7. .212   8. .1844   9. .022

PRACTICE EXERCISE 118, page 320

1. $2.84   2. .24   3. .75
4. 56.06   5. 28.69   6. .032
7. 2.2563   8. $25.48   9. .88

PRACTICE EXERCISE 119, page 321

1. $2.77   2. 5.2   3. 2.9
4. 4.97   5. 26.4   6. 32.79
7. $4.05   8. 5.727   9. $3.09

PRACTICE EXERCISE 120, pages 322–323

1. .032
    –.028
    ⎯⎯⎯
    .004 in.

2. 4,830.4
    –4,543.7
    ⎯⎯⎯⎯⎯
    286.7 mi.

3. $59.95
    12.95
    + 32.95
    ⎯⎯⎯⎯⎯
    $105.85

    $110.00
    – 105.85
    ⎯⎯⎯⎯⎯
    $ 4.15

4. 1.51
    –1.46
    ⎯⎯⎯
    .05 min.

5. 5.850
    – .005
    ⎯⎯⎯
    5.845 in.

6. $564.75
    – 35.95
    ⎯⎯⎯⎯
    $528.80

7. 12.7
    +15.386
    ⎯⎯⎯⎯
    28.086

    34.671
    –28.086
    ⎯⎯⎯⎯
    6.585 ft.

8. a. $P = 138.2 + 170.6 + 92.4$
        $P = 401.2$ yd.
    b. $P = 89.6 + 89.6 + 47.8 + 47.8$
        $P = 274.8$ yd.
    c. Triangular field
    d. 401.2
        –274.8
        ⎯⎯⎯⎯
        126.4 yd.

9. 24.0
    –13.4
    ⎯⎯⎯
    10.6 gal.

10. .704
    + .514
    ⎯⎯⎯
    1.218

    2.080
    –1.218
    ⎯⎯⎯⎯
    .862 in. $= x$

CHAPTER TEST 11B, pages 325–326

1. .9
    –.3
    ⎯⎯
    .6

2. .78
    –.36
    ⎯⎯⎯
    .42

3. 8.4
    –5.0
    ⎯⎯
    3.4

4. .59
    –.55
    ⎯⎯⎯
    .04

5. 8.79
    –5.67
    ⎯⎯⎯
    3.12

6. 9.04
    –1.03
    ⎯⎯⎯
    8.01

7. 5.6
    –2.8
    ⎯⎯
    2.8

8. 85.9
    – 4.6
    ⎯⎯⎯
    81.3

9. 7.0
    – .5
    ⎯⎯
    6.5

10. 6.88
    − .59
    ‾‾‾‾
    6.29

11. 31.1
    −13.9
    ‾‾‾‾
    17.2

12. .407
    −.29
    ‾‾‾‾
    .117

13. 9.53
    −6.49
    ‾‾‾‾
    3.04

14. .048
    −.037
    ‾‾‾‾
    .011

15. 1.345
    − .265
    ‾‾‾‾
    1.08

16. 3.04
    − .79
    ‾‾‾‾
    2.25

17. 28.37
    − 5.6
    ‾‾‾‾
    22.77

18. 7.0
    −3.8
    ‾‾‾‾
    3.2

19. 27.50
    − .38
    ‾‾‾‾
    27.12

20. .830
    −.752
    ‾‾‾‾
    .078

21. 350.0
    −273.2
    ‾‾‾‾
    76.8 gal.

22. 15.75
    − 5.25
    ‾‾‾‾
    10.5 ft.

23. $16.99
    + 8.75
    ‾‾‾‾
    $25.74

24. 7.0
    −6.6
    ‾‾‾‾
    .4 s.

25. 103.2
    − 98.5
    ‾‾‾‾
    4.7 lb.

    $50.00
    − 25.74
    ‾‾‾‾
    $24.26

Chapter 12

PRETEST 12, pages 327–328

1. .8
2. 6.9
3. 5.26
4. .4
5. 8.06
6. .394
7. .092
8. 2.16
9. 8.28
10. 5.628
11. .7488
12. .837
13. .04224
14. .0203
15. a. 40
    b. 12
    c. 3,840
16. 42.38
17. 33.116
18. 3.9298
19. $4,612.14
20. $10.0875 ≈ $10.09
21. 1.5 lb.
22. 2 in.
23. 19.2 mi.

PRACTICE EXERCISE 121, pages 330–331

1. $10.00 = 10$	2. $12.00 = 12$	3. $2.0 = 2$	4. 18.48
5. $3.150 = 3.15$	6. $.4$	7. $2.50 = 2.5$	8. 9.5
9. $.7245$	10. $.36$	11. 62.5	12. 21
13. 174	14. 2	15. 289.92	16. 4.8
17. $.12$	18. $\$100.66$	19. $\$3,218.60$	20. $\$500.00$

PRACTICE EXERCISE 122, page 334

1. 1.848	2. $3.150 = 3.15$	3. $.07245$
4. $.036$	5. 2.8992	6. $.480 = .48$
7. $.0012$	8. $\$1.0066 \approx \$1.01$	9. $\$324.786 \approx \$324.79$
10. $\$68.750 = \$68.75$	11. 1.384	12. 134.712
13. 5.676	14. $.5024$	15. $.00037$
16. $\$2.9136 \approx \$2.91$	17. $\$22.467 \approx \$22.47$	18. $\$176.000 = \$176.00$
19. $\$812.592 \approx \$812.59$	20. $\$3.51750 \approx \$3.52$	

PRACTICE EXERCISE 123, page 336

1. $42. = 42$	2. $1,362.5$	3. $14. = 14$	4. $2400. = 2400$
5. 2.47	6. 314.16	7. $6,200. = 6,200$	8. $4. = 4$
9. $\$12,500. = \$12,500$		10. $\$34.7 = \$34.70$	

PRACTICE EXERCISE 124, page 337

1.
$$\begin{array}{r} \$8.39 \\ \times \quad 3.5 \\ \hline 4195 \\ 2517 \\ \hline \$29.365 \approx \$29.37 \end{array}$$

2.
$$\begin{array}{r} \$57.69 \\ \times \quad 52 \\ \hline 11538 \\ 28845 \\ \hline \$2999.88 \end{array}$$

3.
$$\begin{array}{r} .641 \\ \times \quad 12 \\ \hline 1282 \\ 641 \\ \hline 7.692 \text{ in.} \end{array}$$

4.
$$\begin{array}{r} 22.6 \\ \times \quad 11.8 \\ \hline 1808 \\ 226 \\ 226 \\ \hline 266.68 \text{ mi.} \end{array}$$

5. $\$1.75 \times 1,000 = \$1,750$

6.
$$\begin{array}{r} \$48.45 \\ \times \quad .07 \\ \hline \$3.3915 \approx \$3.39 \end{array}$$

7.
$$\begin{array}{r} \$1.50 \\ \times \quad 6 \\ \hline \$9.00 \end{array}$$

8.
$$\begin{array}{r} 31.37 \\ \times \quad 26.8 \\ \hline 25096 \\ 18822 \\ 6274 \\ \hline 840.716 \text{ ft.}^2 \end{array}$$

9.
$$\begin{array}{r} 3.14 \\ \times \quad 14 \\ \hline 1256 \\ 314 \\ \hline 43.96 \text{ in.} \end{array}$$

10.
$$\begin{array}{r} \$2.99 \\ \times\;\; 3.2 \\ \hline 598 \\ 897 \\ \hline \$9.568 \approx \$9.57 \end{array}$$

CHAPTER TEST 12, pages 339–340

1.
$$\begin{array}{r} .3 \\ \times 3 \\ \hline .9 \end{array}$$

2.
$$\begin{array}{r} 3.2 \\ \times\;3 \\ \hline 9.6 \end{array}$$

3.
$$\begin{array}{r} 3.53 \\ \times\;\;2 \\ \hline 7.06 \end{array}$$

4.
$$\begin{array}{r} .7 \\ \times .6 \\ \hline .42 \end{array}$$

5.
$$\begin{array}{r} 3.04 \\ \times\;\;2 \\ \hline 6.08 \end{array}$$

6.
$$\begin{array}{r} .186 \\ \times\;\;3 \\ \hline .558 \end{array}$$

7.
$$\begin{array}{r} .32 \\ \times\;\;.4 \\ \hline .128 \end{array}$$

8.
$$\begin{array}{r} 34 \\ \times .07 \\ \hline 2.38 \end{array}$$

9.
$$\begin{array}{r} 6.4 \\ \times 1.8 \\ \hline 512 \\ 64\;\; \\ \hline 11.52 \end{array}$$

10.
$$\begin{array}{r} .76 \\ \times\;\;8.4 \\ \hline 304 \\ 608\;\; \\ \hline 6.384 \end{array}$$

11.
$$\begin{array}{r} .87 \\ \times\;\;.69 \\ \hline 783 \\ 522\;\; \\ \hline .6003 \end{array}$$

12.
$$\begin{array}{r} 9.5 \\ \times .07 \\ \hline .665 \end{array}$$

13.
$$\begin{array}{r} .407 \\ \times\;\;\;.06 \\ \hline .02442 \end{array}$$

14.
$$\begin{array}{r} .92 \\ \times\;\;.07 \\ \hline .0644 \end{array}$$

15.
a. $.6 \times 100 = .60. = 60$
b. $9.2 \times 10 = 9.2. = 92$
c. $3.75 \times 1,000 = 3.750. = 3,750$

16.
$$\begin{array}{r} 44.8 \\ \times\;\;1.2 \\ \hline 896 \\ 448\;\; \\ \hline 53.76 \end{array}$$

17.
$$\begin{array}{r} 84.7 \\ \times\;\;.68 \\ \hline 6776 \\ 5082\;\; \\ \hline 57.596 \end{array}$$

18.
$$\begin{array}{r} 9.02 \\ \times\;\;.38 \\ \hline 7216 \\ 2706\;\; \\ \hline 3.4276 \end{array}$$

19.
$$\begin{array}{r} \$39.43 \\ \times\;\;\;\;\;87 \\ \hline 27601 \\ 31544\;\; \\ \hline \$3,430.41 \end{array}$$

20.
$$\begin{array}{r} \$3.29 \\ \times\;\;2.7 \\ \hline 2303 \\ 658\;\; \\ \hline \$8.883 \approx \$8.88 \end{array}$$

21.
$$\begin{array}{r} .5 \\ \times 3 \\ \hline 1.5 \text{ cups} \end{array}$$

22.   .015
     × 150
      750
      15
   2.250 = 2.25 in.

23.   18
     × .6
   10.8 mi.

Chapter 13

PRETEST 13, pages 341–342

1. .21
2. 3.1
3. .6
4. 5.1
5. .04
6. 1.003
7. .021
8. .007
9. a. .08   b. .132   c. .0542
10. .8
11. .44
12. 12.2 mi.
13. .31 oz.
14. 3
15. .5
16. 7. = 7
17. 80. = 80
18. 1.2
19. 520
20. .005
21. 120
22. 2.4
23. 70,050
24. 6 lb.
25. $6.97
26. 12 lb.
27. $398.50
28. 62.8
29. .0625
30. $\dfrac{7}{8}$

PRACTICE EXERCISE 125, page 344

1. .3
2. 4.4
3. .71
4. .83
5. .197
6. .23
7. .27
8. 7.5
9. .183

PRACTICE EXERCISE 126, page 346

1. .03
2. .09
3. .009
4. .07
5. .0007
6. .013
7. .0017
8. .074
9. .006
10. .0808
11. .0543
12. .0092

PRACTICE EXERCISE 127, page 347

1. .4
2. .25
3. .5
4. .125
5. .8
6. .625

PRACTICE EXERCISE 128, page 349

1. .3
2. .6
3. .3
4. .1
5. .44
6. .88
7. .29
8. .83
9. .9
10. .67 ft.

PRACTICE EXERCISE 129, page 352

1. 3
2. .3
3. .3
4. .03
5. .05
6. 5
7. 5
8. .5
9. 367.1
10. 4.25

PRACTICE EXERCISE 130, pages 353–354

1. 2.5
2. 25
3. 250
4. 250
5. 250
6. .025
7. 2,500
8. 2,500
9. 250
10. .025
11. .82
12. 530
13. .87
14. 28
15. 24.9

PRACTICE EXERCISE 131, pages 355–356

1. $1.69\overline{)8.45.}$ 5 lb.

2. $\dfrac{60}{80} = \dfrac{3}{4} = \$.75$

3. $4\overline{)12.8}$ 3.2 ft.

4. $450\overline{)751.50}$ \$1.67

5. $\dfrac{1.30 + 2.56 + .50 + 1.96 + 2.38}{5} = \dfrac{8.70}{5} = \$1.74$

6. $\dfrac{263.22}{21.4} = 12.3$ in.

7. $11.7\overline{)289.2.}$ 24.7 mi.

8. 47.5 m.p.h.

9. $5\overline{)16.7}$ 3.34 in.

10. $.8\overline{)28.8.}$ 36. operations

11. 7.2 gal.

12. a. $\dfrac{63}{135}$ b. .467

13. $.59\overline{)5.31.}$ 9 lb.

PRACTICE EXERCISE 132, page 357

1. 4.2
2. 13.625
3. 1.4
4. 2.4
5. .247
6. .0456
7. 3.1416
8. .62
9. .4
10. \$1.25
11. \$3.47
12. .00456

PRACTICE EXERCISE 133, page 360

4. $\dfrac{1}{5}$
5. $.16\dfrac{2}{3}$
6. $\dfrac{1}{8}$
7. .1
8. $\dfrac{2}{5}$
9. .6
10. $\dfrac{3}{10}$
11. .8
12. $\dfrac{5}{6}$
13. $\dfrac{5}{8}$
14. $\dfrac{7}{8}$
15. .7
16. $\dfrac{9}{10}$
17. $\dfrac{2}{3}$
18. .75
19. .375
20. $\dfrac{3}{7}$

PRACTICE EXERCISE 134, page 362

1. $\dfrac{3}{5}$ 2. $\dfrac{1}{25}$ 3. $\dfrac{3}{8}$ 4. $\dfrac{31}{400}$

5. $\dfrac{111}{500}$ 6. $\dfrac{3}{20}$ 7. $1\dfrac{1}{5}$ 8. $\dfrac{4}{5}$

9. $\dfrac{7}{8}$ 10. $\dfrac{5}{6}$ 11. 12 cans   12. $5\dfrac{1}{3}$ yd.

PRACTICE EXERCISE 135, pages 364–365

1. $2.01   2. .05   3. $4.02   4. .0042 in.
5. .99   6. $5.02   7. 24 ft.   8. 22 hits
9.   10. 180 rivets

	Fraction	Decimal
Wareham	$\dfrac{20}{32}$	.625
Cotuit	$\dfrac{17}{33}$	.515
Harwich	$\dfrac{18}{35}$	.514
Chatham	$\dfrac{17}{35}$	.486
Falmouth	$\dfrac{13}{32}$	.406
Yarmouth	$\dfrac{11}{33}$	.333

CHAPTER TEST 13, pages 368–369

1. $3\overline{).63}$ .21   2. $4\overline{)8.4}$ 2.1   3. $3\overline{)1.2}$ .4   4. $7\overline{)35.7}$ 5.1

5. $6\overline{).18}$ .03   6. $7\overline{)7.007}$ 1.001   7. $2\overline{).084}$ .042   8. $9\overline{).081}$ .009

9. a. .6 ÷ 100 = .00.6 = .006   b. 4.27 ÷ 10 = .4.27 = .427
    c. 63.8 ÷ 1,000 = .063.8 = .0638

10. $\dfrac{3}{5} = 5\overline{)3.0}$ .6   11. $\dfrac{2}{7} = 7\overline{)2.00}$ .285 ≈ .29

12.
$$13\overline{)158.6}$$ = 12.2 mi.
$$\begin{array}{r} 13 \\ \hline 2\,8 \\ 2\,6 \\ \hline 2\,6 \\ 2\,6 \\ \hline \end{array}$$

13.
$$45\overline{)14.0000}$$ = .3111 ≈ .31 oz.
$$\begin{array}{r} 13\,5 \\ \hline 50 \\ 45 \\ \hline 50 \\ 45 \\ \hline 50 \\ 45 \\ \hline 5 \end{array}$$

14. $.4\overline{)\,.8\,.}$ = 2.

15. $.8\overline{)\,.7.2}$ = .9

16. $.6\overline{)\,4.2.}$ = 7.

17. $.5\overline{)\,25.0.}$ = 50.

18. $.7\overline{)\,.8.4}$ = 1.2

19. $.04\overline{)\,16.80.}$ = 4 20.

20. $.6\overline{)\,.0.030}$ = .005

21. $.87\overline{)\,104.40.}$ = 1 20.
$$\begin{array}{r} 87 \\ \hline 17\,4 \\ 17\,4 \\ \hline \end{array}$$

22. $6.3\overline{)\,15.1.2}$ = 2.4
$$\begin{array}{r} 12\,6 \\ \hline 2\,5\,2 \\ 2\,5\,2 \\ \hline \end{array}$$

23. $.69\overline{)\,48,334.50.}$ = 70,050.
$$\begin{array}{r} 48\,3 \\ \hline 34\,5 \\ 34\,5 \\ \hline \end{array}$$

24. $.79\overline{)\,4.74.}$ = 6. = 6 lb.

25. $\$318.55 \div 34\dfrac{5}{8} =$

$\$318.55 \div \dfrac{277}{8} =$

$\dfrac{\overset{1.15}{\cancel{318.55}}}{1} \times \dfrac{8}{\underset{1}{\cancel{277}}} = \$9.20$ per yd.

26. $\$1.47\overline{)\,\$4.41.}$ = 3.

$3 \times 3 = 9$ lb.

27. $52\overline{)\,\$20,982.00}$ = \$403.50
$$\begin{array}{r} 20\,8 \\ \hline 182 \\ 156 \\ \hline 26\,0 \\ 26\,0 \\ \hline \end{array}$$

28.
$$\dfrac{38+41+35+45+49+37}{6}$$
$$= \dfrac{245}{6} = 40.8$$

29. $.08\dfrac{1}{5} = .082$

30. $.62\dfrac{1}{2} = \dfrac{62\dfrac{1}{2}}{100} = 62\dfrac{1}{2} \div 100$

$$= \dfrac{\overset{5}{\cancel{125}}}{2} \times \dfrac{1}{\cancel{100}} = \dfrac{5}{8}$$

Chapter 14

PRETEST 14, pages 371–372

1. 5%

2. $\dfrac{43}{100}$

3. 65%

4. .45

5. $.06\dfrac{1}{4} = .0625$

6. 2.15

7. 6%

8. 125%

9. $\dfrac{1}{2}\% = .5\%$

10. $\dfrac{1}{4}$

11. $\dfrac{11}{200}$

12. $\dfrac{7}{200}$

13. $1\dfrac{1}{4}$

14. 60%

15. 10%

16. $33\dfrac{1}{3}\%$

17. 475%

18. 75%

19. $24

20. $37\dfrac{1}{2}\%$

21. 80%

22. $70

23. a. $34 profit

    b. $\dfrac{\$34}{\$102} = \dfrac{1}{3} = 33\dfrac{1}{3}\%$

24. $\begin{array}{r} \$\ 48.50 \\ -\ \ 38.80 \\ \hline \$\ \ \ 9.70 \end{array}$ ⟶ $\dfrac{\$9.70}{\$48.50} = .20 = 20\%$

25. $\dfrac{100}{300} = 33\dfrac{1}{3}\%$

PRACTICE EXERCISE 136, pages 375–376

1. 6%

2. 50%

3. 70%

4. 28%

5. 5%

6. 215%

7. 90%

8. 72.5%

9. $4\dfrac{1}{2}\%$

10. $\dfrac{12}{100}$

11. $\dfrac{2}{100}$

12. $\dfrac{76}{100}$

13. $\dfrac{6\frac{1}{2}}{100}$

14. $\dfrac{3}{4} = \dfrac{75}{100} = .75 = 75 \text{ percent} = 75\%$

15. $\dfrac{2}{5} = \dfrac{40}{100} = .40 = 40 \text{ percent} = 40\%$

16. $\dfrac{5}{8} = \dfrac{62\frac{1}{2}}{100} = .62\frac{1}{2} = 62\frac{1}{2} \text{ percent} = 62\frac{1}{2}\%$

17. $\dfrac{1}{2} = \dfrac{50}{100} = .05 = 50 \text{ percent} = 50\%$

18.   100%   19.   15%

PRACTICE EXERCISE 137, page 378

1.   .75   2. $.12\frac{1}{4} = .1225$ 3.   .05

4.   .99   5.   1.25   6. $.02\frac{1}{2} = .025$

7. $.00\frac{1}{2} = .005$ 8.   .80   9.   .015

10. $.66\frac{2}{3}$

PRACTICE EXERCISE 138, page 380

1.   38%   2.   6%   3. $37.5\% = 37\frac{1}{2}\%$

4. $6\frac{1}{4}\% = 6.25\%$ 5. $.5\% = \frac{1}{2}\%$ 6. $3.25\% = 3\frac{1}{4}\%$

7.   125%   8. $\frac{1}{4}\% = .25\%$ 9.   312%

10.   60%

PRACTICE EXERCISE 139, page 382

1. $\dfrac{3}{4}$ 2. $\dfrac{1}{20}$ 3. $\dfrac{99}{100}$

4. $\dfrac{546}{10,000} = \dfrac{273}{5,000}$ 5. $\dfrac{222}{1,000} = \dfrac{111}{500}$ 6. $\dfrac{1}{400}$

7. $\dfrac{1}{40}$ 8. $\dfrac{1}{6}$ 9. $\dfrac{3}{40}$

10. $\dfrac{1}{4,000}$

PRACTICE EXERCISE 140, page 385

1. 80%

2. $12\frac{1}{2}\%$

3. 30%

4. $33\frac{1}{3}\%$

5. 5%

6. $42\frac{6}{7}\%$

7. $31\frac{1}{4}\%$

8. 25%

9. $62\frac{1}{2}\%$

10. $22\frac{2}{9}\%$

11. $\frac{60}{100}, \frac{6}{10}, \frac{3}{5}$

12. $.125, \frac{1}{8}$

13. $62\frac{1}{2}\%, .625$

14. $\frac{20}{100}, \frac{1}{5}, .2$

15. $\frac{4}{6}$, about .67

PRACTICE EXERCISE 141, page 388

1. 2.4

2. 6

3. .64

4. 25

5. 18

6. .6

7. 26

8. 2.05

9. $27.98

10. 16.5

11. $.18

12. .25

13. $169.68

14. 39.96 lb. copper
       1.74 lb. lead
       .12 lb. tin
       21.18 lb. zinc

15. $3,067.50

16. $22,500 + $ 13,800
       or $36,300

PRACTICE EXERCISE 142, page 390

1. $\frac{18}{48} = \frac{3}{8}$

2. $\frac{50}{75} = \frac{2}{3}$

3. $\frac{100}{8} = \frac{25}{2} = 12\frac{1}{2}$

4. $\frac{9}{24} = \frac{3}{8}$

5. $\frac{8}{20} = \frac{2}{5}$

6. $\frac{6}{15} = \frac{2}{5}$

PRACTICE EXERCISE 143, page 392

1. 75%

2. 40%

3. $33\frac{1}{3}\%$

4. $66\frac{2}{3}\%$

5. 8%

6. 40%

7. 75%

8. 42%

9. 15%

10. a. A is $55\frac{5}{9}\%.$

     b. B is $33\frac{1}{3}\%.$

     c. C is $11\frac{1}{9}\%.$

     d. $44\frac{4}{9}$ lb. of A

     e. $26\frac{2}{3}$ lb. of B

     f. $8\frac{8}{9}$ lb. of C

PRACTICE EXERCISE 144, pages 395–396

1. 400
2. 400
3. 12
4. 40
5. 200
6. 88
7. 80
8. 300
9. 120
10. 12
11. $125,000

CHAPTER TEST 14, pages 397–398

1. $\dfrac{8}{100} = 8\%$

2. $39\% = \dfrac{39}{100}$

3. $\dfrac{20}{100} = 20\%$

4. $55\% = .55$

5. $5\dfrac{1}{2}\% = 5.5\% = .055$

6. $185\% = 1.85$

7. $.08 = 8\%$

8. $1.35 = 135\%$

9. $.00\dfrac{1}{4} = .0025 = .25\%$

10. $15\% = \dfrac{15}{100} = \dfrac{3}{20}$

11. $4.5\% = \dfrac{45}{1,000} = \dfrac{9}{200}$

12. $4\dfrac{1}{4}\% = .0425 = \dfrac{425}{10,000} = \dfrac{17}{400}$

13. $150\% = 1.5 = 1\dfrac{1}{2}$

14. $\dfrac{2}{5} = .40 = 40\%$

15. $\dfrac{3}{10} = .3 = 30\%$

16. $\dfrac{2}{3} = .66\dfrac{2}{3} = 66\dfrac{2}{3}\%$

17. $3\dfrac{1}{4} = 3.25 = 325\%$

18. $\dfrac{9}{12} = \dfrac{3}{4} = .75 = 75\%$

19.
$$\begin{array}{r} \$80 \\ \times\ .03 \\ \hline \$2.40 \end{array}$$

20. $\dfrac{12}{30} = \dfrac{2}{5} = 40\%$

21. 16% of the number = 24

   1% of the number = $\dfrac{24}{16} = 1.5$

   100% of the number = 150

22. 40% of the original price = $48

   1% of the original price = $\dfrac{48}{40} = 1.2$

   100% of the original price = $120

23. a.
$$\begin{array}{r} 320 \\ -112 \\ \hline 208 \end{array}$$
   b. $\dfrac{112}{320} = .35 = 35\%$

24.
$$\begin{array}{r} \$379.00 \\ -\ 227.40 \\ \hline \$151.60 \end{array}$$
   $\dfrac{\$151.60}{\$379.00} = .40 = 40\%$

25. $\dfrac{19}{57} = .33\dfrac{1}{3} = .333$

Chapter 15

PRETEST 15, pages 399–400

1. 7 da. 16 h.
2. 4 h. 30 min.
3. 5 lb. 11 oz.
4. 9 wk. 2 da.
5. 3 gal. 1 qt.
6. 13 min. 50 s.
7. 1 pt.
8. 5 yr. 24 wk. 6 da.
9. 16 h. 30 min.
10. 2 wk. 2 da. 40 min.
11. 22 yd. 2 ft.
12. 4 ft. 3 in.
13. 2 ft. 3 in.
14. 2 yd. 5 in.
15. 1,440 min.
16. 4,000 lb.
17. 43 h. 30 min.
18. 2 ft. 5 in.
19. 1 qt. 1 pt. 6 oz.
20. 6 gal. 3 qt. 1 pt.
21. 4,200 cm
22. 0.265 m
23. 3,000 m
24. 0.008 km
25. 5,760 mm

PRACTICE EXERCISE 145, page 403

1. 2 wk. 4 da. 14 h.
2. 2 yd. 1 ft. 11 in.
3. 13 wk. 2 da.
4. 8 yd.
5. 2 h. 56 min. 21 s.
6. 7 yd. 2 in.
7. 12 lb. 6 oz.
8. 12 gal. 2 qt.

PRACTICE EXERCISE 146, page 405

1. 6 da. 6 h.
2. 1 ft. 1 in.
3. 1 wk. 4 da.
4. 3 yd.
5. 2 yd. 6 in.
6. 2 h. 30 min. 49 s.
7. 2 lb. 12 oz.
8. 2 qt. 1 pt.

PRACTICE EXERCISE 147, page 407

1. 1 wk. 5 da. 18 h.
2. 13 h. 20 min.
3. 13 lb. 8 oz.
4. 50 wk. 1 da.
5. 38 gal. 2 qt.
6. 1 h. 45 min. 20 s.
7. 1 gal. 1 pt.
8. 5 T. 1,001 lb. 4 oz.

PRACTICE EXERCISE 148, page 409

1. 2 da. 3 h.
2. 1 h. 25 min.
3. 1 lb. 9 oz.
4. 2 wk. 2 da.
5. 2 yd. 1 ft. 4 in.
6. 5 min. 47 s.
7. 1 pt. 7 oz.
8. 1 gal. 1 qt. 1 pt.

PRACTICE EXERCISE 149, pages 409–410

1.
$$\begin{array}{r} 12 \text{ oz.} \\ 3\overline{)2 \text{ lb.} \quad 4 \text{ oz.}} \\ 2 \text{ lb.} = 32 \text{ oz.} \\ \hline 36 \text{ oz.} \\ 36 \text{ oz.} \\ \hline \end{array}$$

2.
$$\begin{array}{r} 3 \text{ lb. 2 oz.} \\ +1 \text{ lb. 6 oz.} \\ \hline 4 \text{ lb. 8 oz.} \end{array}$$

3.   6 lb. 12 oz.
    −2 lb. 3 oz.
    ─────────
     4 lb. 9 oz.

4.   3 ft. 2 in.
      ×4
    ─────────
    ~~12 ft.~~ 8 in.
     4 yds.

5.   1 pt. 2 fl. oz.
         ×20
    ─────────────
    20 pt. ~~40~~ fl. oz. =
     2 pt. 8 fl. oz.
    ─────────────
    ~~22~~ pt. 8 fl. oz. =
    ~~11~~ qt.
    ─────────────
     2 gal. 3 qts. 8 fl. oz.

6.   10 h. 15 min. = 9 h. 75 min.
    − 8 h. 30 min. = 8 h. 30 min.
    ──────────────────────
                1 h. 45 min.

7.   2 yd. 2 ft. 6 in.
     1 yd.   7 in.
     1 yd. 1 ft. 3 in.
    +   2 ft. 11 in.
    ──────────────────
     4 yd. 5 ft. ~~27~~ in.
           2 ft. 3 in.
    ──────────────────
     4 yd. ~~7~~ ft. 3 in.
     2 yd. 1 ft.
    ──────────────────
     6 yd. 1 ft. 3 in.

8.
           5 h.   45 min.
        ──────────────
      4)23 h.
        20 h.
        ─────
         3 h. = 180 min.
                180 min.
                ────────

9.   6 h. 10 min.
      5 h. 30 min.
    + 5 h. 40 min.
    ──────────────
     16 h. ~~80~~ min. =
      1 h. 20 min.
    ──────────────
     17 h. 20 min.

10.   67 ft.   4   in. = 66 ft. 16   in.
      −66 ft. 10$\frac{1}{2}$ in. = 66 ft. 10$\frac{1}{2}$ in.
    ─────────────────────────────────
                           5$\frac{1}{2}$ in.

PRACTICE EXERCISE 150, pages 411–412

1.	500 cm	2.	210 mm	3.	9,000 mm	4.	13,000 m
5.	3 cm	6.	6 mm	7.	4 mm	8.	39 m
9.	3.42 m	10.	79.6 cm	11.	2.5 m	12.	3.8 km
13.	0.82 m	14.	40 cm	15.	0.053 m	16.	0.068 km

CHAPTER TEST 15, pages 414–415

1. 5 da. 3 h.
    +4 da. 11 h.

    9 da. 14 h. = 1 wk. 2 da. 14 h.

2. 2 h. 40 min.
    +1 h. 40 min.

    3 h. 80̸ min. =
    1 h. 20 min.

    4 h. 20 min.

3. 2 T. 150 lb.
    +1 T. 400 lb.

    3 T. 550 lb.

4. 7 wk. 3 da.
    +3 wk. 5 da.

    10 wk. 8̸ da. =
    1 wk. 1 da.

    11 wk. 1 da.

5. 6 gal. 3 qt.
    −2 gal. 2 qt.

    4 gal. 1 qt.

6. 4 h. 38 min. = 3 h. 98 min.
    −2 h. 52 min. = 2 h. 52 min.

    1 h. 46 min.

7. 5 qt.   = 4 qt. 2 pt.
    −2 qt. 1 pt. = 2 qt. 1 pt.

    2 qt. 1 pt.

8. 2 yd. 1 ft. 3 in. = 2 yd. 0 ft. 15 in. = 1 yd. 3 ft. 15 in.
    −1 yd. 2 ft. 9 in. = 1 yd. 2 ft. 9 in. = 1 yd. 2 ft. 9 in.

    1 ft. 6 in.

9. 11 min. 13 s.
          ×5

    55 min. 65 s. = 56 min. 5 s.
        1 min. 5 s.

10. 2 pt. 1 fl. oz.
         ×5

    10 pt. 5 fl. oz. = 5 qt. 5 fl. oz.

11. 6 yd. 1 ft.
         ×4

    24 yd. 4 ft. = 25 yd. 1 ft.

12.   3 ft. 2 in
    5)15 ft. 10 in.

13.   2 ft.   3 in.
    7)15 ft.   9 in.
      14 ft.

        1̸ ft. = 12 in.

             21 in.
             21 in.

14.   1 yd.   10 in.
    4)5 yd.   4 in.
      4 yd.

        1 yd. = 3 ft. = 36 in.

             40 in.
             40 in.

15. $\quad$ 1 h. = 60 min.

$$\begin{array}{r} 60 \\ \times\ 60 \\ \hline 3{,}600\ \text{s.} \end{array}$$

16. $\quad$ 1 qt. = 2 pt.

$2 \times 4 = 8$ pt.

1 pt. = 16 fl. oz.

$8 \times 16 = 128$ fl. oz.

17. 8 h. 15 min.

$$\begin{array}{r} \times 5 \\ \hline 40\ \text{h. } \cancel{75}\ \text{min.} = \\ 1\ \text{h. } 15\ \text{min.} \\ \hline 41\ \text{h. } 15\ \text{min.} \end{array}$$

18.

$$\begin{array}{r} 2\ \text{ft.} \qquad 7\ \text{in.} \\ 5\overline{)12\ \text{ft.} \qquad 11\ \text{in.}} \\ \underline{10\ \text{ft.}} \\ 2\ \text{ft.} = 24\ \text{in.} \\ \underline{35\ \text{in.}} \\ 35\ \text{in.} \\ \hline \end{array}$$

19. $\qquad$ 1 pt. 3 oz.

$$\begin{array}{r} \times 3 \\ \hline \cancel{3}\ \text{pt. } 9\ \text{oz.} = \\ 1\ \text{qt. } 1\ \text{pt.} \\ \hline 1\ \text{qt. } 1\ \text{pt. } 9\ \text{oz.} \end{array}$$

20.

$$\begin{array}{llll} 10\ \text{gal.} & = 9\ \text{gal. } 4\ \text{qt.} & = 9\ \text{gal. } 3\ \text{qt. } 2\ \text{pt.} \\ -\qquad\quad 3\ \text{qt. } 1\ \text{pt.} = & 3\ \text{qt. } 1\ \text{pt.} = & 3\ \text{qt. } 1\ \text{pt.} \\ \hline & & 9\ \text{gal.} \qquad 1\ \text{pt.} \end{array}$$

21. $\quad$ 1 m = 100 cm

36 m = 36 × 100 cm

$\quad\quad\ $ = 3,600 cm

22. $\quad$ 1 m = 0.001 km

5,000 m = 5,000 × 0.001 km

$\quad\quad\quad\ $ = 5 km

23. $\quad$ 1 cm = 10 mm

122 cm = 122 × 10 mm

$\quad\quad\ $ = 1,220 mm

24. $\quad$ 1 mm = 0.001 m

6,800 mm = 6,800 × 0.001 m

$\quad\quad\quad\ $ = 6.8 m

25. $\quad$ 1 km = 1,000 m

14 km = 14 × 1,000 m

$\quad\quad\ $ = 14,000 m

INDEX

Really. This isn't going to hurt at all . . .

BARRON'S

Learning really doesn't hurt when middle school students open any of Barron's *Painless* titles. These books transform subjects many kids consider boring or confusing into fun—emphasizing clear presentation, a touch of humor, and entertaining brain-tickler puzzles that are fun to solve.

Each book: Paperback

Painless Algebra, 3rd Ed.
Lynette Long, Ph.D.
ISBN 978-0-7641-4715-9, $9.99, Can$11.99

Painless American Government
Jeffrey Strausser
ISBN 978-0-7641-2601-7, $9.99, Can$11.99

Painless American History, 2nd Ed.
Curt Lader
ISBN 978-0-7641-4231-4, $9.99, Can$11.99

Painless Biology
Joyce Thornton Barry
ISBN 978-0-7641-4172-0, $9.99, Can$11.99

Painless Chemistry
Loris Chen
ISBN 978-0-7641-4602-2, $9.99, Can$11.99

Painless Earth Science
Edward J. Denecke, Jr.
ISBN 978-0-7641-4601-5, $9.99, Can$11.99

Painless English for Speakers of Other Languages
Jeffrey Strausser and José Paniza
ISBN 978-0-7641-3562-0, $9.99, Can$11.99

Painless Fractions, 2nd Ed.
Alyece Cummings
ISBN 978-0-7641-3439-5, $8.99, Can$10.99

Painless French
Carol Chaitkin and Lynn Gore
ISBN 978-0-7641-3735-8, $8.99, Can$10.99

Painless Geometry, 2nd Ed.
Lynette Long, Ph.D.
ISBN 978-0-7641-4230-7, $9.99, Can$11.99

Painless Grammar, 3rd Ed.
Rebecca S. Elliott, Ph.D.
ISBN 978-0-7641-4712-8, $9.99, Can$11.99

Painless Italian
Marcel Danesi, Ph.D.
ISBN 978-0-7641-3630-6, $8.99, Can$10.99

Painless Math Word Problems, 2nd Ed.
Marcie Abramson, B.S., Ed.M.
ISBN 978-0-7641-4335-9, $9.99, Can$11.99

Painless Poetry, 2nd Ed.
Mary Elizabeth
ISBN 978-0-7641-4591-9, $9.99, Can$11.99

Painless Pre-Algebra
Amy Stahl
ISBN 978-0-7641-4588-9, $9.99, Can$11.99

Painless Reading Comprehension
Darolyn E. Jones
ISBN 978-0-7641-2766-3, $9.99, Can$11.99

Painless Spanish, 2nd Ed.
Carlos B. Vega
ISBN 978-0-7641-4711-1, $9.99, Can$11.99

Painless Speaking
Mary Elizabeth
ISBN 978-0-7641-2147-0, $9.99, Can$11.99

Painless Spelling, 3rd Ed.
Mary Elizabeth
ISBN 978-0-7641-4713-5, $9.99, Can$11.99

Painless Study Techniques
Michael Greenberg
ISBN 978-0-7641-4059-4, $9.99, Can$11.99

Painless Vocabulary, 2nd Ed.
Michael Greenberg
ISBN 978-0-7641-4714-2, $9.99, Can$11.99

Painless Writing, 2nd Ed.
Jeffrey Strausser
ISBN 978-0-7641-4234-5, $9.99, Can$11.99

Barron's Educational Series, Inc.
250 Wireless Blvd.
Hauppauge, N.Y. 11788
Order toll-free: 1-800-645-3476
In Canada:
Georgetown Book Warehouse
34 Armstrong Ave.
Georgetown, Ontario L7G 4R9
Canadian orders: 1-800-247-7160

Prices subject to change without notice.

To order—Available at your local book store or visit **www.barronseduc.com**

(#79) R3/11